MARINE FISHES OF FLORIDA

MARINE FISHES OF FLORIDA

DAVID B. SNYDER AND GEORGE H. BURGESS

JOHNS HOPKINS UNIVERSITY PRESS / BALTIMORE

This book was published with the generous support of CSA Ocean Sciences Inc. and the Florida Museum of Natural History, University of Florida.

Printed in China on acid-free paper
9 8 7 6 5 4 3 2 1

Johns Hopkins University Press
2715 North Charles Street
Baltimore, Maryland 21218-4363
www.press.jhu.edu

Library of Congress Cataloging-in-Publication Data

Snyder, David B., 1955–
Marine fishes of Florida / David B. Snyder and George H. Burgess.
pages cm
Includes bibliographical references and index.
ISBN 978-1-4214-1872-8 (pbk. : alk. paper) — ISBN 978-1-4214-1873-5 (electronic) — ISBN 1-4214-1872-X (pbk. : alk. paper) — ISBN 1-4214-1873-8 (electronic) 1. Marine fishes—Florida. 2. Marine fishes—Florida—Identification. 3. Fishes—Florida. 4. Fishes—Florida—Identification. I. Burgess, George H., 1949– II. Title.
QL628.F6S69 2016
597′.4809759—dc23 2015014320

A catalog record for this book is available from the British Library.

Frontispiece: French grunts (*Haemulon flavolineatum*) swim through a gap in a reef off Jupiter, Florida.

Special discounts are available for bulk purchases of this book. For more information, please contact Special Sales at 410-516-6936 or specialsales@press.jhu.edu.

Johns Hopkins University Press uses environmentally friendly book materials, including recycled text paper that is composed of at least 30 percent post-consumer waste, whenever possible.

Contents

Preface

The initial idea for this book began in the late 1970s in the ichthyology collection of the Florida Museum of Natural History, University of Florida, in Gainesville, Florida. David was an undergraduate student taking Carter Gilbert's ichthyology course. George was a graduate student assisting with the course, his office crammed with fish books, scientific journals, and piles of reprints on taxonomy and the ecology of fishes. Amid all of this was a set of three-ring binders filled with 35 mm color slides of fishes lying on cooler tops, measuring boards, and trawler decks, mostly from North Carolina and Florida. David was impressed, thinking "This guy is a real fish nerd who will surely appreciate my underwater shots," and he later returned with several boxes of slides taken with a Nikonos II underwater camera. The images were grainy, mostly underexposed (with a monochromatic blue cast), and in many the subject was a tiny speck in the frame. Not commenting on their poor quality, George simply held them up to the light, one by one: porkfish, barracuda, gray snapper, snook, manta rays feeding at the surface. One image labeled as an ocean sunfish, taken in the ocean off Jupiter, Florida, caught his eye. He quickly pointed out that it was no ordinary *Mola mola*, it was a sharptail mola (*Mola lanceolatus*), and this was the only underwater photograph he had seen of the fish.

That brief commentary on the images sent David on a quest to document, on film, every fish that swims. A formidable task, to say the least. After a couple of trips to the Indo-Pacific, he decided to scale back the geographical scope to something more attainable: a 100-mile radius from his home in southeastern Florida. Although more realistic, it still was no small undertaking, encompassing some the most diverse fish habitats in the southeastern United States, including Indian River Lagoon, Lake Worth Lagoon, the western edge of the Gulf Stream, and (at the eastern periphery) the Little Bahama Bank.

As time passed, our respective slide and, eventually, digital libraries swelled, but it wasn't entirely clear what we were going to do with all of our images. We discussed writing a book on several occasions, but, as often is the case, there were always more species to find and upgraded photographs to take. Thus, when Vince Burke of Johns Hopkins University Press offered us the opportunity to submit a proposal for a book about marine fishes of Florida, to be copiously illustrated with photographs, we jumped at the chance.

In this volume we use our own observations and photographs, along with published information, to tell a story about Florida marine fishes. The story is one of diversity, borne out of adaptations to Florida's geological history and habitats. Observations compiled over our combined 70-plus years of observing and collecting fishes in virtually all corners of Florida's marine and estuarine environments allow us to provide some context to accompany practical clues to identification. Having spent countless hours underwater and having employed trawls, seines, dipnets, traps, castnets, spears, anesthetics, ichthyocides, electricity, and hooks to collect and study these fishes, we have come to truly love and respect the diverse ichthyological resources found in our wonderful but increasingly weather- and human-modified state waters.

We had to make some hard choices regarding which families and species of fishes to include. As biologists and ichthyologists (fish specialists), we longed to present photographs and text for each and every marine fish that ventures up into freshwater rivers or out to the 100-fathom depth curve, but restrictions on space and time had to be honored. Relying on our field notes, memory banks, and records from the Florida Museum of Natural History fish collection, we settled on 133 families. Making cuts was toughest at the deeper end of our designated depth range, where a number of species living on the upper continental slope (the relatively steep decline from the edge of the continental shelf to the ocean floor) seemingly ignore and cavalierly cross our artificial depth boundary. We grudgingly left out roughies, alfonsinos, and beardfishes but included the wreckfish, admittedly because it is big and charismatic. We also excluded epipelagic fishes (those living in the upper layer of the open ocean) such as opahs, oarfishes, ribbonfishes, crestfishes, lancetfishes, sauries, and pomfrets. Adding to the roster at the other end of our range—the upper reaches of coastal rivers—was challenging for some groups. We included the anadromous (moving upstream to spawn) sea lampreys, shads, and sturgeons, as well as the catadromous (living in freshwater but breeding in saltwater) American eels but excluded the striped bass, an important commercial and sport fish in northerly waters but rare in Florida. We did not include invasive freshwater species, such as blackchin tilapias and Mayan cichlids, and some native freshwater sunfishes that make their way into estuaries in certain areas of the state.

In this day and age, any fundamental treatment of a regional fauna has to acknowledge diversity at the genetic level. DNA techniques now are routinely applied in taxonomic revisions of closely related species, often uncovering what are known as "cryptic species" that have existed right under our noses all along. Usually these involve small fishes (such as gobies, labrisomids, and triplefins), but even larger-sized fishes (including sea chubs, bonefishes, and hammerhead sharks) are among those groups with recently discovered cryptic or misidentified taxa. Cryptic species usually are closely related to more-common, co-occurring, and similarly appearing species; correct identification requires having fresh specimens in hand to fully assess the often intricate or subtle characteristics used to separate out and distinguish among species. We wholeheartedly support this avenue of research, as it sheds light on speciation (the process through which new biological species arise), adaptation, and zoogeography (the geographical distribution of species), but at present its many findings are not practical for field identification. For some groups, such as the goby genus *Bathygobius*, their documented DNA separation is not accompanied by obvious morphological (structural or functional) or color differences, making on-the-spot field (or laboratory) identification infeasible.

We strove, sometimes to an unreasonable extent, to present images taken in Florida waters. Reality set in when deadlines loomed, so it was fairly easy to reconcile occasional exceptions, such as using a photo of a basking shark from the United Kingdom and one of a white shark from Australia (thanks to Jeremy Stafford-Deitsch). Similarly—and by necessity—a portion of the images, particularly of reef fishes, were taken in the northern Bahamas. See the photo locations at the end of the book for those images taken elsewhere than in Florida waters.

Florida's extensive coastline supports the bulk of the state's human population, leading to excessive pressure on nearshore habitat and water quality that are important to fishes. While it would be easy to project gloom and doom and repeat the mantra "It's not like it used be," we follow the lead of the great zoologist Archie Carr, who was on target when he said, in the preface to *A Naturalist in Florida* (p. xv):

> I am especially susceptible to the disease of bitterness over the ruin of Florida—over the partly aimless, partly avaricious ruin of unequaled natural riches of the most nearly tropical state. But in my case I decided simply, "What the hell, you cry the blues and soon nobody listens." And that made me see there was no sense in writing another vanishing Eden book at all. . . . The way to get my point across would be to talk mostly about what joy still remains in the Florida landscape then just sneak in some factual tooth-gnashing every now and then when the readers might really be reading.

We greatly appreciate the efforts and patience of dive partners, ROV (remotely operated vehicle) pilots, anglers, and others who helped us get images of fishes, especially Karen Snyder, Steve Viada, John Thompson, Randy Jordan, Jim Abernethy, T. J. Stewart, Les Crocker, Mike Kendrick, Robin Snyder, Kelly Snyder, Edward Doyle, Rick van Tol, Dave Dutton, Jamie Sherwood, Tony Saucier, Toshi Mikagawa, Ben Hartig, and Keith Spring. Special thanks go to Scott Taylor for showing us how to catch the elusive mangrove rivulus. Ken Lindeman of the Florida Institute of Technology reviewed chapters on grunts and snappers. Jim Abernethy, Noel Burkhead, Don DeMaria, Steve Ross, Wayne Shoemake, and Jeremy Stafford-Deitsch kindly allowed us to use their images. Masa Ushioda of SeaPics.com assisted with the smalltooth sawfish (Doug Perrine), swordfish (Franco Banfi), and bigeye thresher (Jason Arnold) photos. Jay Fleming provided the sea lamprey, American eel, and Atlantic cutlassfish images. All photos without source credits were taken by David Snyder. We thank Kevin Noack and Keith VanGraafeiland for preparing the maps and the sea-surface temperature figure.

Vince Burke, Kathryn Marguy, and Catherine Goldstead of Johns Hopkins University Press offered patience and guidance, as well as graciously granted needed deadline extensions. Kathleen Capels expertly copyedited the entire manuscript and kept us true to our intended audience. We are grateful for the generous financial support provided by Kevin Peterson of CSA Ocean Sciences Inc. and the Florida Museum of Natural History, University of Florida, that made the production of this book possible.

Above all, we extend heartfelt thanks to our patient and understanding wives, Karen Snyder and Linda Burgess, and families—Robin and Kelly Snyder, and Matthew and Nathan Burgess—for allowing us to engage in our passion, sometimes at the expense of family obligations. True fish nerdism requires a strong support system at home.

Introduction

> There is perhaps no State in the Union whose fishes have attracted more general attention than have those of Florida. The interest in the fishes of this State is shared by the commercial fisherman, the angler, and the ichthyologist. The number of species that are sought because of their commercial value is far greater than any other section of America. Those that are of interest to the angler are more numerous than any other can boast, while the richness and peculiarities of the fish fauna of Florida have made this State a fascinating field to the ichthyologist and student of geographic distribution.
>
> —*Barton Warren Evermann, ichthyologist, U.S. Fish Commission, 1898*

Florida marine waters support an incredibly rich array of fishes, from tiny gobies to giant marlins. This richness prompts residents and visitors alike to catch, release, sell, eat, stuff, feed, photograph, or simply watch and ponder Florida fishes. The state is so large—spanning 6.5 degrees of latitude—that one can experience temperate saltmarshes in the north, tropical reefs in the south, and oceanic (blue) water from just about any port in Florida. Potential candidates for encounter are regionally diverse: in the St. Johns River estuary one can catch black drums as thick as your leg; at Satellite Beach the air may have the aroma of schools of menhadens as they pass through a gauntlet of predators; snooks that make popping sounds when feeding on mullets at the water's surface can be heard along the banks of the Indian River Lagoon; huge, barrel-sized Atlantic goliath groupers share waters with divers off the coastline by the town of Jupiter; simply gazing from a car while traveling along the Overseas Highway reveals pods of giant-sized tarpon rolling in the water; the backwaters of the Everglades provide kayakers with the opportunity to encounter an endangered smalltooth sawfish resting under the fringe of mangrove trees along the shoreline; in the gin-clear, freshwater springs of Crystal River, hordes of crevalle jacks, gray snappers, sheepsheads, and other marine fishes mix with freshwater basses and sunfishes; and snorkelers by the jetties flanking the entrance to St. Andrews Bay can observe the vagrant young of tropical reef fishes. Several excellent regional guides (including *A Field Guide to Coastal Fishes from Maine to Texas* by Kells and Carpenter; the *Peterson Field Guide to Atlantic Coast Fishes* by Robins and Ray; *Caribbean Reef Fishes* by Randall: the *National Audubon Society Field Guide to Tropical Marine Fishes* by Smith; *Reef Fish Identification* by Humann and DeLoach; and *Fishes of the Greater Caribbean* [iPhone app] by Robertson and Van Tassell) cover Florida fishes, but all generally have a wider geographical scope. Surprisingly, no book-length summaries dedicated solely to the marine fishes of Florida exist.

This book is written for those who want a little more insight than they may get from a field guide before (or after) exploring marine or estuarine waters of Florida. Our intended audience includes anglers, divers, fish watchers, biologists, and natural-history buffs interested in learning more about marine fishes of this state. Because of space limitations, we restricted our treatment to the 133 most-encountered families, each covered by a single chapter. Every chapter begins with a list of all family representatives known from Florida; here we generally follow the American Fisheries Society's most recent (2013) edition of *Common and Scientific Names of Fishes from the United States, Canada, and Mexico* for common-language (vernacular) family and species names. We only deviated from this reference in cases where more-recently published scientific studies prompted changes in species names or their status as species. We include sections on the species' diversity, geographical distribution and habitat, and natural history. In the species section, we provide cues that will help identify members of the families covered. We focus primarily on the color and shape of the fishes and their behavior and habitat choices as much as possible, rather than going into detail about the more difficult (but diagnostic) conventional characteristics used by fish biologists (ichthyologists), such as counts of scales, fin rays, or gill rakers. In a few cases we had no other choice and include more-technical characters to separate some species. A knowledge of the basic parts of a fish are all that is needed to follow most of the species descriptions. When identifying members of a multispecies family, we start by describing the most conspicuous (and, therefore, easiest) species to identify and work systematically through the group, ending with the most difficult pair to distinguish. This approach sometimes cuts across taxonomic lines and thus does not follow conventional evolutionary relationships. We don't disagree with the evolutionary method, but it often does not lend itself to practical field identifications for the uninitiated. In the geographical distribution and habitat sections, our emphasis is on Florida, and whenever possible we interject our own observations. The natural-history section provides information on reproduction, feeding, and related behaviors as they pertain to Florida. Much of this material comes from our own notebooks, as well as published reports and unrefereed "gray" literature. The breadth of information presented depends on the state of knowledge within the scientific community. In an appendix, we offer a checklist of fishes known from Florida, although we exclude deep-sea and mesopelagic (living at midwater depths) species.

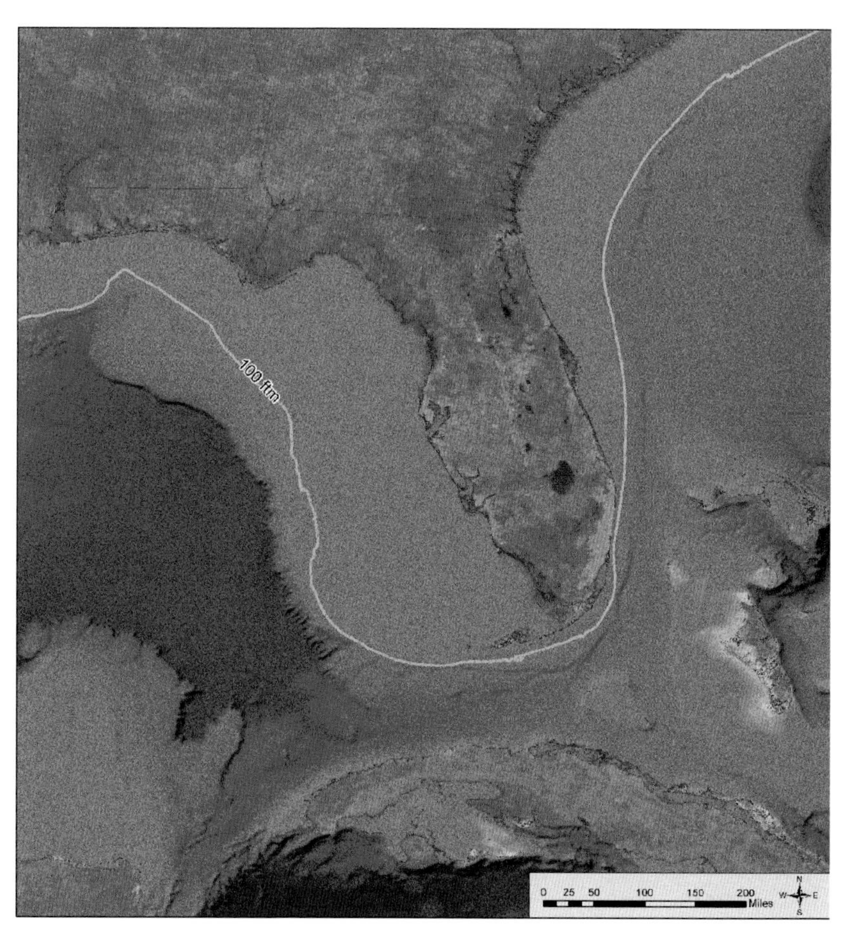

The Florida Platform

The Marine Environment

To properly address Florida's marine environment, we must understand the physical foundation of the region, necessitating a brief discussion of recent geological history. Then there are the characteristics and motion of the overlying water, the medium that supports the fauna and flora of the sea. Knowing the area's geological history and water characteristics helps us understand the distributions of many present-day habitats and the fishes that live in them.

Geological History

The Florida we know today is the emergent portion of a much broader limestone feature called the Florida Platform. Sea level does not have to drop very much to expose great expanses of this platform. During the past 100,000 years, the sea level rose and fell multiple times as massive glaciers sequentially formed and melted over the continents. About 20,000 years ago, sea level was 300 ft lower than at present. If this were the situation today, a resident of Tampa would have to drive 80 mi west for a day at the beach and, upon arrival, would find steep cliffs reminiscent of Big Sur, California, rather than a gradually sloping Florida beach. When the continental shelf was exposed during these glacial periods, sand, bits of shell, and other materials were blown by the wind into linear dunes or shoreline bluffs. Over time the grains gradually cemented together to form hardened ridges, called paleoshorelines by geologists. These ridges underlie many of the reefs and hardbottom habitats we see on the continental shelf and shoreline around Florida today. Other seafloor features—such as solution holes, river beds, and coastal inlets—were formed when past sea levels were lower than the present one. On the other hand, when glaciers melted, the sea level was so high that at times water covered most of the peninsula, leaving only an archipelago of islands to represent Florida. High-water periods permitted temperate (moderate temperature ranges) fishes to freely move between the Atlantic Ocean and the Gulf of Mexico.

Most of the submerged Florida Platform is overlaid by sediments, which exist in broad mosaics of different textures. The coarsest sediments are derived from coralline algal nodules (hard, compact, ping-pong ball–sized masses of limestone formed by red algae), and the finest are river-based muds. Grain size varies, as does the type of sediment found around the peninsula. Quartz sand, eroded over eons from the Appalachian Mountains and carried to Florida by ancient rivers, is prevalent off northeastern Florida north of Cape Canaveral, over much of the panhandle, and in the Big Bend region in the Gulf. Carbonate sediments—derived from the hard shells of sea urchins, barnacles, clams, snails, corals, and various types of algae—cover much of the west Florida shelf and the shelf offshore from Cape Canaveral to the Florida Keys. Quartz sand is most prevalent in the northern Gulf of Mexico and along the beaches of Florida's east coast.

Oceanography

Florida's marine environment is greatly influenced by two major ocean currents that are, in fact, one and the same. Off the west coast there is the Loop Current, which enters the Gulf of Mexico from the Caribbean through the narrow Yucatan Channel and pushes northward toward the Mississippi-Alabama-Louisiana coast. But it never quite gets there, as it takes a hairpin turn to the right near the shelf break and heads back south, generally along the outer margin of the west Florida shelf. It continues through the Straits of Florida between Cuba and the Florida Keys (along this stretch it is called the Florida Current) and becomes the Gulf Stream Current. At this point there is some input from the east as the Antilles Current pushes into the Straits of Florida. The Gulf Stream pushes northward along the eastern Florida shelf edge and occurs progressively farther from shore as the shelf widens to the north. Along this circuitous route around the peninsula, these two currents affect the temperature, salinity, clarity, and movement of water on the adjacent shelves, and

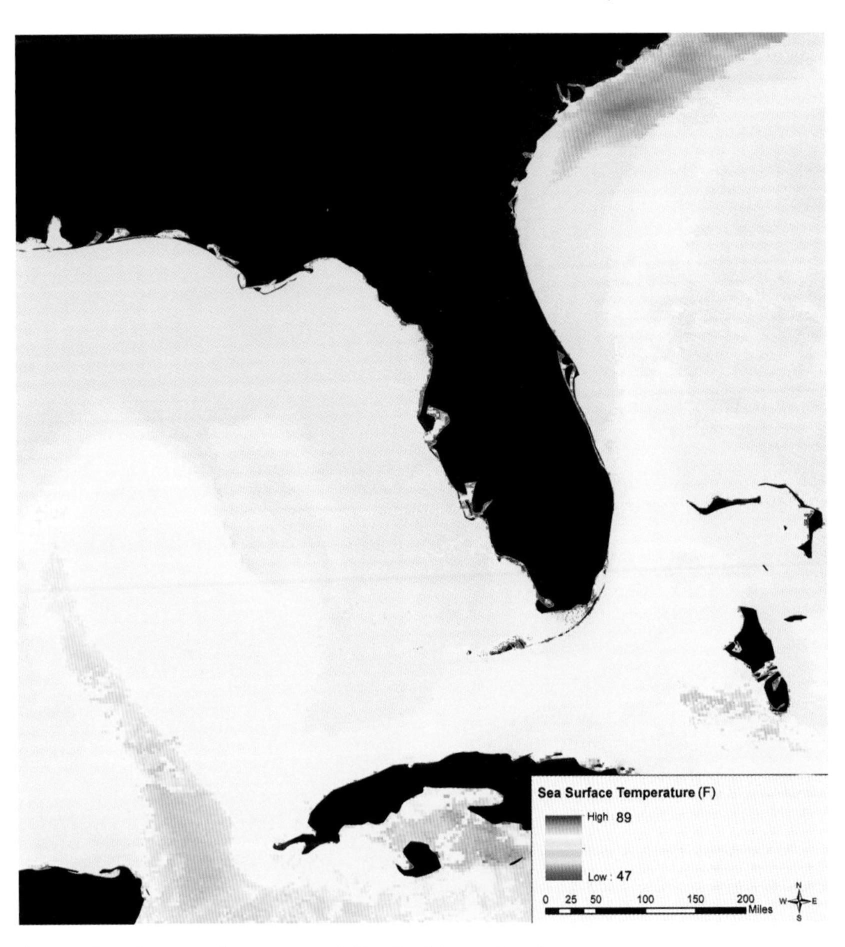

Sea-surface temperatures around the Florida peninsula

even in nearshore waters. Both the Loop and Gulf Stream Currents are dynamic, meandering back and forth along this general track. The presence of clear blue (oceanic) waters offshore of Destin in the panhandle is the result of landward excursions of the Loop Current. The clear waters along the Atlantic side of the Florida Keys are directly related to the movement of the Gulf Stream. Florida's southeast coast benefits most from the presence of the Gulf Stream. It brings warm oceanic waters close to land or pulls cold, nutrient-rich water from the depths and casts it shoreward. Paradoxically, the permanence of the Gulf Stream off southeastern Florida does not mean that things are stable. When the Gulf Stream meanders offshore, typically in the summer months, cold waters from the adjacent depths upwell (flow upward) onto the shelf. Cold-water upwellings are regular occurrences along the southeast coast, from Jupiter to Cape Canaveral. On some occasions we have seen dead or dying fishes that have restricted habitats and cannot go elsewhere, and the more-mobile species frequently move into shallow waters to escape the cold.

Major Habitats

The Florida marine environment is a haven. Many reasons exist for its high diversity of fishes, and certainly the range and extent of habitat types are a huge part of the story. Florida's coastline extends for 1197 mi and its tidal shoreline encompasses 2276 mi. The present-day continental shelf (out to the 600 ft depth contour) is 12,000 mi^2 off the Atlantic coast and 47,000 mi^2 off the Gulf coast. Fishes seek habitats that offer protection from predators, provide abundant food resources, and encourage reproduction. Habitats in the marine environment, like the sediments that are integral to their character, occur in mosaic-like patches across the seascape, from shallow to deep water. Habitats can be built around physical structures like rocks and sand, but some of the most extensive habitats are built on foundations of living organisms: seagrasses, marsh grasses, algae, mangrove trees, oysters, worm rock, and corals. The following sections discuss major habitat types found in Florida's inshore (estuarine and riverine), coastal, and shelf waters.

Seagrass Meadows

Seven seagrass species are found in Florida, with three making up most of the extensive meadows: turtle grass, manatee grass, and shoal grass. These species grow in mixed stands—ranging from small patches to vast meadows—in shallow water where light penetration, nutrient availability, and water flow are favorable. The density of their blades provides excellent hiding places for both small fishes and the invertebrates that are food for fishes. Some herbivorous (plant-eating) fishes feed on directly on seagrasses, as well as the algae on the blades, but

Top: Seagrass meadow, with turtle grass, in the Indian River Lagoon, Florida. *Bottom*: Seagrass meadow composed of manatee grass in Florida Bay.

Marsh grasses, Inglis, Florida

decaying seagrasses also contribute detritus (dead organic matter) and various compounds to local food webs. Seagrass beds serve as highly important nurseries for juveniles of many species, including snappers, grunts, porgies, groupers, and parrotfishes, as well as for all life stages of pipefishes, seahorses, and silversides.

Saltmarshes

While seagrass meadows flourish in subtidal environments, where the grasses almost always are submerged, saltmarshes develop within the intertidal range, where periodic submergence and emergence are the norm. In the northern portions of the state, saltmarshes are predominantly colonized by two emergent grasses: black needlerush and smooth cordgrass. These provide shelter for fishes and invertebrates during flood tides. Marsh grasses contribute material that ultimately is recycled into the food web. Saltmarshes usually are transected (cut across) by tidal creeks, which serve as reservoirs for animals during low tides and are aquatic roads in and out of the marsh grasses. Typical saltmarsh fishes include topminnows, pupfishes, livebearers, silversides, croakers, anchovies, and herrings.

Mangrove Shorelines

Mangrove trees thrive in shallow, warm, salty to brackish water, where sediments range from clean sand to sulfur-laden muck. The air temperature needs to be consistently above 60°F for mangroves to thrive. Three mangrove species (red, white, and black) are found in Florida, but only red and black mangrove trees provide structural habitat for fishes, since white mangroves grow above the high-tide line. Red mangroves have prop roots that emerge from branches or trunks and arch downward toward the substrate(material underlying a feature, such as the bottom of a body of water on which organisms live). Prop roots, combined with shade created by the overhanging branches and bows, make a cozy fish habitat. If you snorkel under a mature stand of red mangroves

Top: Red mangroves in the Indian River Lagoon. *Bottom*: Black mangrove pneumatophores, Key West, Florida.

Top: Nearshore reef, composed of wormrock, Hutchinson Island, Florida. *Bottom*: Inner-shelf reef off Venice, Florida.

in Florida Bay, you will notice a decline in light and temperature levels, as if you were swimming into a cave. The maze-like bundles of roots provide attachment sites for myriad algae and invertebrates and shelter for thousands of small fishes. Black mangroves do not have prop roots. Instead they send out clusters of vertical woody roots, having the diameter of drinking straws, that grow upward, 8–12 in off the seafloor. In addition to offering physical shelter, mangrove leaves, branches, and roots contribute detritus to food webs. Like seagrass beds, mangroves are a fertile nursery for the young of snappers, grunts, Atlantic goliath groupers, barracudas, and a laundry list of other species.

Oyster Bars

Oyster bars are formed by oysters settling and growing on hard substrate (often other oyster shells). The old shells cement together, forming clumps and clusters. Oysters and oyster bars are found in inshore waters where the salinities range between 6 and 12 psu, but because of temperature preferences, they are primarily a north Florida phenomenon. Oyster bars or reefs can be extensive in the lower portions of rivers or shallow, muddy estuaries, especially in northwest Florida. The small nooks and crannies of oyster bars host many small species, such as blennies, gobies, and clingfishes.

Hardbottom Reefs

Florida boasts the only true living coral reefs in North America. The Florida Reef Tract extends from Fowey Rocks (just south of Miami) in a gentle arc that parallels the Florida Keys and goes out to the Dry Tortugas. The Reef Tract is on the seaward edge of the same limestone platform that includes the Florida Keys, a series of emergent islands. Brain, star, elkhorn, and staghorn corals provide structure, modify currents, and process organic material. Add algae, sponges, and sea whips, and a complex fish habitat is created, replete with hiding places and feeding opportunities. The Reef Tract is composed of bank reefs and patch reefs. Bank reefs are semicontinuous structures in the arcing chain that sweeps west-southwest from the peninsula to the Dry Tortugas. Patch reefs, as the name implies, are comparatively small and isolated.

North of the Florida Reef Tract, reefs are linear outcrops and ledges that are actually ancient shorelines or sand dunes hardened over time. These reefs occur in nearshore (0–12 ft), shelf (15–150 ft), and shelf-edge (150–600 ft) depths. Reefs generally parallel the present-day shoreline, and the distance between reef lines is related to the width of the shelf. Depending on local environments (water temperature, depth, clarity, sedimentation, freshwater input), outcrops are colonized by algae, sponges, stony corals, and other organisms. Most reef-forming coral species drop out north of Jupiter on the east coast and north of the Florida Keys on the west coast. The west Florida shelf north of the Dry Tortugas

Top: Midshelf reef, Jupiter, Florida. *Bottom*: Outer-shelf reef, Florida Middle Grounds.

has a broad, relatively low-relief (composed of shallow, rather than taller structures), exposed hard-bottom covered with sponges, sea whips, and less-complex stony coral colonies.

Shelf-edge or deep reefs lie in water depths of 150 to at least 600 ft. Some shelf-edge reefs are low relief, while others are well-developed projections. Shelf-edge reefs lie in the twilight (mesophotic) zone, where limited sunlight reaches the bottom. Algae, a staple of shallower reefs, are scarce on shelf-edge reefs. Bush-like stands of ivory tree coral grow on a rocky spit that extends from Jupiter to north of Cape Canaveral.

The fish fauna on reefs is the most diverse of any Florida marine habitat. Just about every Florida fish family is represented there by at least a species or two. Among the more obvious and charismatic fishes are damselfishes, parrotfishes, wrasses, butterflyfishes, angelfishes, barracudas, snappers, grunts, and groupers, and the mini-fauna includes a plethora of gobies and blennies.

Artificial Habitats

Artificial habitats occur throughout Florida's inshore bays, lagoons, and rivers, in the form of bridges, docks, rip-rap (material lining shorelines, to try to reduce erosion), seawalls, channel markers, sunken vessels, and other debris. Offshore artificial habitats include ships, barges, plane wrecks, and other metal debris. Many of these structures are of accidental origin (e.g., ship and plane wrecks) or are in place for reasons other than attracting fishes (e.g., bridges, seawalls, docks). Artificial reefs are also deployed to compensate for habitat loss due to dredging or other construction. Most Florida coastal counties intentionally deploy artificial reefs to enhance habitat and provide increased fishing and diving resources. Although artificial reefs draw many of the species that occur on natural reefs, including commercially and recreationally important species, they do not duplicate the biodiversity found on natural reefs.

Sunken boat, Lake Worth Lagoon, Florida

Sargassum algae, southeastern Florida

Sargassum Flotsam
Floating sargassum "weed" is one of the few structural habitats found in the pelagic realm (open ocean, as opposed to nearshore or inland water). Sargassum (the general name for multiple species of brown algae in the genus *Sargassum*) drifts in parcels that range from fist-sized clumps to expansive rafts that reach football-field proportions. Each clump or raft provides shade and structure for fishes of all sizes. Sargassum is not the only floating, shadow-producing object in pelagic waters. Logs, trees, boards, shipping pallets, buoys, buckets, hawser lines, and other flotsam also attract fishes. Fishes that grow up in blue water need either wings to escape water-dependent predators or a good place to hide, and sargassum provides the latter. A number of species utilize sargassum as cover in the otherwise vast expanse of open ocean, including young dolphinfishes (mahi-mahi), jacks, triggerfishes, filefishes, sea chubs, driftfishes, and adult and juvenile sargassumfishes. Such adult predators as dolphinfishes, tunas, billfishes, Atlantic tripletails, and silky sharks are well-known foragers that prowl the edges and shadows of sargassum.

Softbottom
Soft, sedimentary seafloor may look barren, desert-like, and devoid of life, but a host of species reside on and within sedimentary bottom. Coarse, shelly, or rubbly sediments attract certain kinds of fishes, as do smooth, fine-grained, muddy bottoms. Fishes may bury themselves directly in the sediment, construct their own burrow, or share a burrow with another critter. Waves and tides can shape sediments into structural features, such as ripples, waves, or bars. Shoals form in areas where thick deposits of sand are perched on the hard, flat substrate that underlies the continental shelf. Open softbottom habitats can be very productive, supporting a considerable biomass of fishes.

High- and Medium-Energy Beaches
Surf is produced in areas where the sea waves hit the shoreline. Such areas in Florida typically have sandy beaches, but there are a few places where the surf smashes

Top: Open sand bottom, Palm Beach, Florida. *Bottom*: Sandy beach, Juno, Florida.

onto hard shoreline. High-energy beaches, such as those occurring on Florida's east coast, bear the brunt of giant ocean waves, while medium-energy beaches, like those along the Gulf coast, are subjected to smaller waves. Fishes inhabiting surf zones encounter significant water movements and must either be very good swimmers (e.g., herrings, bluefishes, mullets, jacks, sharks) or adept at keeping a low profile by hugging the bottom (e.g., drums, flounders, lizardfishes).

Regional Habitat Settings

To summarize how major habitats are distributed in different areas of the state, we use present-day physiography (physical patterns of features) and geology to divide the coastal environment into ecoregions. Each geographical region offers a different combination of habitats. From west to east, these ecoregions are Panhandle, Big Bend, Westcentral, Southwest, Florida Keys, Florida Bay, Southeast, Eastcentral, and Northeast.

Panhandle

The Panhandle ecoregion extends from Perdido Pass (at the Florida-Alabama border) eastward and southward to Cape San Blas. This stretch has a series of inshore bays lying behind barrier islands: Pensacola Bay, East Bay, Santa Rosa Sound, Choctawhatchee Bay, St. Andrews Bay, and St. Joe Bay. The Escambia, Perdido, and Ochlockonee Rivers deliver freshwater into these bays, creating estuarine conditions. The salinity of the coastal ocean depends on local rainfall, freshwater runoff, and river flow. Tides along the Panhandle ecoregion are diurnal (one high and one low per day). Tidal range averages about 1.2 ft. Seagrasses, marsh grasses, and oyster bars provide most of the structural habitat within these estuaries. Uninhabited shorelines are generally lined by marsh grasses. The Panhandle coast has moderate wave energy, and Gulf waters range from blue and clear (when the Loop Current sweeps shoreward) to green-brown (following the passage of storms and inland rain). Panhandle beaches are sugary white, with sandbars paralleling much of the coastline. Hardbottom along the beaches is limited to piers and rock jetties, such as those at St. Andrews Bay, Pensacola, and Destin. Offshore the seafloor is mostly level sand, with some natural (ancient shorelines or dune lines) and artificial reefs. The major topographic feature of the region is De Soto Canyon—remnants of an ancient seaway that once separated Florida from the mainland. The head of De Soto Canyon lies in about 250 ft of water and has stretches of hardbottom, with relief exceeding 30 ft in some segments.

Big Bend

The Big Bend ecoregion lies just east of Cape San Blas. This is a vast, shallow, low-energy corner of the Gulf of Mexico, extending from the Apalachicola River delta to the Anclote Keys. Tides here are diurnal, ranging from about 2 to 4 ft, and can be greatly influenced by local winds. Other than the island that borders Apalachicola Bay, true barrier islands are not to be found. The Apalachicola River creates a major sedimentary delta, but the Ochlocknee, St. Marks, Aucilla, Ecofina, Fenholloway, Steinhatchee, and Suwannee Rivers carry far less suspended sediment into the Gulf. The Big Bend coastline supports extensive saltmarshes, representing 30 percent of the state's entire saltmarsh acreage. Saltmarshes here consist mostly of black needlerush, with lesser amounts of smooth cordgrass. Oyster bars are extensive where salinity and water depth are favorable. Unlike anywhere else in Florida, seagrasses spread well out onto the shelf in the Big Bend. These seagrass beds support spotted seatrouts, red drums, sheepsheads, tarpons, cobias, and sharks. On the adjacent shelf, the sediment is mostly quartz sand overlaying a broad limestone platform, with depressions, springs, solution holes, and sparse hardbottom colonized by macroalgae, sponges, and octocorals. The Florida Middle Grounds, lying in 85–160 ft of water about 85 mi south-southeast of Cape San Blas. is the most conspicuous hardbottom feature of the west Florida shelf. It covers almost 600 mi^2, with irregular bottom that supports algae, sponges, sea whips, stony corals, and other attached organisms. Offshore of the Florida Middle Grounds are prominent shelf-edge reefs—Madison-Swanson, Steamboat Lumps, and the Elbo—lying in water depths of 150–400 ft. Red snappers, speckled hinds, roughtongue basses, scamps, wrasse basslets, tattlers, and bank butterflyfishes inhabit these deeper reefs. Gags aggregate (group together) to spawn at Madison-Swanson and Steamboat Lumps.

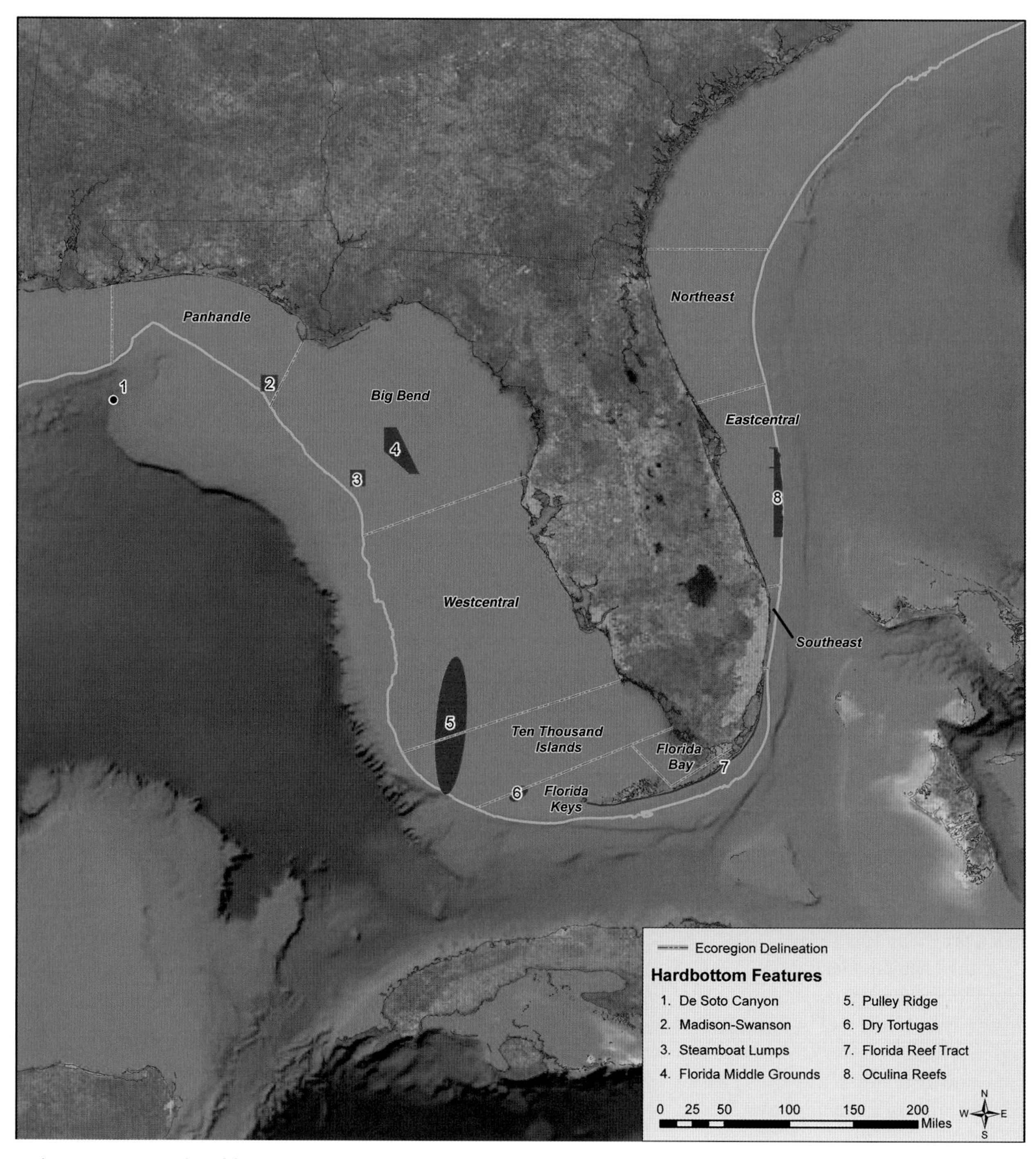

Major ecoregions and reef features around Florida

Westcentral Florida

The Westcentral Florida ecoregion extends from the Anclote Keys south to Cape Romano. Here wave energy is moderate and tides are mixed (two unequal high and low tides per day), with a range of about 2 ft. Major estuaries include Tarpon Springs, the Tampa Bay system and all its tributaries and sub-bays, Sarasota Bay, Charlotte Harbor, Pine Island Sound, and Estero Bay. Rivers emptying into the region include the Anclote, Hillsborough, Alafia, Manatee, Little Manatee, Peace, Myakka, and Caloosahatchee. Barrier islands extend along the coast, from Anna Maria Island southward to Sanibel Island. Seagrasses, marsh grasses, mangroves, and oyster bars flourish behind these islands. Mangroves make their first appearance as the dominant marsh vegetation in this region. On the shelf, extensive hardbottom areas with sponges, stony corals, and algae often are covered by a veneer of sand about 1 in or more thick. Some areas have outcrops or ledges with 5–10 ft relief. Larger sponges and star corals protrude above this mobile sand sheet, providing small islands of structured habitat. Herrings, whitings, Florida pompanos, and red drums occur along the shoreline, and the extensive seagrass meadows contain pinfishes, spots, silver perches, tomtates, white grunts, lane snappers, and toadfishes.

Southwest Florida

The Southwest Florida ecoregion, comprising the Ten Thousand Islands, encompasses a broad, shallow body of water bounded by Cape Romano to the northwest and Cape Sable to the southeast. This is a low-energy coastline, with mixed tides having a range of over 3 ft. The shoreline from Marco Island southward is undeveloped and lined with mangroves. The water is generally tannin stained (dilute brown) inside the mangrove islands but clears along the open Gulf of Mexico, depending on sea conditions. The sediments are sand and mangrove peat. The shelf in this area is mostly coarse carbonate sediment, with areas of exposed hardbottom containing coralline algal nodules. Offshore and south along the shelf edge is Pulley Ridge, a 70 mi long, north–south trending reef feature in 200–275 ft depths. Good water clarity allows lettuce corals and several algae species to be common on this deepwater ridge, which is composed of ancient barrier islands. Inshore of Pulley Ridge are vast areas of level seafloor covered by carbonate sediment, sponges (among them, large loggerhead sponges), and sea whips, including an area known as the Tortugas Shrimp Grounds, an historically important trawling ground for shrimps.

Florida Keys

The Florida Keys are the emergent part of a large limestone bank that arcs gently from the eastern tip of the peninsula to the Dry Tortugas, a distance of over 200 mi. The Keys have no rivers, so, in the absence of outflow, the water is generally blue to aquamarine and clear. Tides are mixed, with a 3 ft range. The sediment is carbonate and tend to be muddier on the Gulf of Mexico side of the Keys (which, from Islamorada north to Key Largo, is part of Florida Bay). Red mangrove trees hug much of the shoreline, and sandy beaches are few and far between. On the seaward edge of the bank, extending in semicontinuous fashion from Fowey Rocks to the Dry Tortugas, lies the Florida Reef Tract, the only living coral reef in the continental United States. Reef-building brain, star, elkhorn, and staghorn corals provide structure, modify currents, and process nutrients and organic material. Sponges, soft corals, sea whips, and algae add structural complexity, offering abundant hiding places and feeding opportunities. Patch reefs, seagrass meadows, and areas of hardbottom with solution holes, sponges, algae, and sea whips are located between the Reef Tract and the emergent Keys. An extensive limestone platform called Pourtales Terrace lies offshore of the Reef Tract, in water depths as shallow as 600 ft. This platform has relief features as well as large, deep solution holes, the latter reflecting an earlier period with lowered sea levels. The best-know topographic feature is the Hump, offshore of Islamorada. The Gulf Stream Current runs along the outside of the Florida Reef Tract, bringing oceanic conditions close to shore.

At the northern end of the Florida Keys ecoregion is Biscayne Bay, a diverse inshore embayment lying in the shadow of downtown Miami. The bay is 32 mi long, with a maximum width of 8 mi and water depths ranging from 6 to 12 ft; its bottom is a mixture of carbonate sands and low-relief hardbottom. Large loggerhead sponges and sea whips grow on the hardbottom. Shorelines predominantly contain mangroves, and extensive seagrass meadows grow in shallow-water areas of Biscayne Bay and Card Sound. The northern portion of the Florida Reef Tract parallels Elliot Key, Old Rhodes Key, and Soldier Key, which collectively enclose Biscayne Bay from the east.

Florida Bay

Florida Bay is a shallow expanse of water (850 mi^2), averaging 5 ft deep, bounded by the Florida Keys to the southeast and the tip of the peninsula to the north. Tides in this ecoregion are mixed and range from 2 to 6 ft. Freshwater input historically was sheet flow (a shallow, slow-moving layer of water) from the Everglades (lying to the north), but currently much of this sheet flow has been diverted or interrupted by human actions. The bottom of Florida Bay has a web-like pattern, with the banks forming basins (sometimes called lakes). Banks provide substrate for mangrove trees, extensive seagrass meadows, and (sometimes) sponges and sea whips. A plethora of mangrove islands dot much of the Bay. Beyond these islands and toward the open Gulf, an expanse of shallow seafloor is covered with low-relief hardbottom that supports sponges and sea whips, and occasional stony corals.

Southeast Florida

The area from Fowey Rocks northward to Jupiter encompasses the Southeast Florida ecoregion. The coastline is considered high-energy, but it is protected from large swells to some extent by the Bahama Banks, lying just 60 mi to the east. The continental shelf is very narrow regionally, and the western edge of the warm Gulf Stream comes closest to shore from Boynton to Palm Beach, giving this region some of the clearest ocean water in the state. The inshore waters are limited to the man-made channel of the Intracoastal Waterway, oriented south to north, which connects Biscayne Bay with Lake Worth Lagoon at Boynton Inlet and ends 15 mi to the north, near Lake Worth Inlet. The tides are semidiurnal (two high and two low tides each day). The shorelines of the inland waterways are lined with mangroves, seawalls, and docks. Limited beds of seagrasses grow where conditions are favorable. Freshwater input is directly related to regional rainfall and inflows from a series of man-made canals and water-control structures draining the western uplands, including the eastern edge of the Everglades. Discontinuous nearshore reefs occur along the shoreline, in intertidal to about 20 ft water depths, from Miami to Jupiter Inlet. In some areas worm rock colonizes the hard substrate, creating more surface area and habitat complexity. Reefs are mostly north–south ledges, representing ancient shorelines or sand dunes hardened over time. These reefs parallel the present-day shoreline, in water depths of 30–150 ft, along the narrow shelf from Miami to Jupiter. North of there, the distance between ledges increases, commensurate with the width of the widening shelf. These outcrops are colonized by stony corals, sponges, algae, and other organisms. Staghorn coral occurs northward to an area off of Ft. Lauderdale. Artificial reefs and wrecks also have been placed along the coast.

Eastcentral Florida

The Eastcentral Florida ecoregion extends from Stuart north to Cape Canaveral. The gem of this region is the Indian River Lagoon, which extends from the Loxahatchee River in Jupiter northward to Mosquito Lagoon in Titusville, a distance of 130 mi. The lagoon has inlets at Jupiter, St. Lucie, Ft. Pierce, and Sebastian. Tides flow in and out of these inlets on a semidiurnal rhythm. Undeveloped shorelines of the lagoon are fringed with mangroves, emergent marsh grasses, and sand beaches. The bottom of the lagoon is mostly sand. The coastline facing the Atlantic Ocean has high-energy beaches, where groundswells from the north or east travel unimpeded by the Bahama Banks. Nearshore reefs run parallel to sandy coastal beaches. These reefs are especially prevalent off Satellite Beach, Vero Beach, Hutchinson Island, and Jupiter Island. Worm-rock reefs are most extensive in this region and are particularly well developed on the shore of Hutchinson Island. The continental shelf is considerably wider here than it is to the south, but it is narrower than the section found in the Northeast ecoregion. The carbonate portion of the sediments changes to mostly quartz and shell. Reefs are mainly paleoshorelines; outcrops follow the depth contours and are farther from shore as one goes north. The Gulf Stream also flows farther offshore, bluewater intrusions close to shore are less common, and the water ordinarily is turbid. The outer-shelf portion supports the unique Oculina coral banks. These structures are formed by *Oculina varicosa*, a pearly-white, branching coral that grows in colonies (thickets) reaching 5–7 ft high and covering up to acres of seafloor. These mounds and thickets grow on an underlying hardbottom ridge, at depths of 200–330 ft, that extends from offshore of St. Lucie Inlet to as far north as Jacksonville.

Northeast Florida

The Northeast Florida ecoregion runs from Cape Canaveral to the Florida-Georgia border. The major inshore features are the natural lagoons and the man-made Intracoastal Waterway, which connects with the southerly Mosquito Lagoon and extends to the northerly St. Johns River estuary. Tomoka River, Moultrie Creek, and Pelicer Creek drain into the Intracoastal Waterway along its route north. Marsh habitats, consisting of smooth cordgrass and black needlerush, and oyster bars dominate the undeveloped shallows. The two major inshore systems in this region are the St. Marys and St. Johns Rivers. When compared with other estuaries in the southeastern United States, Florida's St. Johns River has been labeled "unusual" or "atypical" with respect to its physical, chemical, and ecological characteristics, which include high alkalinity, a low-elevation gradient, low freshwater inflow, an extensive upstream tidal influence, and an abbreviated salinity gradient. Although saline water regularly extends to about 20 mi upstream of the mouth of the St. Johns River, numerous marine fishes are found well above this point. Coastal areas in this ecoregion are mostly high-energy sandy beaches, but rock outcrops occur near Matanzas Inlet and Washington Oaks Gardens. Offshore the shelf is broad, about 75 mi wide, and overlaid with mostly quartz sediments. Hardbottom is sparse, found as isolated ledges and outcrops in depths of about 250 ft.

Fish Biology

Form and Function

The generalized fish shape (called fusiform) is streamlined, with a pointed or rounded snout, terminal mouth, and forked tail. A fish's mouth is terminal when it is at the front of the head, subterminal if it is below the front of the head, and superior (upturned) if the lower jaw extends beyond the upper jaw. Modifications of the basic fish shape produce a body that may be variously flattened (compressed or depressed), truncated (shortened), or attenuated (elongated) to produce the diversity of forms we see in living fishes. Compression results in the side-to-side flattening observed in snappers, butterflyfishes, spadefishes, flounders (flounders are compressed, they just lay on their sides), and many others. Depression produces fishes that are pancaked from top to bottom, such as angel sharks, skates, whiptail stingrays, and batfishes. Truncated species (e.g., triggerfishes, filefishes, butterflyfishes, and ocean sunfishes) are rectangular or disc shaped. Attenuated fishes are stretched into long, cylindrical tubes (trumpetfishes and needlefishes) or serpentine shapes (eels and pipefishes).

Fins help stabilize and orient the body in relation to water movement or to the substrate. Most fishes have one or two dorsal fins on the back, an anal fin on the underside behind the cloaca (vent), and a caudal (tail) fin. Collectively these are called median fins. Pectoral and pelvic fins are paired and may be either widely separated or close together on the body axis, with the pectorals usually positioned higher up on the body. Fins

Parts of a fish (dog snapper)

are fleshy membranes supported by spines and/or rays, which, in turn, are supported by small bones embedded beneath the musculature. Fin spines are stiff, unbranched rods that usually taper to a sharp point. Fin rays are flexible, segmented rods that may be branched near the ends. Primitive fishes (e.g., tarpons, herrings, anchovies, and eels) lack stiff spines. The fins of more-advanced species fins tend to be divided into spiny portions and soft ones (rays). In most fishes the caudal fin is the primary source of propulsion, especially for short bursts of speed. Tail fins come in a range of shapes, including forked, lunate(crescent-shaped), emarginate (shallowly notched), truncate (having even ends), rounded, and lanceolate (tapering to a point).

A fish's skin usually is covered with tough scales and a coating of mucous (slime). Mucous reduces friction, contains antimicrobial substances, and (in some species) even includes toxins. The skin hosts pigment cells and the compounds that give fishes their color. Colors are controlled by the migration of pigment within these cells, called chromatophores. The iridescent silvery color on fishes is caused by a substance called guanine, which is actually a waste product carried from the blood to the skin.

Scales are dead material laid down by living cells deep within the skin. Scales orient toward the tail and overlap like roof shingles. They grow in rows and keep pace with fish growth by getting bigger themselves. This allows biologists to use regular marks on scales to estimate the age of an individual fish. Scales vary greatly in shape and size. Porcupinefish spines, trunkfish plates, batfish bucklers, and surgeonfish spines (scalpels) are all modified scales. Sharks and rays have skin so rough that it feels like sandpaper. This is because the covering on the skin has tiny versions of teeth, called dermal denticles.

Fishes swim by moving the body and the tail back and forth, undulating the median fins, or flapping the pectoral fins like wings. Swimming modes span a range of styles, many of which depend on the body type and genealogy of a particular species or family. For example, eels oscillate (swing back and forth) over the entire length of the body when swimming. Many sharks also swim this way, due to their flexible, cartilaginous vertebral columns. Continuously swimming fishes (e.g., jacks, mackerels, and tunas) rapidly oscillate the tail and stiff posterior third of the body. Trunkfishes (with rigid bodies) move by only use their pectoral and caudal fins. Some species undulate (wave) their median fins (soft dorsal and soft anal fins) as their primary means of propulsion. Triggerfishes and filefishes are the most common purveyors of this swimming mode, but black brotulas are one of its most striking examples.

In several groups of fishes, the pectoral fins are used for locomotion. Eagle rays, mantas, and cownose rays swim by flapping their large pectoral fins in a bird-like fashion. Skates, whiptail stingrays, and round stingrays swim by oscillating the outer margin of their pectoral fins. Other species, most notably wrasses and parrotfishes, use their pectoral fins to scull through the water. These species swing the fins forward, flattened in the horizontal plane, and then turn them through about 90 degrees and pull them back, providing forward propulsion. You can pick parrotfishes and wrasses out of the crowd on a coral reef by watching for individuals swimming with their pectoral fins.

Swimming forward is certainly important, but holding position, feeding, avoiding predators, and other activities require a range of small-scale maneuvers and orientation. These subtle movements are achieved with coordinated adjustments of the body and fins. Fins help with pitching, rolling, and yawing, and they are used to aid fishes in braking during a lunge forward, maintaining position in a current, or moving vertically in the water column (a conceptualized column in a body of water, extending from the water surface to the bottom sediments).

Other adaptations that facilitate swimming are drag reduction (lessening the force of friction) and buoyancy (increasing or reducing density). Fishes reduce drag functionally by coating their bodies with mucous and behaviorally by schooling, where individuals look out for themselves by following their nearest neighbor. Another strategy primarily employed by some jacks (pilotfishes are the masters) is to ride in the slipstream of sharks, rays, or other large-bodied fishes as they swim.

Density is a basic problem for fishes: sinking or floating when they don't want to can mean life or death. Most fishes have a swim bladder that holds gas, making it an internal life preserver of a sort. Gas volume is regulated by moving it in solution through the blood. Some fishes (sharks and rays) lack a swim bladder and have to constantly swim to maintain lift and keep from sinking.

Reproduction

Reproduction is a product of courtship, mating, spawning, and embryonic development. Courtship is a complex behavior that signals mutual interest in mating. Mating is an actual sexual act where at least one female and one male come in close contact and release their gametes (reproductive cells) into the surrounding water or, in the case of internal fertilization (for sharks, rays, and a few other fishes), sperm is transferred from male to female. Spawning is the release of either gametes or developing young into the environment. It is synonymous with mating in species with external fertilization, but where fertilization is internal, the release of young occurs at some postfertilization stage. Embryonic maturation occurs within the female prior to birth in internal fertilizers and in the water column (as larvae) in external fertilizers. During courtship, marine fishes signal prospective mates with species-specific movements, postures, or color patterns. Signaling helps species attract and properly recognize members of their own kind. Attraction is no problem for schooling species, where males and females travel together. But in solitary species it is often the male that establishes a breeding territory or spawning area and then signals females to join him. In most cases potential mates exchange "I'm interested" signals; once these signals are synchronized, mating ensues. Although most fishes use visual signals (e.g., color patterns and postures), sound and smell may come into play in some species. Damselfishes, toadfishes, drums, and croakers are examples of species that use sound to signal prospective mates. Courtship behavior presumably evolved in part to prevent cross-breeding (hybridization) between closely related species, which is a wasted effort, since hybrid offspring are usually not fertile.

External Fertilization

Most fishes engage in external fertilization, where eggs are released into the water column, onto the substrate, in prepared nests, or on other objects. The most common mode for marine fishes is to broadcast buoyant eggs and sperm into the open water column. Even bottom-oriented fishes may spawn upward in the water column, presumably to lessen the impact of bottom-dwelling egg predators. Species that broadcast pelagic eggs include tarpons, herrings, eels, scorpionfishes, snooks, snappers, jacks, tunas, and mackerels. These species have high fecundity (producing many eggs) and low larval survival, and they exhibit no parental care.

Some species have nonbuoyant eggs that sink to the bottom, where they may adhere to vegetation or sediments. Topminnows, pupfishes, and filefishes release eggs onto the substrate or vegetation. The eggs of silversides, flyingfishes, and halfbeaks have wispy, sticky threads that adhere to drifting plants or stationary objects. Other species deposit eggs directly on specially prepared sites (nests); once these eggs are fertilized, one or both parents guard the eggs until hatching. Damselfishes prepare a site by nipping away algae and epibiota (organisms that live on the surface of rocks and other hard substrates) until bare rock is exposed. Usually the male chooses and prepares the spawning site and then vigilantly defends it. Once a receptive female responds and lays eggs, the male fertilizes them and guards the eggs until they hatch and join the plankton as larvae. Male triggerfishes build multiple pit nests in the sand, where the males attract females to mate; the females guard the nests until the eggs hatch. Blennies deposit their eggs in enclosed shelters (e.g., old shells), under rocks, or in empty barnacles. Bicolor damselfishes often lay eggs in vacant snail shells. Toadfishes deposit eggs in enclosed shelters, oyster shells, discarded cans, or other debris that offers security. In a few species, a parent carries the fertilized eggs until hatching. Most notable among these are the pipefishes and seahorses—the males have brood pouches adapted for holding and incubating the eggs. The other prevalent strategy is mouth brooding, where adults loosely hold the developing fertilized eggs in their mouths, thereby protecting them from predators. Marine catfishes, jawfishes, and cardinalfishes are the best-known purveyors of this method.

Mating systems of spawning fishes include pair spawning, group spawning, haremic spawning, and lek formation. Probably the most common mode is pair spawning, which involves only two individuals. Some species form monogamous pairs for life, but in most species reproductive pairings are fleeting encounters. Some species aggregate during a specific spawning period, which may be single or multiple events over days, weeks, or even months.

Although most fishes have discrete sexual identities, some species are hermaphroditic: individuals are both male and female. The way hermaphroditism is manifested depends on the species and on social and environmental factors. Hermaphrodites are either simultaneous or sequential. Simultaneous hermaphrodites have male and female sex cells ripe and ready in their bodies at the same time, while sequential hermaphrodites are born as one sex and change into the other during their lifetime. When mating, individual sequential hermaphrodites take on either the male or female role; simultaneous hermaphrodites, such as hamlets or dwarf sea basses, may even switch gender roles with each mating bout. Sequential hermaphrodites that begin life as females are called protogynous hermaphrodites, and those that begin life as males are known as protandrous hermaphrodites. Examples of protogynous hermaphrodites include groupers, wrasses, and parrotfishes. Protandry is less common, but snooks are a well-known example—all individuals are born as males. Simultaneous hermaphrodites include hamlets, sea basses in the genus *Serranus*, and the ultimate hermaphrodite, the mangrove rivulus—this species mates with itself!

Fertilized eggs metamorphose into larvae that, in most species, are planktonic (drifting in the water column). This pelagic stage lasts for days to months, depending on the species. Early-stage larvae are essentially passive particles transported by prevailing currents until their backbones stiffen and a tail forms; from this point on, larvae can at least swim vertically. Swimming larvae sense chemical, physical, visual, and even sonic cues to help them find a suitable settling habitat. Ambient (generally surrounding) environmental conditions act as sieves, filtering species by their individual requirements and preferences. When the tiny transparent larvae transform into pigmented juveniles, a particular habitat is often chosen. For many, this nursery habitat is structured and is in shallower water than where the adults normally reside. Depending on currents, larvae are either swept away from or circle back to the spawning area. The duration of the pelagic stage may even dictate the geographical distribution of adults.

In a considerable number of species, pelagic larvae transform into juveniles, which are essentially miniature replicas of the adults. In other species, the juveniles are oddly different from adults, which has led to scientists erroneously giving these peculiar larvae a new scientific name. After this transformation, the young settle down to the bottom (if they are bottom dwellers as adults); continue living in the water column (if pelagic as adults); or settle up, as some species do that are attracted to flotsam and sargassum. A common theme for fishes is settlement into shallow nursery areas, usually a structured habitat such as a seagrass meadow, oyster reef, nearshore reef, mangrove shoreline, or estuarine saltmarsh. After a discrete period (usually less than a year) in the nursery area, juveniles gradually move to the deeper-water haunts of the adults. Grunts, snappers, sea basses, groupers, and porgies are among the groups that depend on shallow-water nurseries. Within each of these groups there also are species that settle and grow up near the spawning grounds, rather than jumping between different habitats as if these were stepping stones.

Internal Fertilization

Sharks, skates, rays, livebearers, and brotulas copulate and have internal fertilization. Males have a specialized intromittent organ (a modification of the pelvic fin) that delivers sperm into the female. Brood sizes are small but survival is high, thanks to the mother's protection and nurturing. When offspring are released from the female, they are well-developed and active miniature adults, with good sensory reception, rendering them capable of avoiding predators and finding their own food.

Feeding Ecology

Marine fishes consume a range of items to sustain themselves. Many species are ecologically flexible, opportunistic generalists when it comes to feeding, but others specialize in specific items. It is not unusual for species to change their diets over the course of their lives, as mouths get bigger and swimming prowess increases. In addition to generalists, there are four basic feeding types among Florida's marine fishes: detrivores, scavengers, herbivores, and carnivores.

Detritivores
Detritus is composed of silt and decaying organic material (usually plant matter consisting of bits of cellulous) covered by a film of bacteria, fungi, and microbes that provide nutritive value. Detritus accumulates on the bottom, as a fragile mat or slurry (a more-fluid mixture), in estuaries and areas where water circulation is low. The most notable detritivores are mullets, gobies, and some topminnows.

Scavengers
Many fishes scavenge for food. Scavengers use their senses of smell or hearing to locate dead or dying organisms. Sharks often are thought of as scavengers, but they and many other species popularly dubbed "scavengers" are probably just being opportunistic. Both hardhead and gafftopsail catfishes are noted scavengers. You can find hardhead catfishes by the hundreds at any reasonably busy fishing marina where fish scraps are regularly discarded.

Herbivores
Herbivores do not have to chase or capture their prey. No highly evolved behavior is needed, only structural adaptations: small mouths, nipping teeth, and long alimentary tracts to help the lengthy process required to digest plant material. Parrotfishes, surgeon fishes, sea chubs, and some porgies are prominent members of this group. Most herbivores may be further classified as either grazers or browsers. Grazers feed by scraping, rasping, or biting at algal-covered substrate and then sucking in the loosened materials. With their beak-like teeth, parrotfishes exemplify this feeding mode. Browsing is biting directly on the leaves or stems of plants with various mouth and tooth configurations. Surgeonfishes, damselfishes, blennies, and some porgies fall into this category. The third specialized category of herbivory is phytoplanktivory: feeding on phytoplankton (microalgae that require sunlight to live and grow). Few Florida marine fishes are strict phytoplankton feeders, but gizzard shads and menhadens come close.

Carnivores
Most of Florida's marine fishes are carnivores in some form or another, making their living by eating other creatures. Carnivory spans a range of feeding modes: benthivory (feeding on softbottom invertebrates), planktivory (feeding on zooplankton—tiny animals—suspended in the water column), and piscivory (feeding on fishes). Some species mix and match modes during their lifetimes.

Benthivores consume worms, clams, mollusks, crustaceans, echinoderms, and other invertebrates that usually live embedded in sediments or sheltered in rocks, plants, or other substrates. Bottom feeders employ a variety of tactics to obtain their meals. Small sea bass species, some wrasses, snappers, mojarras, and some grunts simply grab their prey directly from the substrate. Mojarras use their protrusible (able to be pushed out) mouths to pluck organisms off the surface of the sediment without looking down—an advantage when keeping an eye on would-be predators. Pinfishes, numerically dominant in seagrass habitats, pick small invertebrates from the surface of seagrass blades. Some bottom feeders will disturb the sediment by various means to dislodge prey. For example, bonefishes, stingrays, trunkfishes, and some filefishes jet water through their gill openings and then out of the mouth, exposing buried invertebrates with their versions of a waterpik! Feeding whiptail stingrays leave sand flats pocked with pits dug during these jetting sessions. Feeding bonefishes produce telltale muds while jetting into the sediments of marl and seagrass flats in the Florida Keys. The actions of sediment-disturbing foragers benefit a host of followers seeking the spoils. It is fairly common to see a convoy of wrasses, mojarras, yellow jacks, bar jacks, and/or goatfishes following stingrays as they feed over sandbottom. Goatfishes use chin barbels (fleshy filaments) under their mouths to disturb the sediment and displace invertebrates hiding in the surface sediment layers. Some bottom feeders simply shovel in mouthfuls of sediment, sort out the edible fraction by sifting, and eject the sediment through their gill openings.

A specialized form of picking invertebrates from the substrate is parasite cleaning, where small fishes climb around the bodies of larger host (client) species in search of attached parasitic crustaceans or flatworms. Adult yellowprow and neon gobies are the only obligate cleaner species (i.e., restricted to this mode of feeding) found in Florida waters, but juveniles of myriad species—including spottail pinfishes, porkfishes, leatherjacks, Spanish hogfishes, spotfin hogfishes, and blueheads—happily clean parasites from larger fish clients. This phenomenon is most obvious among reef fishes, but it occurs in estuaries and other habitats as well. Hosts appear to benefit from this, in part as tactile stimulation (a massage of sorts) and partly as relief from external parasites. Cleaner fishes also remove necrotic (dead) tissue from wounds, facilitating healing. We include observations of parasite cleaning by juveniles of three species not previously reported to engage in such behavior: black brotulas (page 84), permits (page 166), and high-hats (page 206).

The aforementioned feeding modes pertain to motile (able to move around) invertebrates. A number of bottom feeders, however, focus on invertebrates attached to hard substrate (e.g., sponges, corals, sea whips, hydroids). The specialized long, forceps-like jaws of butterflyfishes allow them to feed on the soft parts (polyps) of corals, soft corals, and hydrozoans. Angelfishes eat sponges; sheepsheads chisel out oysters and barnacles; eagle rays consume oysters attached to the bottom or to pilings; and cownose rays are known to munch on clams.

Depending on their body size, zooplankton feeders pluck individual organisms from the water column or filter vast quantities of plankton. Larvae of many species of fishes start out as zooplankton feeders but then shift to another class of prey (depending on the species) as they grow. Zooplankton feeders generally have small protrusible mouths, slender bodies, and eyes on the sides of the head. These adaptations favor tireless swimming into the current, picking individual plankters (the tiny organisms that constitute plankton) from what can be an infinite cafeteria line of items. Because zooplankton feeders tend to converge in size, shape, and behavior, individuals from different species frequently school together while feeding. For example, round scads and Spanish sardines regularly intermingle in large schools. Juvenile tomtates, striped grunts, school basses, and bonnetmouths typically form mixed aggregations. Jellyfishes are considered plankton, and a few fish species feed exclusively or opportunistically on them. Atlantic spadefishes and ocean sunfishes are examples of species that eat jellyfishes. Other fishes consume plankton by swimming with their mouths open, like a net. The largest of fishes—the whale shark—is a planktivore, well equipped to engulf entire fields of zooplankton, including fish eggs and larvae. Giant mantas also actively feed on plankton.

Fish-eating carnivores have to ambush, lure, stalk, or chase their elusive prey. Ambush (lie-in-wait) predators are some of the most engrossing of fishes. The form and function of this group entails a superior (upward-facing) mouth, equipped with supple jaw muscles capable of lightning-fast reactions, and eyes set on top of the head. Scorpionfishes and lizardfishes are classic examples of ambush predators. Scorpionfishes lie motionless, blending in with algae-covered rocks. Patience is a virtue with these species, and patience is usually rewarded. Some fishes even have algae growing on their skin. Another ambush strategy is for a fish to bury its body in the sediments. Lizardfishes, stargazers, and various flatfishes exemplify this strategy. Some species carry ambushing a little further by actually luring unsuspecting prey into their strike zone. Frogfishes—equipped with a pole and a lure derived from the dorsal fin—are masters of luring behavior. Twitching and waving the lure with the pole—similar to a majorette in a marching band—frogfishes readily attract unsuspecting prey. In case visual attraction is not strong enough to do the trick, recent research suggests that cells in the lure of striated frogfishes also secrete a chemical attractant. Southern stargazers are another of the ambush predators that lie buried up to their eyeballs in the sand. This species produces a weak electrical current from a bank of modified cells behind the eye that may attract curious fishes to their demise.

Rather than patiently waiting for prey to come within striking distance, stalkers actively but stealthily approach their quarry. Fish-eating stalkers move ever so slightly and generally are either colored to match the background or are disruptively colored (patterns that conceal the shape and outline of the body). Some species even mimic the color patterns of less-harmful species to gain closer access to prey. Groupers, great barracudas, snooks, cornetfishes, and Atlantic trumpetfishes are classic stalkers. Some predators do not rely on surprise but simply chase down and eat their prey. This strategy is employed by mackerels, tunas, jacks, needlefishes, and requiem sharks. These species may smell or hear potential prey from a great distance, but vision takes over once they are in close range. Sharks can detect sound, as well as weak electrical signals, and they certainly have a keen sense of smell. Needlefishes often launch themselves into schools of anchovies. Little tunnies herd and chase schools of Spanish sardines, cornering them against structures like piers, wrecks, or reefs. Opportunistic snooks, barracudas, blue runners, crevalle jacks, and other piscivores take advantage of these events, grabbing individuals that get separated from the fleeing pack. Thus predatory activities may cascade into multiple opportunities for many species in local habitats.

Defense from Predators

The discussion regarding feeding tactics suggests that predators have evolved a suite of complex behaviors and structures to help locate, catch, and ingest their prey. Natural selection is a powerful creative force that has given fishes a variety of feeding adaptations, but it also has allowed prey fishes to develop counterstrategies, such as antipredatory morphology, coloration, and behavior.

The most fundamental way to avoid being eaten is to hide. Many species, especially those living on structured habitat, conceal themselves within the habitat matrix, seeking shelter in holes, crevices, caves, or other gap-like hiding places among living layers of encrusting corals, worm rocks, oysters, barnacles, sponges, and hydroids. On featureless sandbottom, fishes either construct their own burrows (jawfishes, tilefishes, garden eels) or share burrows constructed by something else (gobies associating with snapping shrimps). Searobins, some scorpionfishes, cusk-eels, razorfishes, and snake eels bury themselves in the sand or mud. Pearlfishes win the prize for the most secretive hiding place of all—this species resides within the body cavity of a living sea cucumber. In the pelagic realm, many juvenile forms of a number of fishes associate with drifting sargassum algae. Other pelagic dwellers hide among the stinging tentacles of siphonophores and jellyfishes.

Camouflage and countershading are ways in which fishes have evolved to blend in with the water column or the substrate by using variations of color and pattern. Countershading—silvery or white underneath, dark above—is a fundamental pattern utilized by open-water species such as tunas, mackerels, and sharks, as well as herrings, sardines, and mullets. Camouflage (background matching) helps conceal individuals from predators by allowing them to disappear in plain sight. Filefishes, pipefishes, scorpionfishes, and flatfishes exhibit coloration patterns that allow them to fade into the background by replicating the color and texture of their habitat. Seahorses match the coloration of their substrate and even grow fleshy skin projections that resemble small encrusting invertebrates. Planehead filefishes often look like the sponges or algae that are endemic to their home ranges. Juvenile Atlantic tripletails and spadefishes resemble leaves, not only in shape and color but also in their movement patterns. Juvenile spadefishes roll back and forth in shallow water, even when the water's surge is minimal or nonexistent. Many nocturnal species are red in color, especially those residing on reefs. Red color is filtered selectively from the spectrum with increasing water depth, and it is often called the "gray of the sea," as it looks drab gray or brown in water depths greater than about 15 ft. Cardinalfishes, squirrelfishes, and bigeyes are generally nocturnal and predominantly red-colored.

While hiding is an efficient means to avoid being eaten, some fishes live in habitats that offer little refuge, so other strategies have evolved, involving anatomical and structural adaptations. Puffers have the ability to inflate themselves when threatened, and the related porcupinefishes additionally are loaded with sharp body spines, making them floating pin cushions. Puffers also have poisonous compounds in their body mucous, as do soapfishes. Spines, which are the structural basis of many fins, also serve as defensive structures; deep-bodied fishes with stiff spines (e.g., butterflyfishes, angelfishes, porgies, and spadefishes) deter predators simply by having these body parts. Stingrays bear poisonous, serrated tail spines, and poison glands also are associated with the dorsal spines of scorpionfishes. And, of course, nothing beats excellent swimming speed and maneuverability.

Distribution And Zoogeography

Zoogeography refers to the intersection of the geographical ranges and ecology of organisms and the oceanography, physiography, and geological history of the region. Florida's marine waters support a diverse mix of species having warm-temperate and tropical origins. The present-day fish fauna is weighted heavily toward the tropical component, members of which have been emigrating from southerly (Caribbean) regions since the low sea level of the last glacial period, about 18,000 years ago. During this same period, warm-temperate species that may have thrived in a cool-water Florida sea have been steadily retreating back north. Species in the latter group still residing in Florida include sea lampreys, Atlantic and shortnose sturgeons, American and hickory shads, blueback herrings, oyster toadfishes, goosefishes, Atlantic silversides, mummichogs, northern pipefishes, northern and striped searobins, striped basses, weakfishes, seaboard gobies, striped and crested blennies, widowpanes, smallmouth and summer flounders, and northern puffers.

A host of biotic and abiotic factors come into play (e.g., geological history, habitat type, water circulation, salinity regime, and concentrations of dissolved oxygen and minerals) when considering the regional distribution patterns of fishes, but water temperature and its fluctuations generally take center stage. Florida's coastal and shelf waters are thermally influenced by the Loop Current on the west coast and the Gulf Stream on the east coast. Inshore water temperatures are driven by atmospheric temperature, springs, and freshwater input. In this section we discuss distribution within broadly defined habitat categories: hardbottom, softbottom, coastal rivers, coastal pelagic, and oceanic pelagic.

Hardbottom

Warm Gulf Stream waters bathing southeastern Florida (from Jupiter through the Florida Keys to the Dry Tortugas) support the richest assemblage of tropical species—particularly reef fishes—in the state. This region has mostly carbonate sediment, low turbidity, minimal freshwater input, a consistent Gulf Stream influence, and reefs with stony corals, sea whips, and massive sponges. These are classic insular conditions that promote a diverse tropical-fish fauna, because not only are there reefs, but also vast areas containing seagrasses, mangroves, nearshore hardbottom, and the like that provide juvenile habitat for many reef and shore species. The following are just a few of the tropical reef species that reach peak abundances along the this stretch: queen angelfishes; blue hamlets; black groupers; blueheads; high-hats; blue tangs; rainbow and midnight parrotfishes; cottonwicks; French and smallmouth grunts; and mutton, mahogany, and yellowtail snappers.

The distinction between tropical and warm-temperate faunas is often framed as a north–south phenomenon, with a boundary drawn somewhere between Jupiter Inlet and Cape Canaveral, but the real story is a little more complicated than a simple line emanating from shore. The diversity of tropical reef fishes drops considerably within the stretch between Jupiter and St. Lucie Inlets, but many species make their way to Cape Hatteras, North Carolina, or even farther north. Most of these species—as fertilized eggs and larvae—move northward passively in the Gulf Stream Current, eventually settling on the shelf along the western edge of the "river" as it flows northward. A persistent upwelling of cold water (<60°F) along this same stretch affects the northward advance of temperature-sensitive tropical species. Depending on the frequency and severity of upwellings (usually from July to October), sensitive tropical species may gradually move their distribution northward until an extreme cold event knocks everything back. In addition to periodic cold-water upwellings, conditions from St. Lucie Inlet to Jacksonville become more continental, with greater freshwater input, finer sediments containing more quartz and organic matter, higher turbidity, and lower water temperatures.

Shelf-edge reefs oriented parallel to the shore are dominated by macroalgae and encrusting sponges, with only a few stony corals and sea whips. Fish assemblages on these reefs are less diverse but may support a considerable biomass of some species, such as tomtates and black sea basses. Typical assemblages on continental reefs also include red and vermilion snappers, red porgies, sheepsheads, cubbyus, gags, scamps, blue angelfishes, cocoa damselfishes, seaweed blennies, amberjacks, and bigeyes. On the west Florida shelf, the warm Loop Current flows well offshore, leaving the inner-shelf waters much cooler (especially in winter) than they would be at similar latitudes on the east coast of the state. As a result, reef fishes similar to those found on the continental-shelf areas off eastcentral and northeast Florida are more likely to overwinter on deeper-water hardbottom features of the west Florida shelf, from Cape Romano to the panhandle.

In addition to the species inhabiting east-coast continental reefs, west Florida shelf assemblages may include gray snappers, red and black groupers, belted sandfishes, whitespotted soapfishes, zebratail blennies, painted wrasses, hogfishes, leopard toadfishes, short bigeyes, and spotfin gobies. More–thermally stable hardbottom habitats closer to the Loop Current, most notably the Florida Middle Grounds, host a number of tropical species, such as purple reeffishes, neon and yellow-prow gobies, brown chromises, striped parrotfishes, yellowmouth groupers, spotted morays, and Spanish hogfishes.

Water temperatures drop below the preferred range for many reef species in water depths greater than 150 ft, but shelf-edge reefs in these depths attract a less-diverse but unique set of species. Off the east coast, *Oculina* coral thickets provide most of the reef habitat in mesophotic depths. On the Gulf coast, Pulley Ridge, Madison-Swanson, and Steamboat Lumps are well-known mesophotic reefs. These reefs are inhabited by species not found in shallower depths, plus some thermally flexible holdovers from shallower reefs. Sea basses and groupers dominate the species lists from these reefs: roughtongue basses, red barbiers, Warsaw groupers, scamps, snowy groupers, speckled hinds, tattlers, and wrasse basslets. Representatives from other families include reticulate morays, bank butterflyfishes, yellowtail reeffishes, red hogfishes, and greenband wrasses.

Softbottom

Assemblages living on softbottom (mud or sand) also exhibit latitudinal and depth-related distribution patterns around the state, although these species are less static than the reef fishes. Southeastern Florida's narrow shelf has relatively little open bottom and, therefore, low numbers of fishes such as leopard searobins, sand divers, snakefishes, channel and eyed flounders, spotted snake eels, and cusk-eels. On open bottoms north of St. Lucie Inlet—where sediment-grain sizes decrease, turbidity increases, and temperatures decline—demersal (bottom-dwelling) fish abundance (especially for drums and croakers) greatly increases. Just south of Cape Canaveral, butterfly rays, lesser electric rays, clearnose skates, Atlantic croakers, spots, star and banded drums, silver seatrouts, searobins, and sand and lefteye flounders thrive. Scalloped hammerheads and sandbar, dusky, Atlantic sharpnose, finetooth, and bull sharks are happy to turn any of the above into a meal. This productive assemblage continues, in various combinations and abundances, northward to the Carolinas and in similar habitats across the northern Gulf. On the west Florida shelf, longspine porgies, spots, blackwing searobins, dusky flounders, and red goatfishes, among others, typify the softbottom assemblage. Off southwest Florida, we see more of a tropical influence, with additions such as blue croakers, but because almost any hard point—a bit of exposed limestone or a rock or shell—creates hardbottom habitat, hosting sponges, sea whips, or small corals and fishes that seek such cover, it is not unusual to get critters from both habitat types in a single trawl haul.

These open-bottom assemblages refer to outer- and middle-shelf areas, but a group of fishes found nowhere else in abundance live within the surf zone on open, sandy beaches: kingfishes (whitings, not king mackerels), drums, and threadfins. On the east coast of the Gulf, northern and southern kingfishes, sand drums, Atlantic threadfins, barbus, and spotfin mojarras are the main members of this assemblage, along with juvenile Florida pompanos, permits, and palometas. This same list (minus sand drums, palometas, and barbus) would be found on most sandy beaches of the west coast.

Cownose rays and mullets frequent this habitat, and several shark species (blacktips, finetooths, spinners, and Atlantic sharpnoses), snooks, and tarpons are the apex predators of this habitat.

Coastal Pelagic

Coastal pelagic fishes migrate along the inner continental shelf of both coasts, often entering estuaries or lagoons along the way. Migrations are seasonal and generally triggered by water-temperature changes. Many species move southward during winter: either gradually, by tracking seasonal water-temperature declines, or rapidly, following the passage of cold fronts (perhaps because of sudden drops in atmospheric pressure?). These same species work back north during spring, as water temperatures warm. Archetypical coastal pelagic species are Spanish mackerels, bluefishes, Florida pompanos, ladyfishes, Atlantic bumpers, blue runners, crevalle jacks, anchovies, and herrings. King mackerels, cobias, and little tunnies are coastal migratory species that usually stay offshore near the shelf edge, but all three will venture close to shore when searching for food. Cownose rays and many shark species (e.g., bonnetheads and blacktip, finetooth, blacknose, and Atlantic sharpnose sharks), also fall into this group. Coastal migratory species range from the surf zone to the shelf edge but prefer to remain within a mile or so of the shore, depending on wave height and water clarity. Another group of pelagic species (including scaled and redear sardines, false pilchards, plus silver and various other anchovies) prefer to hug the sandy, wave-swept beaches rather than venture over the open shelf.

Coastal Rivers

One of the most distinctive characteristics of Florida's marine fishes is their widespread and pronounced penetration into freshwater reaches of coastal rivers. Marine species regularly ascend well over 100 mi upstream from the mouth of the St. Johns River at Jacksonville. In assessing this unusual pattern of fish distribution in the St. Johns River, renowned University of Florida naturalist Archie Carr once wrote: "In aggregate, the St. Johns fishes make the stream one of the most interesting rivers in the world." Over 50 species of marine fishes regularly invade freshwater systems around the state. You can readily observe this phenomenon on the west coast, in the gin-clear spring "boils" of the Crystal and Homosassa Rivers and in springs that empty into the St. Johns River. Spring water in the Crystal and Homossassa Rivers is a constant 72°F, which beckons a host of marine species (including Atlantic stingrays, Atlantic needlefishes, snooks, gray snappers, pinfishes, sheepsheads, ladyfishes, tarpons, striped and white mullets, clown gobies, spotfin mojarras, and hogchokers) seeking warmth and stability in winter when adjacent marine waters dip well below this temperature level.

How do fishes that are physically adapted to dealing with a salt-water medium stand freshwater? The answer lies in a substance other than sodium chloride (i.e., table salt, which, in dissolved form, is a main ingredient of sea water). Springs and spring-fed rivers are high in calcium, which physiologically can, in a pinch, keep the fishes' internal balance of salt and water at equilibrium. In addition, some limestone strata around the St. Johns River trapped pockets of ancient seawater that gradually seep into the surrounding water table and, ultimately, into the river. Marine fishes regularly enter freshwater reaches of coastal rivers that even have little or no spring-water input. Bull sharks, snooks, tarpons, gray snappers, striped mullets, striped mojarras, Atlantic croakers, red drums, sheepsheads, crevalle jacks, ladyfishes, pipefishes, hogchokers, pinfishes, and Atlantic stingrays all regularly occur in the freshwater reaches of such rivers. A puzzling observation is that the normally euryhaline (able to live in a wide range of salinities) bull sharks have never been reported from any of the spring-fed rivers in Florida but regularly ascend other coastal rivers around the state.

A subgroup of tropical fishes resides mainly in the Indian River Lagoon and the associated St. Sebastian, St. Lucie, and Loxahatchee Rivers. This group—collectively known as tropical peripherals—includes freshwater and river gobies, bigmouth sleepers, opossum pipefishes, burro grunts, striped croakers, Irish pompanos, smallscale fat snooks, and tarpon snooks. The presence of established breeding populations and the abundance of these species only on the eastcentral coast strongly suggests that an influx of pelagic larvae from their natal grounds to the south (e.g.,

Mexico, Cuba, Hispanola) provided the initial inoculation into Florida. Larval immigration probably is still occurring, providing a parachute against a potentially catastrophic cold winter for fish populations.

Oceanic Pelagic Fishes

Great hammerheads, shortfin and longfin makos, and silky sharks, as well as swordfishes, wahoos, sailfishes, blue and white marlins, dolphinfishes, tunas, flyingfishes, and ocean sunfishes, are oceanic pelagic fishes that effectively are drawn along by the blue waters of the Loop Current and Gulf Stream. Oceanic pelagics are found around the state but, depending on your home port, they occur at differing distances from shore—"home by lunch" places off the southeastern coast, but areas requiring a long steam in the Gulf. Movements by oceanic pelagic species tend to be seasonal, with species such as dolphinfishes, sailfishes, and wahoos passing through almost like clockwork at certain times of the year. Most species follow their preferred temperature ranges vertically and horizontally within the larger currents and eddies.

REFERENCES

Gilmore, R. G., Jr. 2001. The Origin of Florida Fish and Fisheries. Proceedings of the 52nd Gulf and Caribbean Fisheries Institute, pp. 713–731.

Helfman, G. S., B. B. Collette, D. E. Facey, and B. W. Bowen. 2009. *The Diversity of Fishes*, 2nd edition, revised and updated. Wiley-Blackwell, Oxford, 720 pp.

Hine, A. C. 2013. *The Geological History of Florida*. University Press of Florida, Gainesville. 229 pp.

Humann, P. and N. DeLoach. 2014. *Reef Fish Identification: Florida, Caribbean, Bahamas*, 4th edition. New World Publications, Jacksonville, FL. 537 pp.

Keenleyside, M. H. A. 1979. *Diversity and Adaptation in Fish Behavior*. Springer-Verlag, Berlin. 208 pp.

Kells, V. and K. Carpenter. 2011. *A Field Guide to Coastal Fishes from Maine to Texas*. Johns Hopkins University Press, Baltimore. 448 pp.

Myers, R. L. and J. J. Ewel. 1990. *Ecosystems of Florida*. University of Central Florida Press, Orlando. 765 pp.

Page, L. M., H. Espinosa-Pérez, L. T. Findley, C. R. Gilbert, R. N. Lea, N. E. Mandrak, R. L. Mayden, and J. S. Nelson. 2013. *Common and Scientific Names of Fishes from the United States, Canada, and Mexico*, 7th edition. Special Publication 34. American Fisheries Society, Bethesda, MD. 384 pp.

Randall, J. E. 1996. *Caribbean Reef Fishes*, 3rd edition, revised and enlarged, T.H.F. Publications, Neptune City, NJ. 368 pp.

Robertson, D. R. and J. Van Tassell, 2012. *Fishes: Greater Caribbean; A Guide to the Shore Fishes of the Caribbean and Adjacent Areas*. Version 1.0. Smithsonian Tropical Research Institute. Available at https://itunes.apple.com.

Robins, C. R., and G. C. Ray. 1996. *A Field Guide to Atlantic Coast Fishes of North America*. Peterson Field Guide Series 32. Houghton Mifflin, Boston, MA. 354 pp.

Seaman, W. 1985. *Florida Aquatic Habitat and Fishery Resources*. Florida Chapter, American Fisheries Society, Kissimmee. 543 pp.

Smith, C. L. *National Audubon Society Field Guide to Tropical Marine Fishes of the Caribbean, the Gulf of Mexico, Florida, the Bahamas, and Bermuda*. Alfred A. Knopf, New York. 720 pp.

Lampreys
Family Petromyzontidae

- **Sea lamprey** (*Petromyzon marinus*)

Background

An eel-like species, the sea lamprey is the sole anadromous (moves upstream to spawn) member of its family recorded from Florida. This highly primitive organism has 7 round gill openings; a round suctorial oral disc (it lacks true jaws); a single nostril on the top of the head; small eyes; and 2 dorsal fins. Coloration is a mottled yellowish-brown.

Distribution and Habitat

Sea lampreys are (or were) uncommon inhabitants (or perhaps just visitors) to extreme northeast Florida. Sea lampreys apparently never were very abundant in the lower St. Johns River, the only place in Florida where they once were seen (a Gulf of Mexico record is questionable). The placement of the Rodman Dam on the Oklawaha River, the major tributary of the St. Johns River, most likely eliminated any potential spawning habitat—it also affected the reproduction and abundance of striped basses (*Morone saxatilis*) and possibly the anadromous herring species of the genus *Alosa*—and records of sea lampreys since the 1950s (and perhaps prior to that date) probably represent strays from viable spawning populations far to the north.

Sea lampreys (*Petromyzon marinus*). *Photo by Jay Fleming Photography.*

Natural History

Sea lampreys are parasitic lampreys (many other species in this family are nonparasitic) that use their sucking oral disc to attach to larger organisms—primarily fishes—and use horny, specialized, cheese-grater-like teeth to rasp and consume flesh and bodily fluids from the still-living host. Reproduction occurs in shallow streams with running water; males and females communally spawn in shallow nests made by moving rocks around. Young are known as ammocoetes, and this nonparasitic, freshwater phase is prolonged. Metamorphosis then advances them into the adult stage, which is spent in the sea.

Nurse Sharks
Family Ginglymostomatidae

- **Nurse shark** (*Ginglymostoma cirratum*)

Background

Nurse sharks are easily distinguished from other sharks by their light-brown color; large, rear-placed, and rounded dorsal fins; blunt snout; tiny eye; and 2 long nasal barbels. The attractive (by shark standards!) juveniles are pale-tan, with faint brown blotches covered with small black spots.

Distribution and Habitat

Nurse sharks are common on structured habitats around the state, and they are most abundant from southeast Florida, throughout the Keys, and north to about Tampa Bay. Primarily a shallow-water species, they inhabit water depths from inches to about 150 ft.

Natural History

Nurse sharks are prominent residents of coral reefs; marl, rubble and sand flats; grass beds; and man-made structures, such as piers and artificial reefs. Unlike other sharks, they happily rest on the bottom and frequently are seen lying in undercuts of ledges and in solution holes in marlbottom, sometimes in pairs or small groups. They have very thick skin and dense scales that almost form an armored plate, ensuring protection against scrapes and stings, including (for females) from their mates. Reproduction occurs during summer months (peak activity in June) in very shallow water (<5 ft) in the Dry Tortugas and the Keys, where multiple males pursue single females in an exercise that resembles an assault more than true love. The successful male holds onto the female's pectoral fin with a prolonged bite, and copulation occurs in a dual inverted position after much paired rolling. Gestation lasts 5–6 months, resulting in the birth of 30–50 live young. Nurse sharks have numerous, closely placed small teeth, arranged in 7–12 functional rows that form a crushing surface, but the teeth also have a small central cutting blade, the net effect being cheese-grater-like. They are suctorial feeders (inhalers, as opposed to biters) that mostly consume fishes; food-habit studies negate the popular notion that nurse sharks are major predators of spiny lobsters. Although not normally dangerous, when aggravated they bite and persistently hold on to humans, even after being removed from the water. Nurse sharks grow to about 10.5 ft long.

Nurse shark (*Ginglymostoma cirratum*), Jupiter, Florida

Whale Sharks
Family Rhincodontidae

- **Whale shark** (*Rhincodon typus*)

Background

Whale sharks are the behemoth of all fishes, growing to 40 ft long; weighing is done by the ton. Easily recognized by their large size, square snouts, and distinctive coloration, whale sharks are a sight to behold. They exhibit countercoloration: the dorsum is brown, gray, or bluish, with an overlaid checkerboard pattern constructed of white or yellowish spots, bars, and stripes; the ventral surface is white.

Distribution and Habitat

Whale sharks occur around the state in offshore oceanic waters, but they occasionally wander closer to shore if the pickin's are good. We have seen individuals near the beach off southeast Florida and in the Keys where deep oceanic (blue) water is not far away. Their abundance in the northeastern Gulf, over continental-slope water, is highest in the summer.

Natural History

Whale sharks are a globally endangered species. Little is known about their life history. Only recently have we learned that females give birth to more than 300 live young in a single litter, but their copulating and birthing locations and reproductive behavior are unknown. Much like the related nurse sharks, young develop in thin, horny egg cases until they hatch in the mother's uterus. The 300-plus young (16–25 in) from a 35.5 ft, 18-ton Taiwan female probably were near birth, as the largest ones were free of their egg cases and a later-dissected individual swam in an aquarium for 17 days before dying from an infection. Whale sharks probably live for more than 50 years.

Whale sharks feed on plankton, jellyfishes, krill, freshly released fish and coral spawn, and small fishes. Adults predictably aggregate on an annual basis in some areas, most notably off Mexico, Belize, and, in the northwest Atlantic, Louisiana. At least two of the aggregation sites coincide with mass spawning of fishes (snappers and groupers), where whale sharks come to feed on the bounty of oil-rich eggs by opening their large mouths and moving, with jaws agape, across the water surface. Another curious feeding behavior involves individuals, oriented heads up, vertically pushing through competing schools of small planktivorous fishes (and their predators) to inhale entire patches of biota. In the Gulf of Mexico, they have been seen erupting through the water surface with accidentally engulfed tunas jumping out of a whale shark's wide-open mouth! But they are not confined to the upper water column—a tagged individual dove to over 3200 ft. Movement patterns are influenced by the predictable seasonal availability of food. Off Belize, deep dives take place mostly by day, and shallower dives occur during fish-spawning season. Whale sharks are always accompanied by free-riding remoras, and often by jacks and cobias.

Whale shark (*Rhincodon typus*). *Photo by James R. Abernethy.*

Sand Tigers
Family Odontaspididae

- **Sand tiger** (*Carcharias taurus*)

Background

Nothing says "shark" like jaws full of ragged-looking teeth that prevent closure of the mouth. Meet the sand tiger. Those teeth lead to the same species' South African moniker, "ragged-tooth sharks." In addition to their impressive recurved dentition, sand tigers have a quite-pointed snout and a body bearing a series of irregular, dark, lateral blotches and spots. The broadly triangular first dorsal fin is located far back on the body; the similarly shaped second dorsal and anal fins are large, not too much smaller than the first dorsal fin. The overall appearance of a sand tiger is that of a plodder rather than a racer, which is borne out by its behavior and ecology.

Distribution and Habitat

A continental-shelf species that occurs off both the east and west coasts of Florida, sand tigers are found worldwide, primarily in temperate waters. They are not coral-reef inhabitants but are rather fond of rocky outcroppings, ledges, and shipwrecks. While found in shallow Florida waters, they more regularly occur over somewhat deeper, midshelf, hard-bottom habitats.

Natural History

The reproductive mode of sand tigers is notable. One embryo in each uterus becomes dominant early in development; it proceeds to eat all other developing embryos and fertilized eggs (ova), and then the unfertilized ova coming down the pipe (oviduct) from the ovary. The net result of this uterine cannibalism is the production of two large-sized (>3 ft at birth) young, one from each uterus. In Florida, birthing occurs during winter months. Sand tigers are voracious piscivores, consuming active schooling fishes like bluefishes and a who's who of bottom dwellers; larger invertebrates (e.g., squids) also are happily eaten.

A peculiar behavior sand tigers engage in is gulping atmospheric air. The air (which is in the shark's stomach) may be utilized as a means of achieving neutral buoyancy, because this species, like all other sharks and rays, lacks the swim bladder (a body part that serves the same purpose) found in most bony fishes. Despite their fierce appearance, sand tigers are not common, unprovoked attackers of humans, but they are active thieves when speared fishes are around and can be a bit aggressive with divers. In tandem with their physique, sand tigers sometimes act as the neighborhood bully.

Owing to their crowd-pleasing appearance and adaptability to an aquarium environment, sand tigers are a must-have species in public aquaria. Aquarists have learned that they need to find a balance between overfeeding (and having pot-bellied, gone-to-seed stars) or underfeeding (and watching a multispecies community tank turn into a solo exhibit). This species has been globally overfished and, because of its limited

Sand tiger (*Carcharias taurus*). *Photo by James R. Abernethy.*

reproductive capacity (2 young per litter), recovery is particularly slow. In Florida, their ferocious appearance, coupled with their lethargic and uncaring behavior, led to many of these sharks being dispatched by bold scuba divers wielding powerheads (a bullet-tipped spear).

Thresher Sharks
Family Alopiidae

- **Bigeye thresher** (*Alopias superciliosus*)
- **Common thresher shark** (*Alopias vulpinus*)

Background

Other than hammerheads, no other sharks are more distinctive than threshers. Reaching great sizes, owing to their outrageously elongated upper caudal lobes, common thresher sharks (20 ft) and bigeye threshers (15 ft) have very long pectoral fins; tiny second dorsal and anal fins; and very short, pointed snouts. Common threshers are brownish- to bluish-gray, with 2 discrete, irregularly edged white patches on the undersides: one ranging from under the jaw backward to midbelly, and the other surrounding the pelvic fins to about the base of the anal fin. The extreme tips of the pectoral fins may also be white. Bigeye threshers are a bit darker on the upper and lateral surfaces, and they have white undersides, from the lower jaw posteriorly to the pelvic fins. Their eyes (some of the biggest in the fish world) are notably larger than those of common threshers, and a pronounced crest extends from top of the head back to the gill slits. This helmet look is most highly developed in adult males.

Distribution and Habitat

Both thresher species are widespread around Florida, but bigeyes are more common. Common thresher sharks tend to occur in shallower and cooler waters, and bigeye threshers inhabit deeper and warmer waters. Bigeyes occur on the edge of the continental shelf and beyond, both at the surface and at great depth.

Bigeye thresher (*Alopias superciliosus*), southern Florida. *Photo © Jason Arnold/SeaPics.com.*

Natural History

The huge tail of both species is used to herd schooling fishes into balls and then smack them senseless before these sharks return to leisurely consume their stunned prey. As a result, many threshers caught by anglers—and most caught on commercial longlines—are foul hooked (hooked on the tail rather than in the mouth). The gigantic eye of a bigeye thresher offers insight (pardon the pun) into its lifestyle—a deep dweller that makes frequent up-and-down yo-yo trips to and from the surface, usually appearing near the surface at night. Females produce litters of two young, one from each uterus. Both pelagic (midwater) and demersal (bottom-dwelling) fishes, as well as squids and pelagic octopi, are popular food items. Common thresher sharks are frequent leapers that acrobatically exit the water while chasing prey and perhaps for other reasons (because they can?). Threshers are among the best tasting of the sharks, and their flesh commands a high price.

Basking Sharks
Family Cetorhinidae

- **Basking shark** (*Cetorhinus maximus*)

Background
Second only to whale sharks in size (32 ft), basking sharks aren't likely to be confused with the former species, as both their morphology and coloration differ markedly. Basking sharks have very long snouts and are uniformly dark-brownish to blackish (vs. a squared-off, wide snout and a unique spotting pattern in whale sharks). In addition, they have the longest gill slits of any shark, their enormous spans almost covering the sides of the body entirely and nearly bisecting the throat. Small basking sharks have been confused with large white sharks by *Jaws*-happy mariners and beachgoers, but a white shark's gill slits are much shorter; its undersides are white; its snout is regal, rather than comical; and, of course, it has very functional and obvious teeth that baskers can only dream about!

Distribution and Habitat
Basking sharks are uncommon visitors to our waters, yet they have been encountered off the northeast and southwest coasts of Florida. We now know from tagging studies that this species undertakes long-distance round-trip jaunts from cooler northerly waters off New England to the Caribbean, suggesting that Florida may not be their final destination—yet another tourist demographic! They appear over varying depths, usually in continental-shelf waters, where there are heavier concentrations of plankton.

Natural History
Basking sharks are seen at the water surface, where they spend a lot of time lazily cruising with their massive mouths wide open, forcing huge masses of water to move over the gills, which filter out plankton for the shark's lunch. When swimming like this, both the dorsal fin and the tip of the upper caudal fin are above the water, often resulting in people "seeing" two sharks swimming in amazing daisy-chain synchrony! That said, basking sharks not uncommonly appear in pairs or small groups outside of Florida. The fins are not as firm as those in some other sharks and sometimes look droopy when they emerge from the water.

Basking shark (*Cetorhinus maximus*). *Photo by Jeremy A. Stafford-Deitsch.*

Mackerel Sharks
Family Lamnidae

- **White shark** (*Carcharodon carcharias*)
- **Shortfin mako** (*Isurus oxyrinchus*)
- **Longfin mako** (*Isurus paucus*)

Background

Mackerel sharks are the ultimate apex predators of the sea (although some might argue that killer whales or sperm whales deserve this title), possessing streamlined bodies, pointed snouts, symmetrical tail fins, and impressive teeth and reaching large sizes. Longfin makos are distinguished from shortfin makos by having longer pectoral fins (duh!) and dusky coloration around the mouth and underside of the snout. Both species of makos are cobalt-blue in coloration. White sharks have a less pointed snout than makos; are greyish, with a black spot in the armpits of the pectoral fins; and have large flat (nonrecurved), triangular, serrated teeth. An additional species, the porbeagle shark (*Lamna nasus*), occurs in deep pelagic waters well offshore of the Florida shelf. Shortfin makos are the most commonly encountered mackerel shark species.

Distribution and Habitat

Mackerel sharks have wide distributions. They are found in temperate and subtropical waters throughout the world's oceans, but some species—such as white, porbeagle, and salmon sharks (*Lamna ditropis*)—prefer the colder end of the spectrum. Makos and whites do occur in tropical waters, but when there they gravitate to cool, deeper waters. White sharks are uncommon, predominantly winter–spring visitors to Florida, while shortfin and longfin makos are primarily offshore species that wander into shelf waters. The latter are encountered by longline fishers hunting tunas and swordfishes and by recreation anglers seeking billfishes and tunas.

Longfin and shortfin makos have the ability to elevate their body temperatures, so they are able to maintain core temperatures of 41°F–52°F. In this sense the makos are somewhat warm blooded, meaning that heat in their blood is conserved within the body and not lost through the gills. Shortfin makos, however, prefer water temperatures between 63°F and 68°F. It has been hypothesized that this species migrates seasonally to warmer waters, and this theory has been supported by some tag-and-release studies.

White shark (*Carcharodon carcharias*). *Photo by Jeremy A. Stafford-Deitsch.*

Natural History

Shortfin mako males mature at about 7 ft, and females at 5.5 ft. They reach sexual maturity between 4 and 6 years of age and have a reproductive cycle of about 2 years. Development of young is ovoviviparous. Embryos hatch in the female's two uteri and are nourished by yolk stored in a yolk sac. There is no placental connection between mother and young. Once the young are hatched in a uterus, uterine cannibalism (known as oophagy) occurs. Oophagy is the ingestion of unfertilized or less-developed eggs by a fetus that is more developed. Young are born upon completion of an approximately 15- to 18-month gestation period. Litters range from 8 to 10 pups, each 2 ft long. Like most other sharks, pregnant females usually abort embryos during or after capture; therefore few specifics about overall shark reproduction are known.

Makos are apex predators that feed on other fast-moving pelagic fishes, such as swordfishes, tunas, squids, and even other sharks. Surprisingly, the stomach contents of shortfin makos caught in gill nets off Natal, South Africa, showed a ratio of 60 percent shark and 40 percent fish, while a study off the northeastern United States found that 77.5 percent of their diet in that area consisted of bluefishes. Marine mammals and sea turtles are rarely ingested by shortfin makos.

Shortfin makos are considered to be one of the premier game fishes in the world, owing to their beauty, power, aggressiveness, grace, and athletic jumping ability. Shortfin makos are recreationally caught with trolled baits and lures, as well as with live or dead baits fished from anchored or drifting boats. A shortfin mako became famous as the protagonist in Ernest Hemingway's *The Old Man and the Sea*. Hemingway also caught a 780 lb mako with a rod and reel off Bimini, in the Bahamas, in 1963.

Cat Sharks
Family Scyliorhinidae

- **Marbled cat shark** (*Galeus arae*)
- **Chain dogfish** (*Scyliorhinus retifer*)

Background

Marbled cat sharks and chain dogfishes are small sharks with short, semidepressed heads and small caudal fins that lack a distinct notch between upper and lower lobes. The 2 almost equally small dorsal fins are located far posteriorly, far on the rear half of the body (farther back in chain dogfishes than in marbled cat sharks). The horizontally elongated, cat-like eyes of both species lack the protective nictatating eyelids found in many other sharks. Both species are yellowish-brown. A reticulated (network-like) chain pattern, alternating with moderately dark, irregularly defined saddles, distinguishes the larger (to 23 in) chain dogfishes from marbled cat sharks (13 in). The latter has a series of ocellated (i.e., a dark area encircled by a band of a different, lighter color) dark-brown spots and saddles on its back and sides, and white-tipped dorsal fins.

Chain dogfish (*Scyliorhinus retifer*). *Photo courtesy of the U.S. Geological Survey.*

Distribution and Habitat

Both species are denizens of continental-slope waters but occasionally are found over the outer shelf. In Florida these bottom dwellers have been captured statewide: marbled cat sharks at depths to 2100 ft, and chain dogfishes to 1425 ft. Chain dogfishes occur in close association with structures over both mudbottom and hardbottom, typically in water temperatures of 47°F–57°F. They are found in deeper water in southern parts (including Florida) of their wide range (from Georges Bank—off the coast of Canada—to Nicaragua).

Natural History

The chain dogfish is the better studied of the two species. Sexual maturity is reached at about 20 in, at 8–9 years of age. Females produce 6–20 ova, solely in the left ovary. They are encased in golden-brown, 2/3″–1″ × 1–2/3″–2–1/2″ egg cases with long sticky corner tendrils that, upon release, attach to hard substrate, which is often associated with soft corals, hydroids, and lost fishing gear. Females in captivity lay egg cases in pairs every 2–3 weeks. Egg cases, which occur singly or in large caches, are eaten by black sea basses. Young break out of their little homes after about 250 days, at lengths of about 4 in. Feeding is suction dominated, favoring squids, fishes, worms, and crustaceans. Marbled cat sharks have two functional ovaries, and each produces a single 1/2″–2″ egg case, with tendrils on the corners. Females are gravid in May–July in the Gulf.

Hound Sharks
Family Triakidae

- **Smooth dogfish** (*Mustelus canis*)
- **Florida smoothhound** (*Mustelus norrisi*)

Background

Members of the genus *Mustelus* are called "gummy sharks" in Australia, a quite sensible name, considering these small sharks have multiple, closely adjacent rows of flattened teeth that form pavement-like upper and lower jaw surfaces. That immediately separates them from all other Florida sharks. Their heads look like someone stepped on them from above, resulting in a somewhat flattened appearance. They have a pair of well-developed spiracles (nasal-like openings) located immediately behind the eyes, as well as a pair of nostrils on the underside of the snout, in advance of the mouth. The second dorsal fin is much larger than and originates much farther forward than the anal fin. Smooth dogfishes are tan to gray in color, not markedly different from Florida smoothhounds, but reach a larger size (up to 5 ft, but usually 3–4 ft vs. about 3 ft). The most obvious difference between the two species is subtle: the lower caudal lobe of smooth dogfishes is pointed, while that of Florida smoothhounds is more curved, with a curved notch above it.

Distribution and Habitat

Distribution helps out a bit if you're on the Florida east coast, where only smooth dogfishes are found. Both species occur in the Gulf of Mexico, where they most commonly inhabit shallow nearshore waters, over mudbottom or sandbottom. In Florida, however, smooth dogfishes have been taken in depths of over 900 ft, and Florida smoothhounds down to about 300 ft. Both are very common and are found in the shallowest of waters in the winter.

Smooth dogfish (*Mustelus canis*). *Photo by George H. Burgess.*

Natural History

The flattened teeth tell a story: these are crushing-type tools, which means Jaws they ain't. The teeth are superb, though, for grabbing and crunching crabs, shrimps, and lots of other bottom-dwelling invertebrates, as well as small fishes. Their olfactory (smell and taste) abilities appear to be fine tuned, as they readily find baited hooks. Hound sharks, in turn, are munched on by many of the larger shark species. Both species are live bearers; litters sizes and size at birth are 10–20 pups at 13–15 in for smooth dogfishes and 7–14 pups at 12 in for Florida smoothhounds. Once of no particular economic importance, smooth dogfishes have become a hot item in recent years, as spiny dogfish populations (not a regular Florida species) became overfished off New England and the Middle Atlantic Bight. Sadly, smooth dogfishes, too, now appear to be under duress.

Requiem Sharks
Family Carcharhinidae

- **Blacknose shark** (*Carcharhinus acronotus*)
- **Bignose shark** (*Carcharhinus altimus*)
- **Spinner shark** (*Carcharhinus brevipinna*)
- **Silky shark** (*Carcharhinus falciformis*)
- **Galapagos shark** (*Carcharhinus galapagensis*)
- **Finetooth shark** (*Carcharhinus isodon*)
- **Bull shark** (*Carcharhinus leucas*)
- **Blacktip shark** (*Carcharhinus limbatus*)
- **Oceanic whitetip shark** (*Carcharhinus longimanus*)
- **Dusky shark** (*Carcharhinus obscurus*)
- **Reef shark** (*Carcharhinus perezii*)
- **Sandbar shark** (*Carcharhinus plumbeus*)
- **Night shark** (*Carcharhinus signatus*)
- **Tiger shark** (*Galeocerdo cuvier*)
- **Lemon shark** (*Negaprion brevirostris*)
- **Blue shark** (*Prionace glauca*)
- **Atlantic sharpnose shark** (*Rhizoprionodon terraenovae*)

Background

Sleek, robust, toothy, and sometimes very large requiem sharks are the critters most folks consider "real" sharks—nurse sharks and dogfishes just don't cut it! At a glance, many requiem sharks, particularly members of the genus *Carcharhinus*, look very similar, but differences in fin placement, coloration, and teeth can be used to separate the species.

Blue sharks (*Prionace glauca*) are elongate, long-snouted pelagic sharks that grow to 12 ft long. They have long, slender pectoral fins and a small first dorsal fin; the back and sides are cobalt-blue, with white undersides (only shortfin and longfin makos are similarly colored). Tiger sharks, the big daddies of the group, grow to 18 ft. They have a squarish snout, large nasal flaps, and large black eyes; and are gray to brown on the back, with a pattern of black bars and spots on the flanks. These markings are more pronounced (as tiger stripes) in the young and somewhat faded in adults. Lemon sharks reach 11 ft; have a rounded snout; are a golden-brown color above and white below; and are unique in having second dorsal fins almost as large as their first dorsal fins. Atlantic sharpnose sharks, the smallest of the requiem sharks in our area (at just over 3 ft long), are separated from other family members by having the origin of the anal fin lying forward of their second dorsal fin and by their coloration: the pectoral fins are distinctly white-edged, and the gray to brownish back is overlaid with faint white spots.

The most distinctive member of the genus *Carcharhinus* is the oceanic whitetip shark, a large (to 12 ft) species readily recognized by its very long, paddle-like pectoral fins and large, rounded first dorsal fin, both bearing distinctive white tips. Other *Carcharhinus* species are more difficult to identify, most being grey, with white bellies, and often having dusky to black fin tips. The presence or absence of an interdorsal ridge (a raised line of skin between the two dorsal fins) helps by separating the various species into two groups.

Those having an interdorsal ridge include bignose, night, Galapagos, sandbar, dusky, silky, and reef sharks. Bignose and night sharks live in deep water and have long snouts. Bignose sharks have a moderately high dorsal fin, long pectoral fins, and the black-colored eyes seen on most sharks; they reach a size of 9.8 ft. The snout of the somewhat smaller (8 ft) night sharks is more pointed; the dorsal and pectoral fins are shorter; the rear tips of the second dorsal fin are long (short on bignoses); and in life the eyes are bright-green. The long pectoral fins of Galapagos sharks can cause confusion with bignose sharks, but Galapagos sharks have a higher first dorsal fin and a shorter snout. A distinctively large, triangular-shaped first dorsal fin that originates above the center of the pectoral fin gives sandbar sharks a bit of a hunchbacked appearance. This species reaches 9 ft long. In contrast, the easily confused dusky and silky sharks have shorter, more-rounded first dorsal fins that originate behind the pectoral fins, and these two species reach larger sizes (13 and 11 ft, respectively). The silky's first dorsal fin is more rounded, however; the free rear tip of the second dorsal fin is elongated; the snout is somewhat more pointed; and the

Tiger shark (*Galeocerdo cuvier*), with a sharksucker (*Echeneis naucrates*)

Lemon shark (*Negaprion brevirostris*), with sharksuckers (*Echeneis naucrates*), Jupiter, Florida

skin has a velvety feel. Reef sharks are most likely to be confused with sandbar sharks, but the reef shark's first dorsal fin is not as large and originates over the rear edge of the pectoral fin; the coloration—black edges on the second dorsal and anal fins and the lower lobe of the caudal fins—differs as well.

Several species of *Carcharhinus* lack an interdorsal ridge: blacknose, bull, finetooth, spinner, and blacktip sharks. Blacknose sharks, a shallow-water species, are often identified by process of elimination because of their generalized appearance. They are small (4 ft) and yellowish-green (fades to grey after death), with a black mustache on the end of the snout. The first dorsal fin originates behind the pectoral fin. Finetooth sharks are the smallest in this genus, barely making 6 ft. The snout is a bit pointed; the pectoral and dorsal fins are relatively small; and the eye is relatively large. If you have one in hand you'll note its very narrow, smooth, erect teeth. Bull sharks are large (11 ft), heavily built, and grayish to brown, with wide, rounded snouts. The eyes are relatively small and are placed forward on the snout. The dorsal fin is broad based and distinctively triangular, but it is not as large as that of a sandbar shark. The pectoral fins are broad, with dark-colored tips. Spinner and blacktip sharks can be difficult to separate. Both species get to about 8 ft long; have narrow, pointed snouts (a bit longer and more pointed in spinners); and exhibit black, dipped-in-ink tips on the pectoral fins. The important distinction is that adult spinners have dark coloration on the anal fin that is absent in blacktips; spinners also have smaller eyes and a slightly more-elongated, pointed snout.

Distribution and Habitat

Requiem sharks prowl coastal, shelf, and offshore waters around the entire state. Notable exceptions to this pattern are oceanic whitetips, which rarely venture into shelf or coastal areas, preferring instead blue (oceanic) water far from land (they seldom are seen off Florida), and reef sharks, reef-dwellers known only from southeastern Florida and the Florida Keys. Bull sharks, in addition to wandering from beyond the shelf edge to the coastline, regularly ascend coastal rivers, well into brackish water and freshwater. Blacknose, blacktip, finetooth, spinner, lemon, and Atlantic sharpnose sharks inhabit shelf waters either by themselves or in small to large aggregations, depending on the density of prey fishes. Blue, night, bignose, Galapagos, silky, and oceanic whitetip sharks are deep-shelf or pelagic, blue water species. Dusky, sandbar, and tiger sharks occur from oceanic waters to near the shoreline but seem to prefer the shelf edge.

Natural History

Requiem sharks reproduce by internal fertilization, but their reproductive behavior is not well known. Mating can be a rough-and-tumble exercise that involves nipping and biting between partners. It is not unusual to see females with bite marks on their flanks and dorsal and pectoral fins. The requiem sharks' life-history patterns follow a life-in-the-slow-lane path, characterized by slow growth, late maturity, and small litters. All species bear live young, and litter sizes vary among species: finetooth (2–6 pups), bignose (3–15), blacknose (3–6), reef (3–6), Atlantic sharpnose (1–7), blacktip (1–10), bull (1–13), sandbar (1–14), silky (2–14), dusky (3–14), oceanic whitetip (1–15), spinner (3–15), Galapagos (6–16), lemon (4–17), night (4–18), tiger (10–82) and blue (to 135, but usually 15–30). Atlantic sharpnose sharks reproduce yearly; most others pup every other year; dusky and sandbar sharks give birth every third year. Like so many other fishes, some requiem sharks seek shallow-water habitats as nursery areas. Tampa Bay, Charlotte Harbor, the Ten Thousand Islands area, Florida Bay, and the Indian River Lagoon support juvenile blacktip, lemon, bull, and blacknose sharks. In the Florida Keys, young lemon sharks are common over seagrass meadows and around mangrove shorelines. Another population

Sandbar shark (*Carcharhinus plumbeus*), Jupiter, Florida

Silky shark (*Carcharhinus falciformis*), Jupiter, Florida

Reef shark (*Carcharhinus perezii*), Jupiter, Florida

Bull shark (*Carcharhinus leucas*), with a sharksucker (*Echeneis naucrates*), Jupiter, Florida

Blacktip shark (*Carcharhinus limbatus*)

of lemon sharks found off the east coast uses the shallow coastal waters just south of Cape Canaveral as its nursery habitat.

Although most requiem sharks migrate, none are more conspicuous in Florida waters than blacktips. Each year, from fall to early winter, thousands of individuals move southward along the U.S. East Coast. In the spring, aided by the clear blue water, Palm Beach County is a perennial producer of aerial images of thousands of blacktips (and some spinners and lemons) roaming the Gold Coast while unwary bathers frolic nearby. At eye level, blacktips and spinners are well known for their habit of jumping and spinning (we have also seen young and old bull sharks jump and spin as well). Because of this jumping behavior, the migrating throngs of sharks are usually referred to as "spinner sharks," but in fact most are blacktips. The blacktip sharks' aggregation off southeastern Florida is an annual phenomenon, one that warrants further scientific study. Although the cause of this migration is unknown, most likely it is related to seasonal drops in coastal water temperatures, with the sharks simply tracking their preferred temperatures to the south each winter. In the spring, the aggregation breaks up as the sharks move northward into warming waters. As an aside, no one knows for sure why they jump, but if not just for the sheer joy in performing a flying axel, they may be communicating to the group, or perhaps just showing off!

Adult lemon sharks also migrate along the east coast of Florida, from Cape Canaveral northward during summer months and southward during winter. This group settles in offshore of Jupiter and, depending on environmental conditions or mysterious calls from deep within their genetic programs, they stay there from November to late March. Individuals congregate near reefs or wrecks, and often over sandy areas, in 60–100 ft of water.

Reef sharks are an exception to the migratory norm exhibited by requiem sharks. Adopting the site-attached mode of existence that so many of the small reef fishes follow, reef sharks reside on particular stretches of the reef for long periods. Favored sites may be transmitted through family groups over time. Reef sharks are capable of lying still on the seafloor, pumping water through their gills using muscular action rather than ram swimming. Thus you might see reef sharks lying under ledges or in caves, seemingly sleeping. Lemon and bull sharks (and the unrelated nurse and cat sharks) also can lie motionless on the bottom.

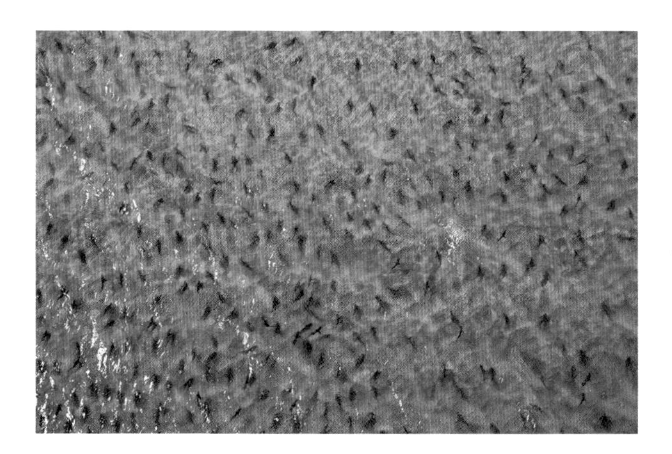

Blacktip sharks (*Carcharhinus limbatus*) schooling off Palm Beach, Florida. *Photo by James R. Abernethy.*

Requiem sharks are generally equipped to feed on fishes, but there are species-specific differences in food habits. Some are active feeders, while others are opportunistic scavengers. For example, spinners and blacktips often are seen herding mullets, sardines, Atlantic bumpers, crevalle jacks, Florida pompanos, and other coastal fishes right up to the shoreline. Bull, sandbar, and dusky sharks may loosely follow migrating schools of Spanish mackerels, king mackerels, little tunnies, and crevalle jacks. These sharks become the scourge of fishers, because they regularly attack hooked fishes. Bull sharks entering inshore waters particularly enjoy eating whiptail stingrays, judging from the number of barbs we have seen embedded in the jaws and faces of large adults. Tiger sharks—often unjustly portrayed as swimming garbage cans—catch and eat large fishes (including other sharks and rays) and adult sea turtles. But they also feed opportunistically on large dead animals, such as whales and dolphins (both of which are mammals), sea turtles, cows, and horses—and yes, human drowning victims—drifting at sea.

Speaking of which, most shark "attacks" on live humans in Florida are best characterized as bites and most likely are perpetrated by blacktip sharks, a moderate-sized species that is primarily a fish eater. Florida has about 20 of these incidents each year, mostly on the east coast and especially in Volusia and Brevard Counties, where abundant numbers of sharks, prey fishes, and surfers regularly share the same surf waters. Florida suffers about one fatality per decade, usually involving bull or tiger sharks, and most of the serious, nonfatal attacks occur in Florida's Gulf waters, where turbidity can be high and bull sharks abound. Despite these numbers, your chances of being bit—far less of dying—are infinitesimally small; a person is more likely to win the lottery in Florida than suffer a shark bite.

Hammerhead Sharks
Family Sphyrnidae

- **Scalloped hammerhead** (*Sphyrna lewini*)
- **Great hammerhead** (*Sphyrna mokarran*)
- **Bonnethead** (*Sphyrna tiburo*)
- **Smooth hammerhead** (*Sphyrna zygaena*)

Background

Hammerheads are the most unmistakable group of sharks. The adaptive function of the odd-shaped head has long been a source of conjecture, and there has been considerable scientific study on this topic. As in other sharks, the snout is covered with electrosensory receptors (ampullae of Lorenzini), but their broad heads give hammerhead species a greater sensory surface (beam angle) for detecting electromagnetic signals emitted by prey—a biological metal detector of sorts—especially for prey that bury their bodies in the sand (e.g., stingrays). The widespread nostrils also enhance their olfactory (smell and taste) senses, and the head shape provides a significant hydrodynamic advantage, making hammerheads very maneuverable. To distinguish the four Florida species, you will need a good look at the front of the head. The head of a great hammerhead is almost perfectly T-shaped, with a relatively

Scalloped hammerhead (*Sphyrna lewini*), Jupiter, Florida

Great hammerhead (*Sphyrna mokarran*), Jupiter, Florida

shallow central indentation, and this species has large, sickle-shaped first dorsal and pectoral fins. They are the largest of hammerheads, reaching 18 ft long. The head of the somewhat smaller (to about 14 ft) scalloped hammerheads is a bit more swept back, and it has a deep central indentation, giving the head a scalloped look. The dorsal and pectoral fins are smaller than those of great hammerheads and are not sickle shaped. Smooth hammerheads have similar fins and a slightly swept-back head that lacks a central indentation. They grow to 13 ft. Bonnetheads, the smallest (to 5 ft) of the hammerheads, are the most recognizable, with a smaller, spade-shaped head; hence the often-used common name "shovelheads."

Distribution and Habitat

Bonnetheads are regular inhabitants in bays and nearshore waters of Florida's Gulf coast, Florida Bay, and the Florida Keys. They are not common in southeast Florida and are more often found off central and northeast Florida. Scalloped hammerheads, the most-abundant large hammerhead species, are found from the surf zone to shelf-edge waters around the state. Great hammerheads occupy the same habitat area, but this species also inhabits oceanic waters well away from land. Smooth hammerheads are more of a cool-water species that are only rarely encountered on the Florida east coast.

Natural History

Hammerheads, like all of Florida's larger sharks, copulate and bear live young. Most hammerheads are oviparous, but two species, great and smooth hammerheads, have a reproductive anatomy that is similar to that of requiem sharks. The litter sizes of the three larger hammerhead species tends to be larger than most requiem sharks, with highs of 56 pups reported for great hammerheads, 44 for scallopeds, and 37 for smooths. The smaller bonnetheads produce litters of only 4–16 pups but, unlike their kin, this species reproduces annually, resulting in about the same number of young generated per year.

All hammerheads have small mouths in relation to their body size, but this does not stop great hammerheads from feeding on a variety of fishes, including such spiny favorites as stingrays and sea catfishes. As with most large sharks, hooked fishes prove irresistible to great hammerheads. Many a disgruntled angler has lost half or more of a trophy tarpon to a large great hammerhead in places such as Boca Grande Pass or Bahia Honda Channel. Locals in these areas name the marauding giants "Hitler" or "King Kong," convinced that a single individual is to blame. "Hitler" apparently is very long lived and well traveled, as he has been cursed at since at least from World War II on and throughout much of coastal Florida! Large great hammerheads often follow blacktip sharks during their annual mass spring migration on the east coast of Florida, ready to extend unprofessional courtesies to their cousins. Stingrays are even more highly regarded as a tasty meal—the face and jaws of large adult great hammerheads often bristle with stingray barbs.

Scalloped hammerheads eat fishes and crustaceans, and smooth hammerheads primarily consume fishes and squids. Bonnetheads feed on crabs and other crustaceans, with the biting pattern of their molar-like teeth specialized for the consumption of hard-shelled prey. Bonnetheads are one of the most abundant coastal sharks in Florida. On the east coast, groups of 5–20 individuals are frequently observed in nearshore waters. During fall and winter, larger groups migrate southward along the coast. Large groups also have been reported along the Gulf coast.

Scalloped hammerheads are known for aggregating around submarine oceanographic features and seamounts in the Gulf of Mexico and eastern Pacific, but no such aggregation points have been identified in Florida waters. Scalloped hammerheads are frequently observed off the eastern coast of Florida, and there appears to be nursery area on the inner shelf, near Cape Canaveral.

Angel Sharks
Family Squatinidae

- **Atlantic angel shark** (*Squatina dumeril*)
- **Disparate angel shark** (*Squatina heteroptera*)

Background

Angel sharks look like sharks that have been run over by a steamroller—they have an angular, stocky appearance, with brown, tan, or gray backs and white undersides. Until recently, scientists recognized only a single species in the northwestern Atlantic, but two additional, similar-looking angel sharks from the western Gulf of Mexico recently were described as a pair of new species. The long-recognized Atlantic angel shark, which reaches 5 ft long, is the only species with a row of raised dermal denticles along the center of the back and lacks distinct markings, other than some small dark spots on its brown-gray back. The newly discovered species—disparate angel sharks and Mexican angel sharks (*Squatina mexicana*)—lack those central thorns. The former is tan, with distinct black spots on the base of the pectoral fins and black spots on the flanks, in line with the first and second dorsal fins; the latter is light-brown, with small, diffuse white spots.

Distribution and Habitat

Atlantic angel sharks are widespread around Florida, mostly in

Atlantic angel shark (*Squatina dumeril*), Pensacola, Florida

outer-shelf and upper-slope waters. This species has been recorded from as deep as 3000 ft. Disparate angel sharks and Mexican angel sharks are recorded from the western Gulf of Mexico in outer-shelf water depths, and both may occur in the eastern Gulf of Mexico offshore of Florida. Precise details of their distribution are unknown, but we have trawled disparate angel sharks offshore of the panhandle, in water depths between 200 and 500 ft. Mexican angel sharks also may eventually appear in Florida's deep waters.

Natural History

Atlantic angel sharks give birth to live young and have a gestation period of about 12 months. Only the left ovary is functional (some angel shark species are like this, while others have two functional ovaries). Average litter size is 7 young, with a range of 4–10. Birthing occurs in the spring (February–June) in the northeastern Gulf of Mexico, after embryos have completely absorbed all the yolk in their external yolk sacs. Mature males develop spines on the outer tips of their wing-like pectoral fins. Fertilization is internal, and copulation occurs in the summer (June–October). Atlantic angel sharks are primarily fish eaters and, to a lesser degree, consume squids and crustaceans. In the Gulf, this species' favorite fish prey are Atlantic croakers, spots, Gulf butterfishes, longspine porgies, and red goatfishes. Atlantic angel sharks are classic lie-in-wait predators that rapidly erupt from the bottom to grab prey that unknowingly pass too close. Nothing is known about the life history of the two newly named species.

Torpedo Electric Rays
Family Torpedinidae

- **Atlantic torpedo** (*Torpedo nobiliana*)

Background

When looked at from above, Atlantic torpedos are almost circular, but their head profile is nearly flat, like the leading edge of a balloon pushed against a wall. The head has two highly developed electric organs on its upper surface that appear as ovoid bulges, and the tail lacks barbs, characteristics that easily distinguished this species from the only other circular ray in Florida, yellow stingrays. Atlantic torpedos have rounded pelvic fins, and the first dorsal fin is larger than the second. Coloration in Atlantic torpedos is dark-brown to grey, sometimes with darker spotting. They reach sizes of 6 ft and 200 lb.

Distribution and Habitat

Although the species often occurs near shore in more northerly locations, Atlantic torpedos are a rare, deeper water form in Florida. We have Florida records of this species from 90 to more than 1600 ft. This pattern of submergence of cool-water species into deeper water in the tropics and subtropics is a common one among many fishes.

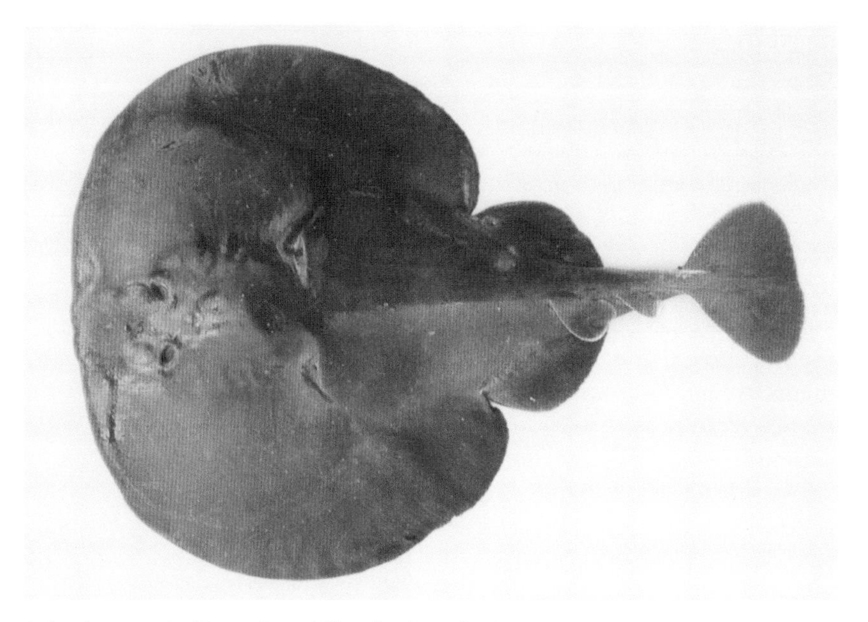

Atlantic torpedo (*Torpedo nobiliana*). *Photo by George H. Burgess.*

Natural History

The electric organs are discharged when an unsuspecting food item swims too close to the partially buried Atlantic torpedo. It then grabs and consumes the stunned or dead prey, usually bottom-dwelling fishes. The electrical discharge is powerful enough (220 V) to put a man on the deck. Atlantic torpedos are live bearers.

Electric Rays
Family Narcinidae

- **Lesser electric ray** (*Narcine bancroftii*)

Background

Lesser electric rays (formerly referred to by the scientific name *Narcine brasiliensis*) have a dorsal surface that varies from tan to reddish-brown, with rectangular blotches that sometimes are encircled with small dark spots. Two dark, rectangular patches under the eyes are typical in adults. Maximum total length is about 2 ft. It differs from the related Atlantic torpedo in having a more elongate body; a broadly rounded head profile; straight-edged pectoral fins; and roughly equal-sized dorsal fins; and only reaching a length of 2 ft.

Distribution and Habitat

Lesser electric rays are found uncommonly in Florida lagoons and inshore waters, out to depths of 150 ft. Off southeastern Florida we have observed them primarily around jetties and piers and inside inlets at night. In winter, individuals move into middle- and outer-shelf waters.

Natural History

Lesser electric rays, like all rays, copulate to fertilize embryos. Lesser electric rays on the coast of Florida commonly mature at lengths of 22–33 in (males) and 20–26 in (females). Developing embryos are retained in the uterus—the ovoviviparous reproductive mode. The embryos are first nourished with yolk and then with histotroph (uterine milk), a protein-rich liquid secreted from the mother ray's uterine lining. Females give birth to 4–16 young after a minimum gestation period of 3 months. Some researchers believe that development may be delayed in response to environmental conditions, resulting in a longer gestation period of 11–12 months. The young are born already able to give off an electrical charge.

During the day, lesser electric rays bury themselves in the sand and emerge at night to forage on benthic invertebrates (living on or in the seafloor), primarily polychaete worms. A feeding lesser electric ray protrudes its mouth into the sediment like a vacuum cleaner attachment and sucks up sediment and infauna, then lifts its snout to spit out nonfood items. The electric organs, derived from modified muscle tissue, are bean shaped and located posterior to the eyes on each side of the body. Lesser electric rays can deliver shocks up to 56 volts; when discharging, they arch their backs. It is thought that the electric organ discharge serves to deter predators rather than stun prey.

Lesser electric ray (*Narcine bancroftii*), Jupiter, Florida

Sawfishes
Family Pristidae

- **Smalltooth sawfish** (*Pristis pectinata*)
- **Largetooth sawfish** (*Pristis pristis*)

Smalltooth sawfish (*Pristis pectinata*), Florida Everglades. *Photo © Doug Perrine / SeaPics.com.*

Background

Despite their elongate, shark-like appearance, sawfishes are more closely related to skates and rays, the key clue being the gill openings, which are located on the underside of the flattened head and body. Smalltooth and largetooth sawfishes historically occurred in Florida and elsewhere the United States, and both are listed as federally endangered species; the smalltooth is the only sawfish species currently remaining in Florida. The two sawfishes can be definitively identified by examining the saw-like snout (rostrum). Largetooths have 16–20 pairs of stout teeth on a relatively wide rostrum; smalltooths have more numerous (24–32), smaller teeth on a narrower rostrum. In addition, the first dorsal fin of largetooth sawfishes originates well in advance of the pelvic fins, versus over or slightly behind the pelvics in smalltooths. Both species once regularly grew to 20 ft long, and occasionally to 25 ft, but we seldom see them over about 15 ft today.

Distribution and Habitat

Largetooth sawfishes are now spatially extinct in U.S. waters—the last capture was recorded half a century ago—and probably they were a rare species in Florida even in their heyday, as they are known only from a few old records. Elsewhere in the United States, largetooth sawfishes were historically most common in the Gulf of Mexico off Texas, but the heart of the species' distribution then and now is in riverine estuaries around the Caribbean southward to the Amazon River. Similarly, smalltooth sawfishes once ranged from New York and New Jersey to Texas and southward to Brazil, but their U.S. range has contracted considerably over the past 200 years, leaving but a remnant population surviving in southwestern Florida, Florida Bay, and the Florida Keys.

Populations of both species are greatly reduced throughout their entire ranges, as are the three other sawfish species worldwide, hence their global endangered status. Declines have occurred because their preferred nearshore habitats are rife with human activities, leading to incidental capture of sawfishes by gillnets, trawl nets, and other fishing gear, including that for recreational fishing. Habitat modification and loss have also been contributing factors to their losses. Key habitat features for young sawfishes include shallow water, mangrove shorelines, and estuarine conditions. Although juveniles have an affinity for brackish water, subadults (6–10 ft) and adults move into more saline waters, utilizing shallow mud banks, flats, seagrass meadows, and mangrove habitats, as well as nearby coral reefs.

Natural History

Our understanding of the reproductive biology of smalltooth sawfishes is very limited. Males have claspers and fertilization is internal. Females give birth to live young and presumably produce a litter every other year. Litter sizes are thought to range from 1 to 12 young, with an average of 7. The gestation period is about 5 months, and size at birth is around 32 in. A breeding season has not been identified; the presence of young of the year during March and April, however, indicates that at least some mating occurs during summer. Several shark species are known to enter shallow water during particular times of the year to release their pups, and it is very likely that smalltooth sawfishes follow a similar pattern. From the available information, it is not clear if smalltooth sawfishes, like many other sharks and rays, are philopatric (where females return to their birthplace to release their young). A recent DNA study reveals that parthenogenesis (virgin birth by females) uncommonly occurs in Florida, an evolutionary strategy probably influenced by the reduced availability of male partners.

Smalltooth sawfish food habits have not been directly studied, but observations indicate that they consume benthic and pelagic fishes, as well as invertebrates. The rostrum is used to herd, stun, and even impale shallow-water fishes like tenpounders, herrings, and mullets, and it may also be utilized to rake the seafloor to uncover partially buried invertebrates, such as shrimps and crabs. The presence of fresh rostrum teeth–wounds on the sides of larger sawfishes suggests that the saw is also used in intraspecific interactions, most likely between males. When hooked, sawfishes aggressively swing their rostra defensively. Young sawfishes probably are on the menu of bull and lemon sharks, which prowl shallow brackish waters in search of easy meals. In Florida Bay, another potential predator is the endangered saltwater crocodile (*Crocodylus acutus*). Adult sawfishes probably are immune from predators—other than humans.

The toothed rostra of sawfishes of all sizes readily entangle in nets, ropes, monofilament line, discarded pipe sections, and other debris. We have often seen individuals with raw wounds from monofilament line or netting buried into the flesh of the head. Even worse, some captures have documented the illegal and cruel practice of removing the rostrum for its curio value, leaving these individuals short changed in food gathering and defense for the rest of their lives. Some sawfishes are caught incidentally on hook and line by anglers seeking sharks, tarpons, or groupers, and they occasionally appear in commercial shark-longline and shrimp-trawl fisheries. As federally listed endangered species, sawfishes may not be harmed, so if one is accidentally captured, it should be kept in the water (no posing with the catch!) while fishing gear is removed, and then the fish should immediately be released. Be quite careful of the rostrum, which is very quickly and accurately swung at whomever and whatever is close; the rostral teeth are sharp and pointed, as documented by scars on one of the author's (GHB) hands.

Guitarfishes
Family Rhinobatidae

- **Atlantic guitarfish** (*Rhinobatos lentiginosus*)

Background

Atlantic guitarfishes are easily distinguished from other rays by their elongate, guitar-like appearance and from sawfishes by the lack of a saw. Atlantic guitarfishes reaches a length of 2.5 ft.

Atlantic guitarfish (*Rhinobatos lentiginosus*), Singer Island, Florida

Distribution and Habitat

Atlantic guitarfishes are found around the state in coastal and shelf waters and are usually associated with hardbottom or softbottom where sponges and soft corals are present, in depths less than 70 ft.

Natural History

Little is known of the life history of Atlantic guitarfishes. The young are born live at about 8 in long. Litter sizes range from 3 to 6 individuals. Adults and young frequent the Florida east coast during winter months; there may be some inshore movement into nearshore waters in the fall and winter. Guitarfishes lie motionless on the bottom, frequently buried in the sand. They feed, often suction-like, on benthic invertebrates, including worms, mollusks, shrimps, and crabs.

Skates
Family Rajidae

- **Spreadfin skate** (*Dipturus olseni*)
- **Rosette skate** (*Leucoraja garmani*)
- **Freckled skate** (*Leucoraja lentiginosa*)
- **Clearnose skate** (*Raja eglanteria*)
- **Roundel skate** (*Raja texana*)

Background

Skates are a diverse group of batoid (ray-like) fishes in the region, but only five species regularly occur in water depths shallower than 600 ft off Florida. Spreadfin skates, the largest (in disc width) of the skates found off Florida, have a pointed snout (which distinguishes them from the others); thorn-like spines on the tail, without hooked points; and 2 separate dorsal fins. Rosette skates are small (10 in disc width), with a rounded snout and a heart-shaped disc. The body is light-brown, with small dark spots that form small rosette patterns around larger central spots. Freckled skates are shaped much like rosette skates but are gray-brown, with a rounded snout and small dark spots interspersed with larger pale spots. Clearnose skates, the most common skate species found in Florida waters, are golden-brown, covered with dark spots and short, dark, wavy lines. Roundel skates are similar in shape to clearnose skates but have a distinctive ocellated spot—black with a yellow ring—at the base of each pectoral fin.

Distribution and Habitat

Skates live on mud-, sand-, or shellbottom in outer-shelf waters. Spreadfin skates occur from nearshore to 1240 ft deep along the west coast of Florida. Rosette and freckled skates have complementary distribution patterns: rosette skate occurs along the east coast in depths ranging from 120 to 1200 ft; freckled skates are found only in the Gulf in a similar depth range (173–1930 ft). Clearnose skates live in 16–390 ft depths statewide and are often seen in coastal and inshore waters of the panhandle and northeastern Florida. Roundel skates are found only off the west coast, in 50–400 ft depths.

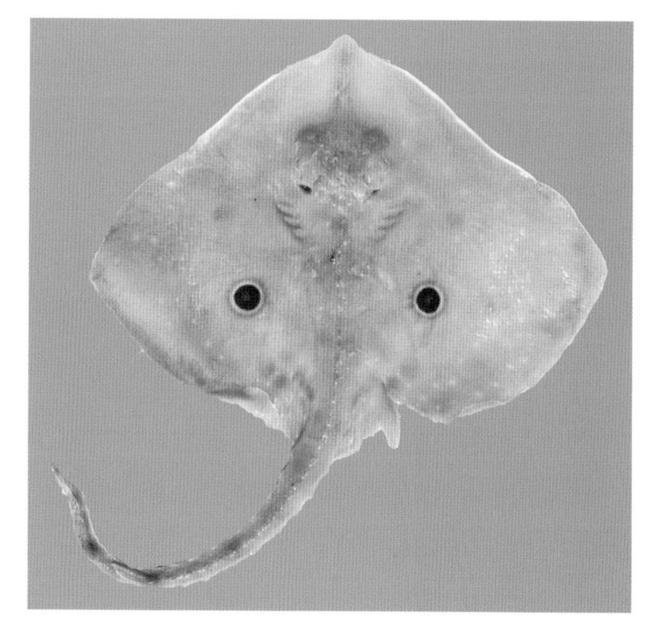

Spreadfin skate (*Dipturus olseni*), southwest Florida. *Photo by George H. Burgess.*

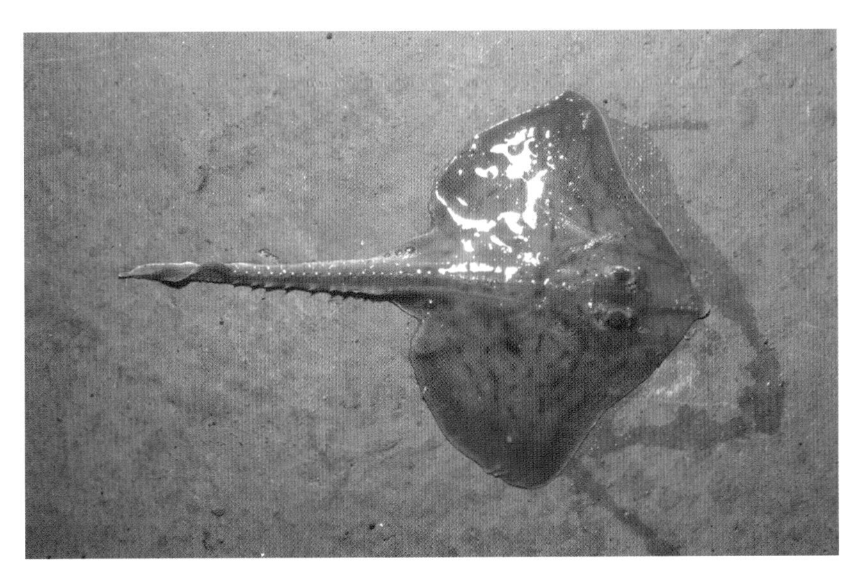

Clearnose skate (*Raja eglanteria*)

Natural History

Skates, like other elasmobranchs (a subclass of cartilaginous fish), reproduce by internal fertilization. Mating by pairs of captive clearnose skates is reported to last for 1–3 hours. During copulation, the male bites down on the female's pectoral fin near her tail while arching his body around to insert the clasper and transfer sperm. Females can store sperm for up to 3 months. Fertilized eggs are produced in multiple pairs over the course of 1–13 days. Eggs are encased in 2.5–4 in, oblong, leathery shells, with curly tendrils on each corner. A single female can lay as many as 66 egg cases during the season (December–May in Florida). Empty egg cases drift ashore, where they are well known to beachcombers as "mermaids' purses." Depending on the water temperature, eggs can incubate in egg cases for up to 3 months. Disc widths of newly hatched young range from 3.3 to 4.1 in. Little is known about reproduction in the other skate species, but roundel skates produce the smallest egg cases, measuring 1.3–1.7 in. Skates that have been studied use their electrical sensitivity to locate prey—bottom-dwelling shrimps, crabs, worms, and fishes—buried in the sand. Clearnoses are generalists, but they show a preference for crustaceans.

American Round Stingrays
Family Urotrygonidae

- **Yellow stingray** (*Urobatis jamaicensis*)

Background

Yellow stingrays are the sole member of this family in the northwest Atlantic. They differ from whiptail stingrays in their rounded caudal fin, round disc, and pretty coloration. This small stingray reaches a disc width of a little over 2 ft. The background color is brown or greenish, with small yellow, golden, or white spots. Yellow stingrays have a stout, serrated, venomous spine on the tail.

Distribution and Habitat

Yellow stingrays are usually found near seagrass beds, reefs, mangroves, and other structured bottom in high-salinity waters. They are most common off southeast Florida and the Florida Keys, in water depths less than 100 ft.

Natural History

American round stingrays are live bearers that reproduce by copulation. Their reproductive behavior is essentially identical to that observed in southern stingrays and, like that species, the teeth of males change shape from molariform to recurved during breeding season, facilitating the male's ability to grab and hold onto a "lucky" female. In Florida there are two mating seasons: January–April and August–September. Mating involves one to several males courting a single gravid female. Embryos initially depend on yolk, and then, when it is depleted, they are nourished by uterine milk (histotroph) secreted by trophonemata (finger-like extensions of the uterus). Gestation time is 5–7 months. Yellow stingrays give

Yellow stingrays (*Urobatis jamaicensis*), Lake Worth Lagoon, Florida

birth to litters of 3 or 4 (up to 7) live young.

Yellow stingrays find their food—amphipods, mantis shrimps, shrimps, crabs, worms, and mollusks—by jetting water into the sediment. They, in turn, are preyed on by various shark species. Waders beware! The envenomation (injection of venom) from a yellow stingray spine is quite painful, so always employ the "stingray shuffle" while wading in stingray waters.

Whiptail Stingrays
Family Dasyatidae

- **Southern stingray** (*Dasyatis americana*)
- **Roughtail stingray** (*Dasyatis centroura*)
- **Atlantic stingray** (*Dasyatis sabina*)
- **Bluntnose stingray** (*Dasyatis say*)

Background

Whiptail stingrays are a common sight in shallow waters around the state and present a constant concern for waders, owing to the venomous serrated barb(s) protruding from the base of the tail. The four whiptail stingray species known from Florida waters look very similar, but careful scrutiny of their disc shape can help separate the various species. Atlantic and bluntnose stingrays have rounded wings tips. Atlantics are the smallest (to 2 ft across the disc); have an almost-round disc (the tips are broadly rounded), with concave anterior edges leading to a relatively long, pointed snout; and are caffe-latte in color. Bluntnose stingrays are larger (a bit over 3 ft); have a diamond-shaped disc (the tips are not as rounded), with straight anterior edges; a shorter snout; and are dark-brown–gray-green. The similarly colored roughtail and

Bluntnose stingray (*Dasyatis say*), Loxahatchee River, Florida

Roughtail stingray (*Dasyatis centroura*), Hutchinson Island, Florida

southern stingrays have more-angular wing tips, and both reach larger sizes (8 ft and 5 ft, respectively). The two are separated by the presence or absence of tail tubercles. Stout dermal bucklers, reminiscent of rose thorns, cover the tail of the appropriately named roughtail stingray; these are absent in bluntnose, Atlantic, and southern stingrays.

Distribution and Habitat

All of the whiptail stingrays occur around the state, differing only in their preferred habitats and water depths. Two species are found on opposite ends of the habitat spectrum. Atlantic stingrays prefer shallow estuarine waters and regularly venture into freshwater. There is a permanent population in the St. Johns River and its tributary springs. In contrast, roughtail stingrays most often are found in deeper shelf waters (rarely to about 900 ft), but large females appear off the beaches of southeastern Florida during the winter and cold, offshore upwelling periods. Southern and bluntnose stingrays inhabit intermediate-depth coastal and shelf waters, to about 200–300 ft. Although both species of these species occur in estuaries, they are encountered more commonly in higher salinities.

Natural History

In Florida, most roughtail stingrays are large adult females (males and young are found northward of Florida off the U.S. East Coast), while the sex ratios of the other whiptail stingray species are about 1:1. Females of all four species grow to and mature at larger sizes than males. Whiptail stingrays reproduce by copulation and, after a 3–8 month gestation period, give birth to live young. Disc widths at maturity (male + female), litter sizes, and disc widths at birth are as follows: Atlantic stingrays (7.9 in + 8.6 in; 1–4; 3.9 in), bluntnose stingrays (1 ft + 1.6 ft; 1–6; 5.9 in), southern stingrays (1.7 ft + 2.5 ft; 2–10; 6.7 in), and roughtail stingrays (4.5 ft + 5 ft; 4–6; 13.4 in). Southern stingray mating behavior has been observed and takes place after one or more (up to seven or eight) males closely pursue a female. The successful male bites the rear third of the female's disc, holding on while he arches his body ventrally, and then flips his

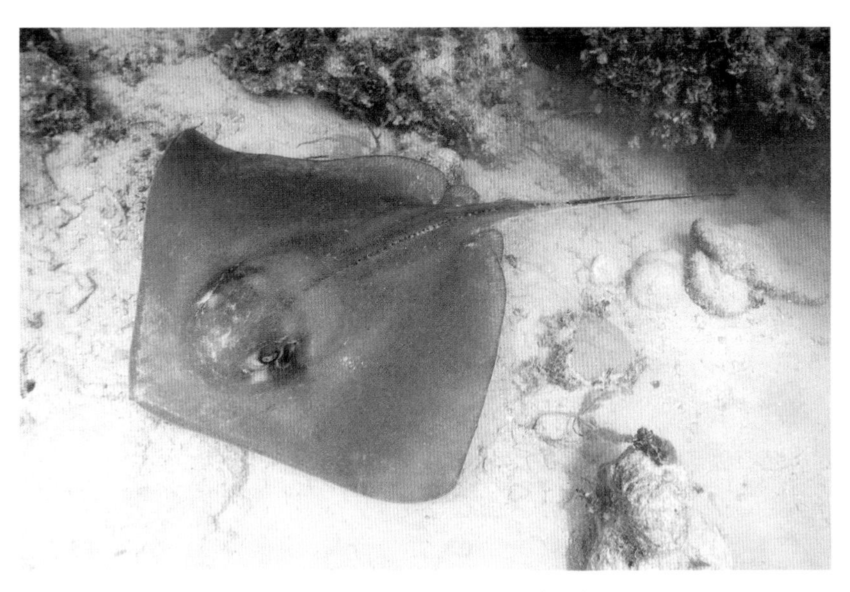

Southern stingray (*Dasyatis americana*), Islamorada, Florida

Atlantic stingray (*Dasyatis sabina*), St. Johns River, Florida

body so the two are belly to belly while inserting his clasper into her cloaca. While *in copulo*, the female continues to swim slowly, dragging along the still-biting and thrusting male. Two males have been observed simultaneously copulating with a single female, each holding on to the female with his mouth. For female stingrays, it truly hurts to be in love! Precopulatory activity of roughtail stingrays observed from a submersible at 262 ft mirrors this behavior, which has also been seen in mantas and American round stingrays. Happily, Mother Nature helps female Atlantic stingrays (and probably the other stingray species) by giving them a 50 percent thicker skin than males to help counter their mate's teeth, which change from a molar shape to recurved cusps during mating season. Mating-bite injuries and scars on females (and males) let biologists know when those seasons are underway.

Whiptail stingrays feed mainly on bottom-dwelling invertebrates that they dislodge from the sediment by using their flattened bodies in creative ways. One feeding mode involves wing flapping, but the most spectacular is jetting water out of the ventrally oriented mouth like a hydraulic dredge. Some lagoons, sand flats, or other shallow sandbottoms are pocked with feeding pits made by whiptail stingrays. If you snorkel in the vicinity of an actively feeding southern stingray, expect the visibility to be that of skim milk. These distinctive pits appear in fossil deposits, allowing us to deduce that such feeding behavior has been going on for millions of years (since the Cretaceous period).

Large roughtail and southern stingrays often are accompanied by remoras and groups of cobias that gain some feeding and movement advantages in the one-sided association. Southern stingrays attract a variety of opportunists when they are feeding. Bar jacks, yellow jacks, blue runners, eyed flounders, spotted goatfishes, and various mojarras can be seen following foraging rays. In most instances the followers pick up motile invertebrates and small fishes flushed from the sand by the ray.

Stingrays of all sizes are sought by sharks, especially great hammerheads and bull and tiger sharks. Stingray barbs protruding from the faces and jaws of these predators is a common sight. A great hammerhead was observed attacking a southern stingray by methodically biting off the pectoral fins, prior to consuming the helpless remainder. Great blue herons are known to capture and eat Atlantic stingrays. But sometimes stingrays get the last laugh, when a spine pierces the esophagus and migrates into vital organs, killing such predators as killer whales and bottlenose dolphins (sharks don't seem to be as vulnerable).

Butterfly Rays
Family Gymnuridae

- **Spiny butterfly ray** (*Gymnura altavela*)
- **smooth butterfly ray** (*Gymnura micrura*)

Background

Butterfly rays have massively wide wings—the pectoral fins—and tiny tails, distinguishing them from whiptail and American round stingrays (diamond- and round-shaped bodies, respectively, with long tails). Their brown to tan base coloration, with small dark and light spots, closely resembles that of flounders. Spiny butterfly rays have 1 or more serrated barbs at the base of the uniformly colored tail, while smooth butterfly rays lack these venomous spines and have a banded tail. The wing tips of spiny butterfly rays also are more rounded than those of smooth butterfly rays. Both species attain large sizes, with smooth butterfly rays growing to about 4 ft in width and spiny butterfly rays reaching almost 7 ft.

Distribution and Habitat

Smooth butterfly rays are found statewide, while the rarely encountered spiny butterfly rays are confined to Florida's east-coast waters. Both species occupy inner-shelf waters, to depths of about 200 ft, but only smooth butterfly juveniles inhabit inshore lagoons and estuaries.

Natural History

Butterfly rays give birth to live young. Sexual maturity is reached at about a 40 in disc width for spiny butterfly rays and at about 20 in for smooth butterfly rays. Only the left ovary of a spiny butterfly ray is functional, and the litter size is 4–7 young, shared by the two uteri. The pups (6–7 in wide) have the barb covered by a fleshy sheath (also found in other barb-bearing rays), keeping Mom from a very painful birthing. After a 4–9 month gestation period, smooth butterfly rays usually give birth to 5 young, each 6–8 in wide. Preliminary tagging studies in Florida show that smooth butterfly rays migrate from inshore waters (Indian River Lagoon) to the adjacent shelf during winter months. Butterfly rays feed primarily on small fishes.

Smooth butterfly ray (*Gymnura micrura*), southwest Florida shelf. *Photo by George H. Burgess.*

Eagle Rays
Family Myliobatidae

- **Spotted eagle ray** (*Aetobatus narinari*)
- **Bullnose ray** (*Myliobatis freminvillei*)
- **Southern eagle ray** (*Myliobatis goodei*)

Background
Some ichthyologists combine eagle rays, cownose rays, and mantas into a single family, the Myliobatidae, but here we treat the three groups as separate families. Eagle rays differ from cownose rays and mantas in having elongated, duck-like snouts. Spotted eagle rays are the largest of our three species, growing to 8–9 ft wide. The back is black to brown, with distinctive white spots and rings, and white underneath; the very long (up to 15 ft) tail has 1 or more serrated spines at its base. The smaller (to about 3 ft) and less common bullnose rays are light-brown, often with very pale, white spots. The dorsal fin is located on the tail, past but close to the pelvic fins (between the pelvics in spotted eagle rays). Southern eagle rays are similar in size and shape to bullnose rays but are uniformly brown to gray in coloration and have a dorsal fin located well back from the pelvic fins.

Distribution and Habitat
Both eagle ray species occur around Florida in shelf and inshore waters, but southern eagle rays are uncommon. Spotted eagle rays are known to enter the low-salinity reaches of rivers, frequent seagrass beds, and commonly occur over reefs and wrecks all the way out to the shelf edge. They travel singly or in small groups. Bullnose rays usually travel in aggregations in nearshore and coastal-shelf waters around Florida but are most common in north Florida.

Natural History
Spotted eagle rays copulate often, with multiple males pursuing a

Spotted eagle ray (*Aetobatus narinari*), Lake Worth Lagoon, Florida

Bullnose ray (*Myliobatis freminvillei*), Lake Worth Lagoon, Florida

single female. Development is ovoviviparous: embryos develop within a yolk sac inside the mother and are born alive. Litter sizes range from 1 to 4 pups that measure between 7and 14 in wide at birth. Bullnose rays have a similar reproductive pattern, but their litter size is 8 pups, measuring 10 in.

Eagle rays are bottom feeders that often bury their shovel-like snouts in the sand to ferret out their favorite prey—bivalves. They appear to employ some water jetting, as well. We have observed spotted eagle rays scraping mussels, oysters, and other mollusks from dock pilings. Eagle rays are well adapted for feeding on hard-shelled invertebrates, as they are equipped with a single, wide row of flat tooth plates on the upper and lower jaws, capable of pulverizing snail and clam shells. The forward extent of these plates protrudes beyond the mouth opening, to help secure prey items. At the back of the throat, fleshy papillae help strain bits of shell, allowing only the soft parts to go down. Bullnose rays have 3 wide rows of tooth plates that serve a similar crushing function.

Spotted eagle rays are well-known jumpers, completely exiting the water and landing with a resounding smack. The motivation for this behavior is unknown, but speculation includes ridding themselves of external parasites, communicating with other rays, or simply for the pure joy of it (as do mullets, billfishes, and requiem sharks). We've noticed that leaping occurs when the rays are spooked, so it may simply be a defensive behavior. Leaping spotted eagle rays have even landed on unsuspecting and unfortunate boaters in southern Florida, causing major injuries and even death.

Cownose Rays
Family Rhinopteridae

- **Cownose ray** (*Rhinoptera bonasus*)

Background

Cownose rays have a distinctively blunt snout that is indented in the center, resulting in 2 lobes. The back is chocolate brown, with white undersides. They have a long, cylindrical tail, with 1 or 2 serrated spines at its base. They grow to a maximum disc width of 3 ft.

Cownose ray (*Rhinoptera bonasus*), Pine Island Sound, Florida. *Photo by Jay Fleming Photography.*

Distribution and Habitat

Cownose rays travel in schools, ranging from a few individuals to aggregations of thousands. Migrations are not well known, but individuals or schools may traverse the entire U.S. East Coast or move around the Gulf of Mexico from Florida to Mexico and back again. Groups enter estuaries and lagoons—such as Charlotte Harbor, Tampa Bay, and Indian River Lagoon—but they also frequent the wash zones of high-, medium-. and low-energy beaches.

Natural History

Cownose rays give birth to live young. Only the left female reproductive tract is functional, and single young are produced. The young develop within the female's body, feeding on a uterine substance known as histotroph. Gestation lasts between 5 and 12 months. Foraging cownose rays are known to eat valuable species of bivalves, so they are not well liked by commercial fishers engaged in clam mariculture. The threat of cownose rays to commercial interests is probably overstated, however, as they feed primarily on crustaceans and polychaete worms. They locate these items, which are often buried in the sand, with their electrosensory system and then dislodge them from sediment by pumping water through the gills and out the mouth in a jet-pump action. One group of researchers have suggested that cownose ray populations may be increasing, due to the decline in major predators (e.g., bull and dusky sharks), but support for this premise is not particularly convincing.

Mantas
Family Mobulidae

- **Giant manta** (*Manta birostris*)
- **Devil ray** (*Mobula hypostoma*)

Background

Mantas can be distinguished from all other rays by the pair of prominent, inward-pointing cephalic lobes (sometimes called "horns" by fishers) that project forward from the head and flank the mouth. The eyes are set on the outside base of these fins. Giant mantas are the largest of the rays. They have a diamond shape when viewed from above, and the pectoral fins have pointed tips. The back is dark-brown and the undersides are white, often with patches of dark pigment. The caudal spine is present as an indistinct, calcified mass on the top of tail, immediately behind the dorsal fin. Devil rays are smaller (maximum of ca. 4 ft wide); have stiff, forward-pointing cephalic fins; and have a subterminal mouth. They are uniformly dark-brown to light-brown on the back and grayish underneath, and they lack a spine at the base of the tail.

Distribution and Habitat

Giant mantas occur around the world, in tropical and warm-temperate coastal and oceanic waters. Giant mantas habitually return to particular locations, but such areas are few and far between in Florida. Devil rays are observed singly or in small groups around the entire state. We have seen this species migrating in schools along the east coast during winter months. Individuals come close to shore in response to cold-water upwellings.

Giant manta (*Manta birostris*), Jupiter, Florida

Natural History

Giant mantas and devil rays reproduce through copulation; only the left reproductive tract is functional, and females bear single pups. At birth, young giant mantas are about 50 in wide and weigh 20 lb. Young are born in shallow coastal water, where they may remain for a few years. Mantas consume plankton, and they frequent areas where currents and plankton blooms are favorable for this type of feeding. On occasion we have observed giant mantas with their backs out of the water as they skim plankton from the water surface. When mantas swim just under the surface, their wingtips rhythmically appear and disappear, looking enough like a shark's dorsal fins to strike fear into the hearts of beachgoers or swimmers, who inevitably report seeing a pair of sharks engaged in synchronized swimming. Adult and subadult giant mantas appear off eastern Florida primarily during spring and summer months. Devil rays are not seen as often as giant mantas and tend travel in groups over the shelf, usually during winter. The wings of mantas cruising in nearshore waters are often adorned with lures left behind by "a little too close" casts from anglers, who are cognizant that big cobias are usually are found slipstreaming underneath these swimming Christmas trees.

Sturgeons
Family Acipenseridae

- **Shortnose sturgeon** (*Acipenser brevirostris*)
- **Atlantic sturgeon** (*Acipenser oxyrinchus*)

Background

Sturgeons are archaic bony fishes easily recognized by the rows of large scales (scutes) along their backs, sides, and belly; an asymmetrical (heterocercal) tail; an inferior mouth, with barbels; and 1 dorsal and 2 ventral fins set way back on the body. Sturgeons are anadromous fishes, living much of their lives in the coastal ocean and then ascending coastal rivers to spawn in freshwater. Two species are known from Florida: shortnose sturgeons and Atlantic sturgeons, the latter represented in Florida by two geographically separated subspecies. One of them, referred to as the Gulf sturgeon (*Acipenser oxyrinchus desotoi*), is confined to the Gulf of Mexico, and the other (*Acipenser oxyrinchus oxyrinchus*) is found in northeast Florida. The more common shortnose sturgeons grow to about 4.5 ft, while the larger Atlantic sturgeons top out at 14 ft. Shortnoses, appropriately enough, have a short snout, while Atlantic sturgeons have a longer snout. The lengths of the 4 mouth barbels also distinguish the two species, those of shortnoses being shorter than those of Atlantics. These barbels

Atlantic sturgeon (*Acipenser oxyrinchus*), Suwanee River, Florida. *Photo by Noel Burkhead, U.S. Geological Survey.*

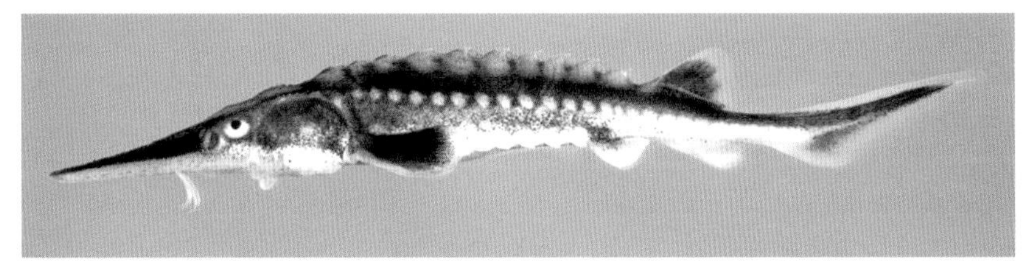

Juvenile Atlantic sturgeon (*Acipenser oxyrinchus*), Suwanee River, Florida. *Photo by Noel Burkhead, U.S. Geological Survey.*

function as sensory feelers to help locate food items on the bottom.

Distribution and Habitat

Shortnose and Atlantic sturgeons occur in larger rivers and estuaries of the North American eastern seaboard, from the St. Johns River in Florida to eastern Canada. Both are rarely encountered in the St. Johns River, but the Gulf sturgeon population is more robust. It has been separated from the parent Atlantic sturgeon populations since the Pleistocene epoch and occurs primarily from the Suwannee River westward to the Mississippi River; there are a few historical records of the species down as far as southwest Florida during really cold winters. The largest extant Florida population is thought to be in the Suwannee River. Sturgeons require freshwater rivers for spawning and, because there are no adequate riverine habitats in southern Florida, this portion of the peninsula acts as a barrier to interchange between the Atlantic and Gulf of Mexico populations. Both shortnose and Atlantic sturgeon populations have been depleted or even extirpated throughout much of their historical ranges by fishing, shoreline development, dam construction, and other human-generated factors. Both species are listed as endangered in Florida and the United States.

Natural History

Shortnose sturgeons ascend rivers to spawn during February–April. Females deposit between 27,000 and 208,000 eggs over hardbottom in shallow, fast-moving water. They grow slowly; females mature in 6–7 years, while males mature in 3–5 years. Individuals can live to at least 67 years old, with an average lifespan of 30–40 years. Gulf sturgeons can achieve an age of 42 years. These females reach sexual maturity between 8 and 17 years, while for males it is between 7 and 21 years.

In the Suwannee River, Gulf sturgeons spawn from March to May, with a peak in April. Females lay up to 3 million eggs, usually in deep areas or in holes with hardbottom where there is some current. Eggs are adhesive and attach to rocks, vegetation, or other objects. They hatch in about 1 week, depending on the temperature of the water. The young fishes remain in freshwater reaches of the rivers for about 2 years, and then they begin to migrate back downstream to feed in estuarine and marine waters. Adults spend from March through October of the year in the rivers and from November through February in estuarine or shelf waters. Changes in water temperature trigger upstream and downstream migrations. While upstream (mostly in summer), adults freely jump out of the water—mostly early in the morning or near dusk—for unknown reasons. In doing so they often hit boaters and sometimes have caused severe injuries.

Young Gulf sturgeons feed on larger planktonic crustaceans and insect larvae, and adults eat bottom-dwelling clams and snails. Near river mouths and on the inner continental shelf, adults generally consume clams and snails but also include other items, such as crabs, shrimps, worms, brachiopods, amphipods, isopods, and small fishes.

Gars
Family Lepisosteidae

- **Alligator gar** (*Atractosteus spatula*)
- **Longnose gar** (*Lepisosteus osseus*)

Background

Gars, looking every bit like living fossils—they have been around since the Cretaceous period (180 million years ago)—have a special type of heavy, diamond-shaped scales (ganoid scales) arranged in diagonal rows that provides a defensive external armor (exoskeleton). Their bodies are elongated and are round in cross-section. The head is long, owing to extremely prolonged jaws that bear prominent and numerous sharp teeth. Pectoral fins are located almost on the belly, and the single dorsal fin is located far back on the body, just in front of the rounded heterocercal (the upper lobe larger than the lower lobe) caudal fin. The spacing between the pectoral–pelvic and pelvic–anal fins is about the same. In alligator gars, the pectoral and pelvic fins are around

Longnose gar (*Lepisosteus osseus*), Rainbow River, Florida

the same size, only a bit smaller than the equal-sized dorsal and anal fins. Compared with longnose gars, the body is not as slender; the stouter jaws are less elongated; and the upper jaw slightly overhangs the lower jaw (vs. a longer but not overhanging jaw). Alligator gars may have some spotting on the dorsal, anal, and caudal fins but otherwise are gray to greenish-gray dorsally and white laterally and on the belly. Longnose gars are dark-grey or greenish-gray above, fading on the sides to whitish underneath; the pectoral, pelvic, anal, and caudal fins and the throat area under the operculum often are yellowish-orange. They have large, obvious spots laterally and on the dorsal, anal, and (especially) caudal fins.

Distribution and Habitat

Both species of gars are primarily freshwater fishes, frequenting the upper, middle, and even lower reaches of riverine estuaries (longnose gars enter salinities of 31.0 ppt). The large (to 6.6 ft) longnose gars occur naturally in rivers and lakes throughout the state, except for the extreme tip of the peninsula south of Lake Okeechobee. They now are immigrating into the region, however, through the maze of man-made water-control canals that crisscross that area. Alligator gars get even larger (just under 10 ft) and barely come into Florida, occurring irregularly on the western half of the panhandle.

Natural History

Gars have a distinct advantage in being able to survive in warm, low-oxygen (hypoxic) waters by being able to gulp atmospheric air (physostomous), which is stored in their swim bladders and provides 70–80 percent of their oxygen needs. In cooler-temperature, higher-oxygenated waters, respiration occurs through the use of gills. Respiration rates are highest during periods of darkness, which correspond with a period of greater activity.

Differences in the shape of the head and anal fin separate the two sexes of longnose gars; these difference presumably aid in the procurement of food and of mates during reproduction. Females grow longer, weigh more, and live longer (32 years) than males (29 years). Sexual maturity is reached at 2–3 years in males and at 6–7 years in females. Females have high fecundity, producing 30,000 toxic eggs that are each 0.1 in wide. Individuals remain at a spawning site for about 2 weeks to 3 months during the late April–early July spawning period; fidelity to a given spawning site may occur year after year. Reproduction typically takes place by day, although night spawning is known. One female works with about five males; both sexes release their gametes while swimming near the bottom, with heads angled downward. Eggs, bearing an adhesive coating, adhere to rocks, stones, ledges, and vegetation in the shallows and in channels, including in the spawning nests of other fishes. The young, at sizes of 1–1.5 in, hatch after 7–9 days in 67°F–86°F water, and they associate with vegetation during their first year. Growth is rapid (in the first year to 16 in) for 2 years in males and 4 years in females, but it then slows down to a level pace after they reach maturity.

Tagging reveals movements of up to 48 mi. In the Gulf, longnose gars move downstream at night into salinities of 3–10 ppt to pursue Gulf menhadens, returning to freshwater by day. Longnose gars are among the apex piscine predators of riverine estuaries, consuming herrings (menhadens are the bomb), killifishes, sunfishes, bullhead catfishes, pickerels, croakers, spots, and other fishes; in estuaries, freshwater and estuarine prey are consumed at a ratio of about 1:1.

Alligator gars have been shown to move up to 14.4 mi. Female alligator gars are longer and heavier and have greater girths than males. At 10 years of age, females reach 3.3 ft long, and after 30 years they attain 6.6 ft. Their maximum age is less than 50 years. Males mature by age 5–6 (at 37 in) and females achieve maturity by age 10–14 (at 55 in). Spawning occurs from March through June in rising-water, flooded backwater areas. Highly fecund females generate 157,000 toxic eggs. Young develop in shallow, protected embayment and tributary nursery areas, growing rapidly (2.4 lb by early August, 6.4 lb by early October). Mullets, croakers, herrings, pinfishes, tonguefishes, eels, sea catfishes, catfishes, sunfishes, minnows, suckers, and gars, as well as lesser amounts of crustaceans, other invertebrates, and birds constitute their diet. Alligator and longnose gars have hybridized in a large public aquarium.

Tenpounders
Family Elopidae

- **Ladyfish** (*Elops saurus*)
- **Malacho** (*Elops smithi*)

Background

Ladyfishes and malachos are silvery, slender, have large eyes and mouths, and reach 3 ft long. They are also called "chiros" and "poor man's tarpons." The name "tenpounder" is a misnomer, as the world record for these species is about 7 lb.

Ladyfishes in the western Atlantic traditionally were considered a single species, but recent study reveals the presence of a second, cryptic species: malachos. Adults of the two species are indistinguishable externally, but there are significant differences in characteristics at the larval stages, in their DNA, and in some countable characters (e.g., the number of gill rakers and vertebrae). Scientific evidence indicates that malachos numerically represent only 1–2 percent of Florida tenpounders.

Distribution and Habitat

Malachos are a tropical peripheral species, reaching Florida and the United States via larval dispersal. The center of abundance for malachos is in the Caribbean and northern South America, and the ladyfishes' center of abundance is in North American waters. The two species (primarily the larval stages) overlap in the southeastern United States and Gulf of Mexico, including Florida. Most malacho larvae probably do not survive to maturity; thus any adult or subadult encountered in Florida most likely is a ladyfish. Ladyfishes occur throughout inshore, coastal, and shelf (<70 ft) waters of Florida, including well upstream in freshwaters of coastal rivers, such as the St. Johns River. Adult ladyfishes usually travel in schools while in the coastal ocean and more singly while in lagoons and estuaries.

Ladyfish (*Elops saurus*), St. Johns River, Florida

Natural History

Spawning in ladyfishes is not well known, but it probably takes place in coastal and shelf waters. Based on the occurrence of larvae, malachos spawn in summer–fall and ladyfishes in winter–spring. In many parts of the state, ladyfish aggregations occur offshore. We have observed large schools of ladyfishes moving southward along the eastcentral coast of Florida during the fall. Like eels and the related tarpons, ladyfishes have a laterally compressed and transparent larval stage. The prolonged pelagic duration of this leptocephalus stage varies, but it lasts at least 2 months.

Ladyfishes consume worms, shrimps, crabs, and small fishes in the water column. They actively feed at night, especially in the glow of dock or bridge lights. They are eaten by bottlenose dolphins, as well as sharks, sawfishes, and other larger, predatory fishes. Along the east coast, ladyfish schools appear inshore during winter months, where they feed in lagoons, estuaries, and coastal rivers, and then move offshore in spring and summer, presumably to spawn.

Tarpons
Family Megalopidae

- **Tarpon** (*Megalops atlanticus*)

Background

A tarpon—his majesty, the silver king—grows to 8 ft long and weighs in at over 300 lb. Small tarpons might be confused with ladyfishes and malachos, because of their silver bodies, large eyes, and deeply forked tails. Tarpons, however, are readily identified by having much larger scales, a deeper body, and a dorsal fin with an elongated fin ray (making it look a bit like an overgrown Atlantic thread herring).

Distribution and Habitat

Tarpons are widely distributed in the tropical–subtropical western Atlantic and found throughout Florida's inshore and coastal zones, occasionally moving into deeper waters. Tarpons are euryhaline and freely move into freshwater. In some areas, tarpons live in landlocked freshwaters, particularly in golf-course ponds.

Natural History

The curious tarpon behavior known as daisy chaining—single-file, follow-the-leader aggregations—seen on the shallow flats of the Florida Keys long was thought to indicate courtship or spawning, but biological research suggests otherwise. Tarpons reproduce in deep offshore waters, with known spawning sites located offshore of western Florida and Trinidad. Adults mature at around age 5 or 6, at a length of 4 ft. Females produce 4–20 million eggs in a season, with spawning peaks in Florida waters occurring from May to July. Fertilized eggs are broadcast into the water column, where they gradually transform into leptocephalus larvae. Leptocephali are only found in eels, tenpounders, tarpons, and bonefishes. Leptocephalus tarpon larvae transform through several stages before finally resembling miniature adults. They actually shrink in size during the transition to the juvenile stage. During this 2- to 3-month transformation, tarpon larvae undergo a multienvironmental migration toward shallow water.

After their time in the plankton, tarpon larvae settle in inshore brackish water or even freshwater. Juveniles (4–18 in) prefer brackish inshore waters, mostly in the southern portion of the peninsula, including coastal rivers, mosquito impoundments, and the freshwater reaches of the Everglades. Large adults occur around the state in

Tarpons (*Megalops atlanticus*), Singer Island, Florida

coastal and nearshore waters. The center of abundance, though, is the Florida Keys. Although some individuals are found there year round, the greatest accumulations of adults occur from April through August, when people can see large numbers of big tarpons from bridges, such as Seven-Mile Bridge, while travelling on the Overseas Highway. A similar run takes place on the Gulf coast, at Boca Grande Pass at the mouth of Charlotte Harbor. Some researchers have speculated that a distinct population is found farther north, in the Big Bend ecoregion off the Homosassa River. It was long thought that this subpopulation, which yields fishes weighing over 200 lb, differed from those in the Florida Keys. Whether there are population differences has yet to be resolved. In either case, schools of tarpons can be found cruising nearshore waters along the entire Atlantic coast, as well as the panhandle, in summer months. The mosquito impoundments in the Indian River Lagoon support large numbers of juveniles.

Tarpons have large, inferior, bony mouths, and the inside of the jaw is covered with tiny teeth. They gulp air, and their stomachs serve as makeshift lungs. For this reason, tarpons habitually roll at the surface of the water, belching and then gulping fresh air. Rolling tarpons push their heads and mouths out of the water and then arch their backs on descent, which rolls the dorsal fin and then the tail. On quiet mornings you can hear them breathing in air while rolling.

Although tarpons possess large mouths, they feed on a variety of relatively small prey items. Adult and subadult tarpons eat shrimps, crabs, and fishes. In the Florida Keys, where vast schools of tarpons appear every spring, there is a feeding phenomenon known as the palolo worm spawning. "Palolo" is the name given to a type of worm that normally lives buried in the sandy bottom. At certain times of the year (late May and early June), they emerge by the thousands to spawn at the surface, and tarpons are well known to be the first in line for this bounty, a phenomenon that is legendary among anglers in the Keys. Along both coasts of Florida, tarpons can be seen showering (spooking schools right out of the water) groups of mullets, Atlantic thread herrings, menhadens, and anchovies. At various marinas in the Florida Keys, large tarpons, which love to hang out under docks, have been conditioned to take handouts from people. Here the fish species that routinely flummoxes the best fly fishers in the world readily takes stinking, filleted carcasses and over-the-hill bait from sunburned tourists, creating cottage industries in the Keys.

Bonefishes
Family Albulidae

- **Bonefish** (*Albula vulpes*)
- **Bigeye bonefish** (*Albula* sp.)

Background

Bonefishes, one of the most revered sport fishes, attract anglers from around the world to the Florida Keys. Until recently, there was only one recognized bonefish species in Florida waters, but recent genetic studies have revealed a second cryptic species, the bigeye bonefish, that is virtually indistinguishable from *Albula vulpes*. This new species, which has not yet been given a scientific name, is fairly common in somewhat deeper Florida Keys waters than those inhabited by the bonefish, and thus is not targeted by fishers seeking the "ghost of the flats." Yet another undescribed bonefish lacking a common or scientific name and identified only by DNA is reported to occur rarely in Florida. Bonefishes are readily distinguished from their cousins, tenpounders and tarpons, by their thick body, subterminal mouth, and gray lateral body bars. They grow to almost 3 ft and 18 lb.

Distribution and Habitat

Bonefishes (hereafter applied only to *Albula vulpes*) are normally found as far north as Indian River Lagoon on the east coast and Cape Romano on the west coast, but primarily they are a Keys and Biscayne Bay species. They are most conspicuous on shallow flats, where they have an affinity for seagrass beds and carbonate sandbottom in clear,

Bonefish (*Albula vulpes*)

shallow water. In general, most of the adults move onto Atlantic-side flats during winter months and then spread into the backcountry during summer. We have seen individual or small groups of adults as far north as Jupiter Inlet. On occasion we have observed large, tight schools of adult bonefishes moving along the east coast, from near Palm Beach to Jupiter, during April–May. Young adults and juveniles also occur along the coast, as far north as Stuart, during fall and winter months. These young are frequently caught by anglers fishing for Florida pompanos.

Natural History

Bonefishes spawn in Florida from November to May. Although they spend most of their time on shallow flats and in adjacent creeks or channels, evidence suggests that adults move into offshore waters to spawn. Like the related tarpons and tenpounders, bonefishes have leptocephalus larvae. These larvae are pelagic for at least 2 months. Juveniles are not well known, but they may be captured in shallow, sandy littoral zones on leeward or windward sides of barrier islands or keys. Bonefishes are well adapted for their feeding style, which involves rooting around in soft sandbottom, marlbottom, and seagrass bottom. The narrow, bony snout is pushed into the bottom at an angle, which causes the upper lobe of the tail to break the water surface—a phenomenon known as tailing. When feeding in soft sediment, individuals generate milky plumes that are referred to as "muds." Bonefish eyes are covered by a hard, clear lens that protects them while feeding in the sediment. Like triggerfishes, filefishes, and some rays, bonefishes can forcefully jet water through the mouth to displace buried prey. Studies conducted in the Florida Keys showed that bonefishes feed primarily on mud crabs, Gulf toadfishes, swimming crabs, and shrimps, but they also consume snails, clams, and worms.

Freshwater Eels
Family Anguillidae

- **American eel** (*Anguilla rostrata*)

Background

American eels are long, slender, and round in cross-section. Adults can achieve a length of 5 ft but rarely surpass 3 ft. The head is short; the snout is pointed, with a small mouth that has prominent lips; the lower jaw is noticeably longer than the upper jaw. Moderately long pectoral fins and body scales are present. The dorsal fin originates much farther back on the body than in other inshore eels: morays, snake eels, and conger eels. This fin is continuous with the anal and caudal fins. The upper body is greenish-gray, brownish-gray, or silvery-gray, often quite dark; the lower body is whitish to yellowish.

Distribution and Habitat

American eels usually are encountered in riverine estuaries and in freshwater streams and rivers. They occur statewide but most commonly are found on both sides of the northern Florida peninsula and panhandle. When at sea, they can occur in quite deep water during their spawning migration.

Natural History

In the late winter and early spring, American eels—the most famous of catadromous (living in freshwater but breeding in saltwater) species—spawn offshore in the Sargasso Sea, south of Bermuda, near where European eels (a separate species) also spawn. American eels go out with a bang, since the spawning event is their last accomplishment before dying. The developing lanceolate (leaf-shaped) larvae, known as leptocephali, miraculously make their way in the correct direction for their particular species—left (westward) or right (eastward)—largely through passive, ocean-current transport. Eventually (250 days to 1 year later), during the winter, they make landfall—at the mouths of rivers once occupied by their ancestors—after metamorphosing into the colorless glass-eel stage while over the edge of the continental shelf. They then turn into pigmented elvers, mini-eels if you will, shrinking in size during the whole transformation process. The little elvers head upstream in the spring, often crawling over dams and other impediments to reach streams, where they spend most of their life as yellow-colored eels, and then gradually move downstream into the estuary as they grow larger. After years in freshwater or estuarine waters, when their age / size is right, in the late winter and spring they enter the sea, after maturing sexually into silver-colored eels. Their eyes grow notably larger, and in the autumn they begin the long trek back to the Sargasso Sea, where it all began. All in all, an absolutely amazing feat of timing and endurance!

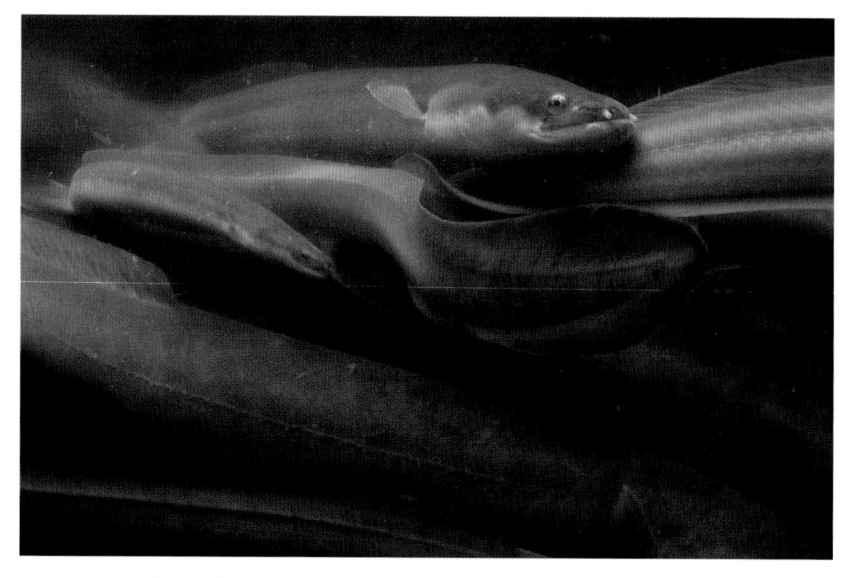

American eel (*Anguilla rostrata*). *Photo by Jay Fleming Photography.*

American eels are nocturnally active, hiding by day and emerging by night to forage for food. Omnivorous and voracious, they consume almost anything they can fit into their mouths, chiefly bottom-dwelling invertebrates and small fishes. Females reach much larger sizes than males. Eels are favorite targets for a host of marine predators and are highly favored by striped basses, a species that shares an affinity for estuarine waters.

Morays
Family Muraenidae

- **Pygmy moray** (*Anarchias similis*)
- **Chain moray** (*Echidna catenata*)
- **Fangtooth moray** (*Enchelycore anatina*)
- **Chestnut moray** (*Enchelycore carychroa*)
- **Viper moray** (*Enchelycore nigricans*)
- **Green moray** (*Gymnothorax funebris*)
- **Lichen moray** (*Gymnothorax hubbsi*)

- **Blacktail moray** (*Gymnothorax kolpos*)
- **Sharktooth moray** (*Gymnothorax maderensis*)
- **Goldentail moray** (*Gymnothorax miliaris*)
- **Spotted moray** (*Gymnothorax moringa*)
- **Blackedge moray** (*Gymnothorax nigromarginatus*)
- **Polygon moray** (*Gymnothorax polygonius*)
- **Honeycomb moray** (*Gymnothorax saxicola*)
- **Purplemouth moray** (*Gymnothorax vicinus*)
- **Redface moray** (*Monopenchelys acuta*)
- **Reticulate moray** (*Muraena retifera*)
- **Stout moray** (*Muraena robusta*)
- **Marbled moray** (*Uropterygius macularius*)

Background

Moray eels are well adapted for life in the constricted crawl spaces of the reef, possessing narrow heads and sinuous, scaleless, mucous-rich bodies lacking pectoral and pelvic fins. Florida's moray species can be distinguished by their coloration, dentition, and size. Two species have small dorsal and anal fins that are found only at the rear tip of the body. The rarely observed pygmy morays, the second smallest (to about 8 in) of the morays, have a tan to brown body that is covered with white stellate (star-shaped) spots and white markings on the chin and head. The similarly colored marbled morays, which reach 1 ft long, have a single small opening (the rear nostril) above each eye, while the pygmys have 2 (the rear nostril and a pore).

All other morays have longer dorsal and anal fins, the dorsal fin originating above or in advance of the gill opening. Redface morays, the smallest (to 5 in) and rarest of the morays, have a brownish-orange body, with a reddish-orange face. The dorsal fin originates approximately over the origin of the anal fin. Chain morays are moderate in size (to 20 in) and differ from other morays in having rounded molariform (vs. sharp and pointed) teeth; a stout body; rounded tail; stubby snout; and high forehead. The body has distinctive alternating, yellow and black, chain-like markings along the flanks. Ferocious-looking chestnut, viper, and fangtooth morays have long snouts and markedly curved jaws that cannot be closed completely, revealing large, pointed teeth. The easily confused chestnut and viper morays are uniformly dark-brown in color; fangtooths are brown, but with pale blotches. The smaller (to 13 in) chestnut morays have a small, horizontally oriented series of white spots on the lower jaw, each surrounding a pore, while the larger vipers (to 2 ft) and fangtooths (to 3.6 ft) lack that spotting but retain the pores.

The species of *Gymnothorax* and *Muraena* have prominent rows of teeth in their jaws and down the center of the roof of the mouths. Many of these morays are similar

Chain moray (*Echidna catenata*), Palm Beach, Florida

Viper moray (*Enchelycore nigricans*), Jupiter, Florida

in shape, but color patterns are useful in species recognition. Green morays are the big daddy of the clan, reaching 8 ft long. They are colored uniformly dark-green to greenish-yellow (actually, the body is bluish and the enveloping mucous is yellow, producing the green color). The coloration of the large (to 4.5 ft) purplemouth morays is variable, ranging from dark-brown to purplish, with dark speckling; in adults there is a submarginal dark stripe on the median fins. The inside of the mouth is dusky purple, and the eye is golden-yellow. Purplemouth morays can be confused with spotted morays, which have a white-cream-yellow base coloration with a dense overlay of dark-brown spots, or a dark base coloration with reticulating light markings—it's all a matter of your perspective. Spotted morays often have a dark margin on the front portion of the dorsal fin, leading to confusion with purplemouths when only the front part of the eel is seen poking out of a hole. Spotted morays grow to about 3.5 ft, the same size as sharktooth morays, which have a large head, with a high forehead. The latter are chocolate-brown to green, with light, round or worm-like patches posteriorly that meld into brown, worm-like patches at the front of the body. A dark spot surrounds the gill opening, and the dorsal and anal fins have light edges, characters spotted morays share with reticulate morays. The latter species is dark-brown, with moderately large pale spots or rosettes on the body and fins.

Stout morays are large (to 5.5 ft) and have a massive head, with a short snout; a pumpkin-yellow inner mouth; and reddish-brown coloration over the rear two-thirds of the robust body, with distinctive large, dark, giraffe-like spots—they are a sight to behold. Their spotting pattern fades to an almost uniform brown in older individuals. Goldentail morays are small (<18 in) and dark-brown to caffe-latte in color, with very small yellow spots covering the body; a yellow ring around the pupil of the eye; and, usually, a golden tip to the tail. Sometimes individuals show a reversal in their background coloration and the colors of spots. Lichen morays, even smaller at 12–14 in, are dark-brown, covered with light dashes arranged in a lichen-like pattern.

The remaining species have longer, more pointed tails. Honeycomb and blackedge morays are very similar appearing and can be confusing to separate. Both are about 2 ft long; are caffe-latte in color, with pale-white–yellow spots on the bodies; and have either dark and light spotting or undulating margins to the fins. Blackedge morays, however, have a continuous black-edged margin on the fins, interacting in a ying-yang pattern with a white, undulating, submarginal band, and more-uniform and smaller-sized body spotting. In contrast, the fin margins of honeycomb morays are mostly a series of half-ocellated spots, black within white, leading to a noncontinuous

Green moray (*Gymnothorax funebris*), Jupiter, Florida

Purplemouth moray (*Gymnothorax vicinus*), Palm Beach, Florida

dark margin to the fins. They also have greater diversity in body-spot sizes, with some notably larger than others (especially toward the tail) that achieve sizes much larger than the eyes. Blacktail morays grow a bit bigger (3.3 ft) and are readily identified by their coloration, ranging from caffe-latte anteriorly to dark-brown (almost black) on the rear quarter of the body, near the tail. On this are very small, pale, ocellated spots that begin to get bigger posteriorly, ending in only 3–4 very large, oblong spots caudally. The color pattern of polygon morays consists of a series of small dark spots on pale reticulations, filling in a large, dark, polygonal web.

Spotted moray (*Gymnothorax moringa*), Lake Worth Lagoon, Florida

Reticulate moray (*Muraena retifera*), Florida Middle Grounds

Distribution and Habitat

Morays are tropical forms and many species—marbled, redface, chain, viper, chestnut, green, purplemouth, and goldentail morays—are pretty much confined to extreme southern Florida, essentially to the Keys. Reticulate, honeycomb, and blacktail morays are the sole species found statewide, and pygmy morays occur in all peninsular and Keys waters. Blackedge morays are a northwestern Gulf species, known from off Mobile Bay, that may wander across the border into Florida waters. Spotted, polygon, lichen, and sharktooth morays occur only off the east coast and in the Keys; fangtooth and stout morays are restricted to eastcentral Florida. Three of these species—fangtooth, sharktooth, and stout morays—are eastern Atlantic species that no doubt have floated across the pond as leptocephalus larvae; sharktooth and stout morays may now be established (reproducing) in the northwest Atlantic.

Chain morays are residents of clear, shallow waters, where they inhabit rocky and sandy shorelines; although recorded to depths of about 40 ft, they seldom are found in waters that deep. Chestnut morays share that depth zone and inhabit rocky shorelines and shallow patch reefs. Green morays, a highly adaptable species, live wherever there is structure—tide pools, dock pilings, tidal (even brackish) creeks, mangroves, and reefs—from the shoreline to depths of about 110 ft. Spotted and purplemouth morays range from depths of 30 to 300 ft or more. They are shallow-water species when on rocky shorelines, seagrass beds, and reefs in south Florida, but they are said to be found in 150–350 ft hardbottom habitats in the northern parts of their ranges. Viper morays also occur in this depth stratum; they are rocky shoreline–patch, reef–sandy bottom eels found to about

Stout morays (*Muraena robusta*), Jupiter, Florida

100 ft deep. Redface morays are a rare, rocky rubble and coral-head dweller, in 35–150 ft of water. Goldentail morays are other shallow to mid-depth (5–164 ft) residents.

Versatile reticulate morays are found on reefs, rocky ledges, sand, and mud, in depths of 65–300 ft. Marbled morays are rocky rubble–reef inhabitants, most often seen at depths of about 100 ft but occurring as deep as 450 ft. Pygmy morays also are rocky rubble–reef residents, a wide-ranging (15–600 ft depths) species normally found at 100–300 ft. Also occurring in mid-depths (100–225 ft) are reef and rocky ledge–dwelling stout morays.

A deeper-water contingent includes lichen morays (200–300 ft), polygon morays (300–850 ft), fangtooth morays (>500 ft), and sharktooth morays (500–925 ft). Three related species typically frequent low-relief habitats, such as offshore sand-mud banks, calico scallop beds, jetties, and seagrasses: blackedge and honeycomb morays, most common in depths of 30–150 ft; and blacktail morays, found deeper, at 150–750 ft.

Natural History

A scant number of direct observations suggest that morays pair-spawn. Males of some moray species fight by locking jaws. Occasionally a spotted moray is seen with cuts on its head, possibly from fights with other spotted morays. Spawning locations and timing are largely unknown, but spotted and reticulate morays apparently are summer spawners, and some other moray species may spawn near their cousins (conger eels and American eels) in the Sargasso Sea. Following presumed pelagic spawning, fertilized eggs drift for up to 100 days. The larval stage, known as leptocephalus, is so bizarre that early naturalists originally placed eel larvae into their own genus, *Leptocephalus*. Leptocephalus larvae usually are longer than juvenile eels, a reduction in size being the norm during metamorphosis.

Morays generally leave their daytime hiding places to move around the reef at night, searching for prey, often in nearby seagrass beds. They feed on fishes (including other morays) and invertebrates. Chain morays are crab specialists that even leave the water on occasion when chasing their prey. In a study performed in an area of co-occurrence, spotted morays foraged twice as often as purplemouths and consumed more crabs than fishes, while purplemouths ate more fishes; surprisingly, both species foraged more often and were more successful on stormy

Goldentail moray (*Gymnothorax miliaris*), Palm Beach, Florida

nights than on calm ones. Morays have large heads, thanks in part to the massive muscles that operate their jaws. The sharp teeth rimming the jaws and crossing the palate are effective at snagging prey, but the small mouth does not generate sufficient suction to ingest the prey. To accomplish that, morays have a remarkable adaptation—a second set of pharyngeal jaws that thrust forward when the oral jaw closes, helping to pull the prey into the throat. Night-feeding spotted and goldentail morays have been observed with an entourage of coneys, graysbies, and greater soapfishes that are eager to grab startled prey items spooked by foraging morays.

A recent study suggests that spotted and reticulate morays live at least 23–24 years. Morays appear to be pretty parochial. Limited tagging reveals that green and purplemouth morays stay on their same reefs over several months; anecdotally, divers have reported seeing the same green moray on a reef for as long as 10 years. Occupation of individual shelters by chain, viper, purplemouth, and spotted morays have been shown to be temporary in nature, however, lasting a few days at a time, but good places are recycled—more than a single species use the same safe houses at one time or another, even being there together. Goldentail morays, in contrast, are more faithful to their homes, residing for at least a 6-week period in the same spot. Many morays are secretive, and most are will not cause harm unless provoked, but spotted and purplemouth morays can be uncommonly aggressive both in and out of the water. Recently a 5 ft green moray, used to being fed by tourists, attacked a nonfeeding diver in Cuba. Lobster-hunting divers frequently are bitten when blindly reaching into a hole occupied by two tenants: one tasty, and the other toothy. One of the authors (DBS) was bitten by a spotted moray when reaching under a ledge to grab an attractive seashell.

Snake Eels
Family Ophichthidae

- **Key worm eel** (*Ahlia egmontis*)
- **Tusky eel** (*Aplatophis chauliodus*)
- **Stripe eel** (*Aprognathodon platyventris*)
- **Academy eel** (*Apterichtus ansp*)
- **Finless eel** (*Apterichtus kendalli*)
- **Sooty eel** (*Bascanichthys bascanium*)
- **Whip eel** (*Bascanichthys scuticaris*)
- **Shorttail snake eel** (*Callechelys guineensis*)
- **Blotched snake eel** (*Callechelys muraena*)
- **Ridgefin eel** (*Callechelys springeri*)
- **Slantlip eel** (*Caralophia loxochila*)
- **Spotted spoon-nose eel** (*Echiophis intertinctus*)
- **Snapper eel** (*Echiophis punctifer*)
- **Irksome eel** (*Gordiichthys ergodes*)
- **Horsehair eel** (*Gordiichthys irretitus*)
- **String eel** (*Gordiichthys leibyi*)
- **Sailfin eel** (*Letharchus velifer*)
- **Surf eel** (*Ichthyapus ophioneus*)
- **Sharptail eel** (*Myrichthys breviceps*)
- **Goldspotted eel** (*Myrichthys ocellatus*)
- **Broadnose worm eel** (*Myrophis platyrhynchus*)
- **Speckled worm eel** (*Myrophis punctatus*)
- **Margined snake eel** (*Ophichthus cruentifer*)
- **Shrimp eel** (*Ophichthus gomesii*)
- **Blackpored eel** (*Ophichthus melanoporus*)
- **Spotted snake eel** (*Ophichthus ophis*)
- **Palespotted eel** (*Ophichthus puncticeps*)
- **King snake eel** (*Ophichthus rex*)
- **Diminutive worm eel** (*Pseudomyrophis fugesae*)
- **Blackspotted snake eel** (*Quassiremus ascensionis*)

Background

The Ophichthidae is most speciose of the eel families. Snake and worm eels are seldom seen, but some of their species are relatively common. Once placed in separate families, current thinking unites the two groups into a single family. Like morays, they have muscular jaws, abundant teeth, and tubular nostrils, but snake eels have stiff, pointed tails (most lack a caudal fin entirely). Other eels, including worm eels, have a more flexible tail, with a caudal fin.

Owing to the rarity of capture

and difficulty of identification for many species, we confine our discussion to the species that are most frequently observed. King snake eels are the largest (reaching 7 ft) snake eel and rival conger eels and green morays as the largest eel in the western Atlantic. They have 14–15 dark saddles on the upper body. Goldspotted, sharptail, palespotted, and spotted eels, as well as snapper eels and spotted spoon-nose eels, have dorsolateral body spotting; differences in spot sizes, colors, and patterns facilitate species recognition. Most other species in this family are more uniformly colored, including Key and speckled worm eels and shrimp eels. Shrimp eels are dark-brown above, pale below, and have dusky fins. Key and speckled worm eels are brown to tannish, with dark, pepper spots (more visible in speckled worm eels); the latter's dorsal fin originates well in advance of the beginning (origin) of the anal fin (posterior to the anal fin in Key worm eels).

Shrimp eel (*Ophichthus gomesii*), Lake Worth Lagoon, Florida

Distribution and Habitat

Snake eels live under rocks and in mud-, sand-, and rubble bottom, emerging only to forage or mate. Although they are bottom dwellers, many snake eels surprisingly make treks to the water surface at night, attracted to lights on docks or ships, as well as to full and near-full moons on cloudless nights. In Florida, species that are associated with hardbottom include sharptail, goldspotted, and blackspotted snake eels, while other species—such as king snake eels, sooty and snapper eels, and spotted spoon-nose eels—inhabit soft sediment. Most of the species are found statewide if their preferred habitat is available. Exceptions are margined snake eels and blackpored eels (east coast only); horsehair and snapper eels, and blotched, shorttail, and king snake eels (essentially confined to the Gulf); and some tropical species that are pretty much south Florida residents, including stripe, slantlip, sharptail, goldspotted, and surf eels, spotted snake eels, and Key worm eels.

Natural History

Hard, pointed tails allow snake eels to burrow tail-first into the sediment, and the pointed snouts of some species probably aid in head-first eruption from the substrate. Like their moray cousins, snake eels aggressively bite when threatened, so care should be taken when attempting to dehook or pick up a writhing eel. The supersized king snake eel happily lunges at any human body part that gets too close, performing a passible

Sharptail eel (*Myrichthys breviceps*), Palm Beach, Florida

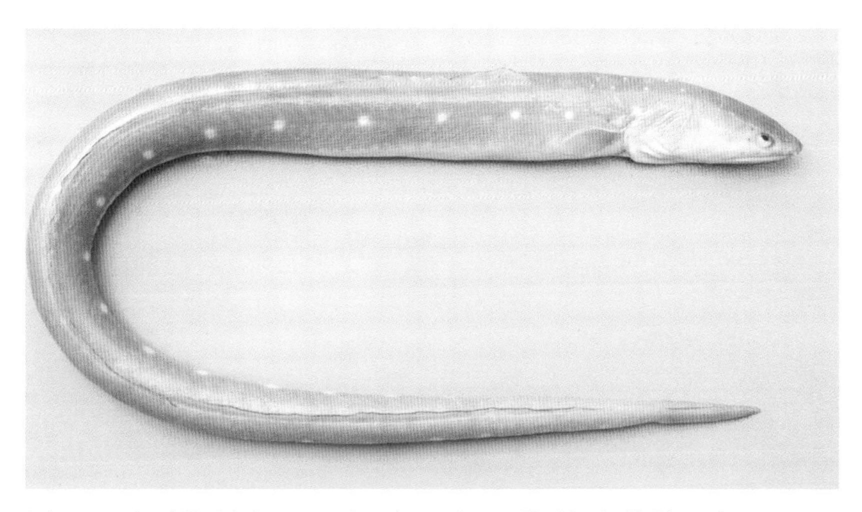

Palespotted eel (*Ophicthus puncticeps*), southwest Florida shelf. *Photo by George H. Burgess.*

imitation of a cobra when rearing off a deck. Although most species appear to be nocturnally active, sharptail eels forage out in the open, on reefs and hardbottom areas, during the day. As they move from place to place, looking under rocks and in crevices for food, they often are accompanied by a shadow fleet of small fishes, hoping to gain access to any startled prey items.

Snake and worm eels are broadcast spawners. Many of the species in this family that normally inhabit inshore waters move offshore to spawn, migrating at night. As with the other eel species, snake and worm eels have leptocephalus larvae, which favor wide-scale dispersal of the young. Speckled worm eels, the best-studied species of the group, spawn in the ocean near the Bahamas and off North Carolina, and their larvae are transported by the Gulf Stream and other currents. After 50–100 pelagic days, they arrive in estuarine inlets and river mouths, shrinking—yes, shrinking—to lengths of 2–4 in as part of the metamorphosis process. They then settle in estuarine sediment and grow into the adult stage, consuming amphipods, worms, and other small, benthic critters. Young often feed in seagrass beds.

Conger Eels
Family Congridae

- **Bandtooth conger** (*Ariosoma balearicum*)
- **Bullish conger** (*Bathycongrus bullisi*)
- **Conger eel** (*Conger oceanicus*)
- **Manytooth conger** (*Conger triporiceps*)
- **Blackgut conger** (*Gnathophis bathytopos*)
- **Longeye conger** (*Gnathophis bracheatopos*)
- **Brown garden eel** (*Heteroconger longissimus*)
- **Yellow garden eel** (*Heteroconger luteolus*)
- **Margintail conger** (*Paraconger caudilimbatus*)
- **Splendid conger** (*Pseudophichthys splendens*)
- **Yellow conger** (*Rhynchoconger flavus*)
- **Whiptail conger** (*Rhynchoconger gracilior*)
- **Guppy's conger** (*Rhynchoconger guppyi*)
- **Threadtail conger** (*Uroconger syringinus*)

Background

As with snake eels, secretive, largely solitary conger eels rarely are encountered by fishers or divers. The most obvious species is the conger eel, the largest of its family, which grows to 7.5 ft and close to 90 lb. Like other congers, they have pectoral fins and, similar to most other congers, they are dark-grey to brown in coloration. The continuous dorsal-caudal-anal fin is edged with a dark margin, and the upper jaw projects out past the lower jaw, separating this species from manytooth congers, which are similarly colored but have equal-sized jaws and are smaller (to 3 ft). The next-largest congers (to a bit under 2 ft) are margintails—which are brownish, with a dusky tip on the jaw, and fins with light margins—and

Left: Yellow garden eel (*Heteroconger luteolus*), Lake Worth Lagoon, Florida. *Right*: brown garden eel (*Heteroconger longissimus*), Lake Worth Lagoon, Florida.

yellow congers. Yellow congers are yellowish-brown, with black margins on the rear portions of the dorsal-caudal-anal fin; lack a dusky jaw tip; and have a shorter trunk (the tip of the snout to the anus is about one-third its total length, just under 0.5 in) than margintails. Bandtooth congers reach only about 1 ft long; are golden-brown, with dark median fin margins; have large eyes; and have a long snout that notably overhangs the lower jaw. Aside from garden eels, which once were placed in their own family, other congers in the area look a lot alike and specimens require careful examination. Brown and yellow garden eels are very long and thin, with small heads, tiny snouts, protruding lower jaws, and large eyes; they are separated out from the others by their body color. Yellow garden eels are golden-brown to light-tan anteriorly, with a white belly; the body gradually turns pale posteriorly (towards the tail). Brown garden eels are dark-brown, with tiny yellow flecks anteriorly, fading to pale-yellow posteriorly. They have white throats and bellies, and the margin of dorsal fin is white.

Distribution and Habitat

Most conger eels live in the sand that covers much the continental shelf, but some species prefer softbottom, near reefs. Brown and yellow garden eels, and bandtooth, margintail, and manytooth congers live near structured habitats; of these, only garden eels are social. Bandtooth, yellow, and margintail congers are found statewide. Bandtooths occupy softbottom, from the shoreline to 1500 ft deep; yellows prefer sandy bottoms, in depths of 50–600 ft; and margintails live near reefs in nearshore waters, to about 150 ft. In the northern part of their range, conger eels are seen on nearshore jetties, down to 1500 ft, but off Florida they are encountered only on deep (>200 ft) reefs and wrecks. They also occupy burrows constructed by tilefishes, in depths of 400–800 ft (see chapter 75). Manytooth conger eels are found from the panhandle and around the peninsula to southeastern Florida, in 10–500 ft depths. Brown garden eels are only known from

Margintail conger (*Paraconger caudilimbatus*), Lake Worth Lagoon, Florida

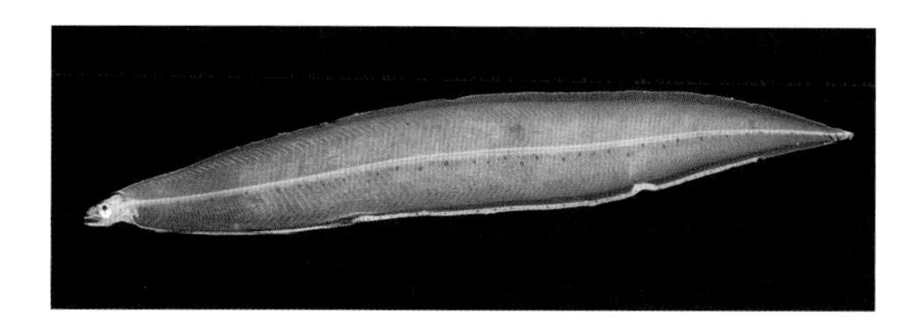
Leptocephalus larva from an unidentified conger eel

the Florida Keys and southeast Florida, and the closely related yellow garden eels occur from the west Florida shelf around to at least Stuart on the east coast, in depths ranging from 10 to 150 ft.

Natural History

Except for conger eels (*Conger oceanicus*) themselves, very little is known about reproduction in most of the conger eel family. All are secretive; many burrow in the sand, while others retreat into crevices in hardbottom or reefs. We have seen manytooth and yellow congers only during night dives, suggesting nocturnal activity.

Brown and yellow garden eels live in tight-fitting burrows in the sand. It is not unusual to see dozens or even hundreds of garden eels arching out of their burrows and facing into the current, picking out plankters. A cluster of individuals resembles a field of plants, hence their vernacular name, garden eel. The large, close-set eyes of garden eels allow for good binocular vision, facilitating plankton feeding. Related garden eels from the Red Sea secrete mucous from special skin cells in the tail. Presumably the mucous helps bind sand grains lining the burrows, making a smooth tube. The Atlantic species probably also produce mucous, as their burrows are smooth, looking like soda straws buried in the sand.

Conger eels are the best-studied member of the Congridae, in part because they are the only real fishery species in this family. Like American eels, conger eels and manytooth conger eels travel to the Sargasso Sea to reproduce in the fall and winter, a massive migration for some individuals that leaves the adults as mere shadows of themselves by the time they reach their destination. The large, greatly specialized leptocephalus larvae travel up to tens of thousands of miles from the Sargasso Sea before they transform into glass eels, then elvers, and finally into early juveniles in estuaries and nearshore waters (leptocephali have even been found in tide pools); they then settle in shallow water. Most larvae and transforming stages of conger eels are found from North Carolina northward, suggesting that subadult and adult congers encountered in Florida probably are Yankee migrators that headed south with growth, eels that initially had passed right by the state as leptocephali while riding the northward-flowing Gulf Stream.

Anchovies
Family Engraulidae

- **Key anchovy** (*Anchoa cayorum*)
- **Cuban anchovy** (*Anchoa cubana*)
- **Striped anchovy** (*Anchoa hepsetus*)
- **Bigeye anchovy** (*Anchoa lamprotaenia*)
- **Dusky anchovy** (*Anchoa lyolepis*)
- **Bay anchovy** (*Anchoa mitchilli*)
- **Flat anchovy** (*Anchoviella perfasciata*)
- **Silver anchovy** (*Engraulis eurystole*)

Background

Anchovies, widely known as "glass minnows," are the quintessential forage fishes or "baitfishes" in shallow waters around the state. At least eight species live in Florida waters. At a glance, all of these fishes look the same—big eyes; big mouth; and dusky gray coloration, with a wide, silver, lateral stripe that looks like mylar. While identification is difficult, a few subtle differences can be used to distinguish the various species. Flat and silver anchovies have relatively short maxillae (upper jaw bones) that do not extend posteriorly to the preopercular margin, separating them from the *Anchoa* species. The latter have long maxillae that extend to the preopercular margin. Relative snout length and relative placement of the anal and dorsal fins can be useful in separating the *Anchoa* species. For instance, the relatively short-snouted bay anchovies have dorsal and anal fins originating over each other. Cuban anchovies look very similar to bay anchovies, but the former's dorsal fin begins above a point forward of the anal fin. The anal-fin origin in long-snouted dusky anchovies is located under the rear base of the dorsal fin. Key, bigeye, and striped anchovies have their anal fins originating under the rear half of the base of their dorsal fins. Key and bigeye anchovies have medium-length snouts, while the snouts of striped anchovies are long. Other characters, such as the width of the silver stripe, also help distinguish the various species.

Dusky anchovy (*Anchoa lyolepis*), Indian River Lagoon, Florida

Bay anchovies (*Anchoa mitchilli*), Indian River Lagoon, Florida

Distribution and Habitat

Anchovies are variously distributed around Florida in inshore, coastal, and shelf waters less than 100 ft deep. Bay, Cuban, silver, and flat anchovies are found around the state, mostly in estuarine and coastal waters. Striped anchovies also occur around the state, but they are uncommon to rare in the Southeast and Southwest ecoregions and absent from the Florida Keys. As their name implies, Key anchovies are restricted to the Florida Keys and southeastern Florida, as are bigeye anchovies. Dusky anchovies are found off eastern Florida but are rare on the west coast.

Natural History

Anchovies are water-column spawners, producing distinctive, elliptically shaped eggs that transform into larvae in a matter of hours. Bay anchovies spawn in coastal and shelf waters, with the females releasing 500 eggs per day over a multimonth spawning season. The early juveniles—transparent slivers with eyes—move into shallow coastal or inshore waters. Little is known about reproduction in the other anchovy species.

In the shallows of southeastern Florida, anchovies often school with juvenile Spanish sardines, round scads, false pilchards, and bigeye scads. Schools disband at night and foraging occurs individually or in smaller groups, presumably doing so to avoid being eaten by the multitudes of fishes and birds that are active by day. Anchovies feed on plankton throughout the water column, their large mouths being well suited for plankton consumption. Some species also eat tiny, bottom-dwelling invertebrates. On calm days, feeding anchovies look like raindrops hitting the water surface.

Anchovies collectively form an important link between plankton and larger fishes in coastal and estuarine food webs. All the notable predators—such as jacks, needlefishes, barracudas, mackerels, snooks, tarpons, tenpounders, and flounders—feed on anchovies. Individuals or schools are attracted to lights at night. In southeastern Florida, it is possible to encounter multiple species of anchovies in a single throw of a cast net or pull of a seine.

Herrings
Family Clupeidae

- **Blueback herring** (*Alosa aestivalis*)
- **Alabama shad** (*Alosa alabamae*)
- **Skipjack shad** (*Alosa chrysochloris*)
- **Hickory shad** (*Alosa mediocris*)
- **American shad** (*Alosa sapidissima*)
- **Gulf menhaden** (*Brevoortia patronus*)
- **Yellowfin menhaden** (*Brevoortia smithi*)
- **Atlantic menhaden** (*Brevoortia tyrannus*)
- **Gizzard shad** (*Dorosoma cepedianum*)
- **Round herring** (*Etrumeus teres*)
- **False pilchard** (*Harengula clupeola*)
- **Redear sardine** (*Harengula humeralis*)
- **Scaled sardine** (*Harengula jaguana*)
- **Dwarf herring** (*Jenkinsia lamprotaenia*)
- **Little-eye herring** (*Jenkinsia majua*)
- **Shortband herring** (*Jenkinsia stolifera*)
- **Atlantic thread herring** (*Opisthonema oglinum*)
- **Spanish sardine** (*Sardinella aurita*)

Background

Herrings are small- to medium-sized schooling fishes, with big mouths designed for wide-open plankton feeding. They are known to most Floridians simply as "baitfishes." All but round, dwarf, little-eye, and shortband herrings have sharp scutes (modified scales) along the belly, and most have blue-green backs, silver sides, and a laterally compressed body. The family is diverse, and the various species often are tough to identify. Two herring species jump out by having an elongate, thread-like ray in the last dorsal fin: Atlantic thread herrings (a.k.a. "greenbacks" or "greenies") and gizzard shads. That ray is longer in Atlantic thread herrings. Both species are silvery, with faint horizontal lines on the body and a black shoulder spot behind the gills (much larger in gizzard shads). The two are easily identified by looking at the head (round and bulbous in gizzard shads) and tips of the dorsal and caudal fins (dipped-in-ink coloration in Atlantic thread herrings).

Members of the genus *Alosa* can also be difficult to separate, but geography helps. Two species, Alabama shads and skipjack shads, occur only in the panhandle and are usually seen in rivers. The other three—American shads, hickory shads, and blueback herrings—occur only on the extreme northeast coast and are usually encountered only in the St. Johns River estuary. Alabama shads grow to almost 20 in and have a shoulder spot behind the gills; the middle of the upper jaw is deeply notched. Skipjack shads reach a similar size but have a more slender jaw notch, a more protruding lower jaw, and small teeth; they also lack the black spot. The best way to confidently separate these two species is by counting gill rakers: Alabama shads have 41–48 on the lower gill arch, while skipjack shads have 20–24. The largest of our herrings—hickory and American shads—grow to 2 ft long and have well-developed, notched jaws; prominent shoulder spots; and about 5 smaller, diffuse, trailing spots. Hickory shads have a more protruding lower jaw and small teeth. The appropriately named blueback herrings have a pretty blue back, a single shoulder spot, grow to about 15 in, and are the most slender of the shads.

The three species of menhadens, often collectively referred to as "pogies" or "bunkers," have especially big heads and can be distinguished by their color patterns. Gulf menhadens are confined to the Gulf, and Atlantic menhadens are found only on the east coast, facilitating the identification process. All three species have dark shoulder spots, but only Gulf and Atlantic menhadens have trailing spots: Gulfs have 1 or 2 rows and Atlantics have about 6 rows, which make the middle part of the body look like it

Yellowfin menhadens (*Brevoortia smithi*), Satellite Beach, Florida

has been hit with birdshot. In life, yellowfin menhadens are distinguished by having golden-yellow dorsal and tail fins, but these colors fade after death. Gulf menhadens have dusky, pale-yellow fins, and often the sides are brassy in color (silver to slightly brassy in Atlantic menhadens, and silver in yellowfin menhadens).

The genus *Harengula* is represented by false pilchards, redear sardines, and scaled sardines. The latter two names are unfortunate, as these are not really sardines; all should probably be called pilchards. Redear sardines are the most distinctive of these three species: they have a dark shoulder spot, faint orange stripes, and an orange spot behind the gill in life; unfortunately, these colors fade after death. False pilchards and scaled sardines are very difficult to tell apart, both being silvery, with green backs, and having a single black shoulder spot. False pilchards are slightly more slender and have a bit larger eye; the dorsal and caudal fins of scaled sardines often are more yellow than those of false pilchards. Spanish sardines, the true sardines of the bunch, are easily recognized by their slender body and green back.

Round herrings—a moderate-sized (to 10 in), slender, small-headed, round-bodied sardine, lacking in scutes—are found only in deep water. In life, the upper body has a brassy sheen (the color again is lost after death). The three *Jenkinsia* species are small (up to 2.5–3 in) and very slender and, in having a central silver body stripe, may be mistaken for anchovies or silversides. These species are almost impossible to identify unless a specimen is in hand.

Distribution and Habitat

Alabama and skipjack shads range from the panhandle to the Suwanee River. American shads, hickory shads, and blueback herrings are found off east Florida to Cape Canaveral, mostly in the St. Johns River. Yellowfin menhadens occur around the state in coastal and shelf waters. Gulf menhadens are only found on the west coast, from the panhandle to about Tampa Bay. Atlantic menhadens occur on the east coast, only to about Cape Canaveral. Young menhadens reside in estuaries, bays, and

Scaled sardines (*Harengula jaguana*), Boca Raton, Florida

False pilchard (*Harengula clupeola*), Lake Worth Lagoon, Florida

Redear sardines (*Harengula humeralis*)

Spanish sardine (*Sardinella aurita*), Lake Worth Lagoon, Florida

lagoons. Scaled sardines are found around the state, in shallow water (<20 ft). False pilchards and redear sardines occur in southeast Florida through the Florida Keys and, on the west coast, to about Tampa, in water depths of less than 50 ft. Dwarf herrings only occur off the east coast, and little-eye and shortband herrings only in the Florida Keys. Round herrings are found around the state but, unlike all other members of this family, live exclusively in deep water, out to the shelf and beyond. Spanish sardines and Atlantic thread herrings occur around the state, in shelf and coastal waters. Young of both species are found in lagoons and estuaries. Gizzard shads occur in coastal rivers around the state. Although they prefer freshwater reaches, they will occasionally venture into the saline lower segments.

Natural History

Herrings are water-column spawners, with prolific fecundity. Alabama, hickory, and American shads, as well as blueback herrings, are anadromous species that spend most of their lives in the ocean, but each year they ascend coastal rivers to spawn in freshwater. Florida's shad populations are, on average, smaller, younger, and produce more eggs than the same species do in river systems north of Florida on the U.S. East Coast. Shad spawning runs have declined from historical levels in all rivers of the southeastern United States; this is particularly true with American shad and hickory shad runs in Florida's St. Johns River. Hickory shad still enter the river in December and remain through March. Menhadens spawn over the continental shelf, but their larvae settle in inshore waters, where they spend part of their first year. False pilchards, redear and scaled sardines, Atlantic thread herrings, and Spanish sardines spawn in coastal or shelf waters, and their young move into shallow coastal or inshore waters. Little is known about spawning in dwarf and round herrings.

Herrings feed on plankton, either by ram filtering or a visually-oriented plucking of individuals from the water column. Menhaden species are ram filter feeders, swimming with their mouths agape, which act like tiny nets to scoop in plankters. Herrings consuming individual plankters often create a surface disturbance; you may see single flicks, or a mass of fishes feeding in unison. Atlantic thread herrings, Spanish sardines, and pilchards eat constantly during the day. The three *Harengula* species school in very shallow water in daytime but disband at night to feed on plankton under the guise of darkness. Many of the species in the herring family are attracted to dock or boat lights at night. In addition, many of the species school together, or with species of similar shape. Spanish sardines regularly school with similarly sized round scads.

The abundant herrings—with their oily, rich flesh and lack of spines—are on the menu of just about every predatory fish that swims. This is why most of them are esteemed as bait and are collectively called "minnows," "whitebaits," "pilchards," and "hadens" by fishers.

Sea Catfishes
Family Ariidae

- **Hardhead catfish** (*Ariopsis felis*)
- **Gafftopsail catfish** (*Bagre marinus*)

Background

Catfishes are mostly freshwater inhabitants, but sea catfishes prefer marine and estuarine waters. The two Florida species are colored steely gray to blue dorsally, with white undersides, but they are easily distinguished by comparing the dorsal and pectoral fins. Gafftopsail catfishes have very long, fleshy filaments, extending from the dorsal and pectoral spines, which are lacking on hardhead catfishes. Hardheads have 3 pairs of rounded barbels on the chin and upper jaw, while gafftopsails have 2 pairs of flattened, ribbon-like chin and upper-jaw barbels, the latter much longer than those of hardheads. Gafftopsail catfishes grow to a larger size (3 ft) than hardhead catfishes (2.2 ft).

Distribution and Habitat

Sea catfishes are continental species that occur in estuaries, lagoons, and shallow coastal waters in all areas of the state except the Keys. We have observed large schools of hardhead catfishes in coastal waters off eastern Florida during January, February, and March of some years. Such individuals may have left cooler inshore waters in search of warmer water. Those left behind in lagoons and bays are readily killed in winter cold snaps.

Natural History

Both species have similar life-history patterns, highlighted by the mouth brooding of eggs and young. Each species produces small numbers of relatively large eggs that are incubated in the mouth of the male. Spawning occurs from May to August, usually in estuarine waters. Mature females develop fleshy tabs on their pelvic fins. Mature eggs look like minigrapes, as they are very large (between 0.5 and 0.75 in) and usually yellow or green in coloration. Because of their size, only 20–65 eggs are produced per female. After release by the female and fertilization, the eggs are gently picked up by the male. Males brood the eggs in their mouths, where the eggs are kept clean and aerated. Developing young, 2–3 in long, still with an attached yolk sac, live for about 2–4 weeks in the mouth of the brooding male. Amazingly, the males still manage to eat while doing this, and occasionally a hooked male will spew eggs or little catfishes on deck.

Sea catfishes are opportunistic foragers that favor amphipods, shrimps, and crabs while young, and small fishes as adults. They are mainstays around docks, particularly where fishes are regularly filleted or gutted and scraps are tossed over the side. Sea catfishes are well-known followers of shrimp trawlers, both in inshore and shelf waters, where they reap the bounty of bycatch (nontarget species) tossed over the side. We have noted that hardhead catfishes learn the particular sounds of certain boats

Hardhead catfish (*Ariopsis felis*), Loxahatchee River, Florida

and follow them into the slip for an easy meal. Their ability to differentiate the low-frequency sound of an engine is not surprising, as they possess directional hearing. This allows them to acoustically detect obstacles by producing short bursts of low-frequency sound pulses that reflect back to themselves—a primitive form of echolocation. Like all catfishes, their sense of smell is acute, and hardhead catfishes can even detect an alarm substance given off by one of their mates, which results in "I'm outta here" behavior.

When caught, both species will sound off by croaking and stridulating (sounds produced by vibrating a small bone on the swim bladder). Gafftopsail catfishes produce prodigious amounts of thick white slime. For recreational fishers, the telltale sign that a gafftopsail catfish has been tampering with your bait is a bead of slime running along the line, well above the terminal tackle. Commercial fishers are not appreciative of how they slime up nets. Both species have stout, serrated pectoral and dorsal spines, with toxin or venom associated with these spines. Anglers punctured by catfish spines experience a great deal of pain. Apparently sharks don't mind, as a variety of species readily munch on sea catfishes, often bearing spines in their jaws and throats as goodbye gifts from the cats. Bottlenose dolphins love them, too, but the spines may pierce the dolphin's esophagus or stomach wall, resulting in fatal infections; one dolphin died after a spine migrated to its lung, leading to pneumonia. In brackish waters, sea catfishes can become prey for water snakes.

Lizardfishes
Family Synodontidae

- **Largescale lizardfish** (*Saurida brasiliensis*)
- **Smallscale lizardfish** (*Saurida caribbaea*)
- **Shortjaw lizardfish** (*Saurida normani*)
- **Inshore lizardfish** (*Synodus foetens*)
- **Sand diver** (*Synodus intermedius*)
- **Largespot lizardfish** (*Synodus macrostigmus*)
- **Offshore lizardfish** (*Synodus poeyi*)
- **Bluestripe lizardfish** (*Synodus saurus*)
- **Red lizardfish** (*Synodus synodus*)
- **Snakefish** (*Trachinocephalus myops*)

Background

Lizardfishes are very elongate, round-bodied bottom dwellers, with sharp teeth; a small, rubbery adipose (lacking rays) fin on the back; and, with one exception, pointed snouts. Snakefishes are the blunt-snouted exception. They grow to 10 in long and have eyes located forward of the center of the mouth, an upturned mouth, and distinctive coloration: alternating yellow-gold and blue stripes on the body and fins, and a dark scapular spot at the upper corner of the gill cover.

The pelvic rays of members

Snakefish (*Trachinocephalus myops*), Lake Worth Lagoon, Florida

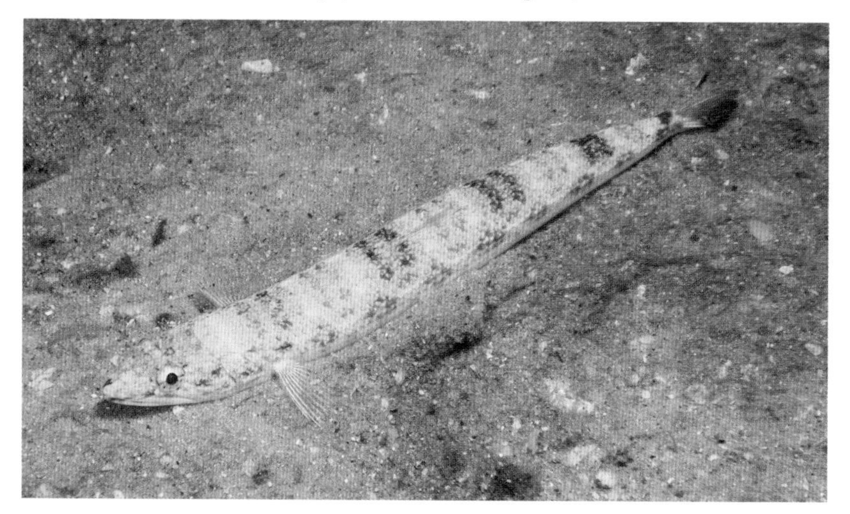

Inshore lizardfish (*Synodus foetens*), Jupiter, Florida

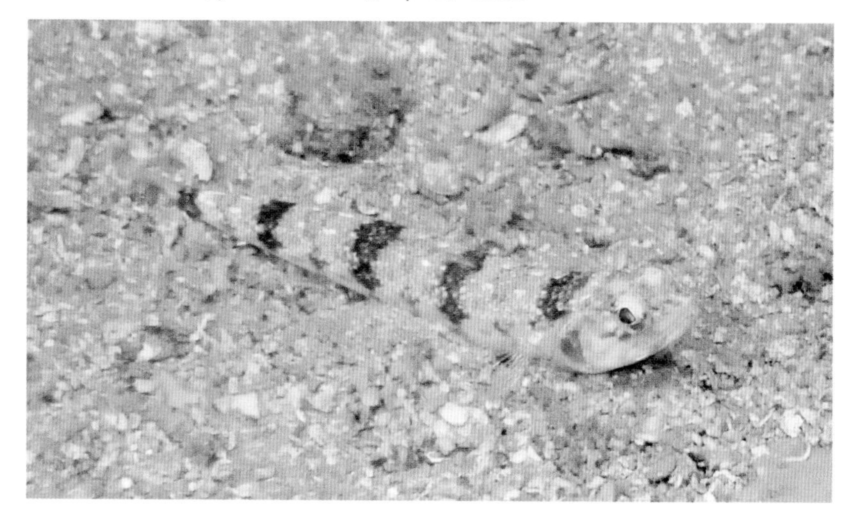

Offshore lizardfish (*Synodus poeyi*)

of the genus *Saurida* are of equal length (variable lengths in other lizardfishes). Largescale and smallscale lizardfishes are small, reaching about 6–7 in and 3–4 in, respectively. Largescales have a dark diagonal band on the dorsal fin; a lower jaw that is longer than the upper jaw; and 3 scale rows between the lateral line and the dorsal fin. Smallscales, in contrast, have 4 rows of scales between the lateral line and the dorsal fin. Shortjaw lizardfishes are especially skinny, with a slightly rounded snout; large eyes; and a silvery-grey body, with 5–6 small, midline blotches.

The most-often-seen lizardfishes are in the genus *Synodus*. Small (7 in) bluestripe lizardfishes have thin, horizontal blue lines on the upper body and 8 dark, diamond-shaped blotches on the flanks. The larger (16 in) inshore lizardfishes are tan to brown, with up to 8 dark, diamond-shaped flank blotches. Offshore lizardfishes (12 in) also have 8 diamond-shaped blotches, but they have larger scales than other *Synodus* species, and a fleshy knob on the lower jaw. Red lizardfishes (10 in) have reddish-brown bars on the dorsal and caudal fins, 4 similarly colored bands on the back, and a tiny black spot near the tip of the snout. Sand divers (16 in) have a small scapular spot, 3–6 dark bands on the tail, and diamond markings that meld with the bars on the back. The similar largespot lizardfishes (12 in) have a large scapular spot and lack bands on the tail (the lower lobe of the tail is dark).

Distribution and Habitat

Most lizardfish species sit directly on or lie buried in sand- or mud-

Red lizardfish (*Synodus synodus*)

Sand diver (*Synodus intermedius*), Venice, Florida

Inshore lizardfish (*Synodus foetens*), with recently caught bandtail puffer (*Sphoeroides spengleri*), Lake Worth Lagoon, Florida

bottom in inshore, coastal, and shelf waters. Bluestripe lizardfishes occur along the east coast, including the Florida Keys, in water depths of <60 ft. *Saurida* species generally occur in deeper (60–1800 ft) waters. Members of another species, the inshore lizardfish, lie partially buried under sand in estuaries, near reefs, in hardbottom, or in seagrass meadows, in waters depths of 1–80 ft, but they have been reported from as deep as 600 ft. Sand divers live over sand, usually on the periphery of reefs or hardbottom, from nearshore to 855 ft, but most commonly in depths of less than 180 ft. Unlike the other sand-loving species, red lizardfishes prefer reefs and hardbottom habitats to depths of almost 300 ft, where they lie on open hardbottom, sponges, or live corals. Three species are found in intermediate depths: largespot lizardfishes (91–655 ft), offshore lizardfishes (88–1050 ft), and snakefishes (to 1300 ft).

Natural History

Little is known about reproduction in lizardfishes, except that they engage in external fertilization; some species apparently scatter eggs over the substrate. The transparent young settle onto sandbottom, where they blend in with the background. Larvae and juveniles differ in coloration from adults and are difficult the identify. Lizardfishes are voracious predators, quietly lying in ambush on or in the substrate, then explosively launching out at and snagging prey fishes and crustaceans in their toothy mouths.

Codlets
Family Bregmacerotidae

- **Antenna codlet** (*Bregmaceros atlanticus*)
- **Striped codlet** (*Bregmaceros cantori*)
- **Keys codlet** (*Bregmaceros cayorum*)
- **Stellate codlet** (*Bregmaceros houdei*)
- **Spotted codlet** (*Bregmaceros mcclellandi*)

Antenna codlet (*Bregmaceros atlanticus*). *Photo by George H. Burgess.*

Background

Codlets are tiny (<4 in), elongated fishes, most easily recognized by the presence of extremely long, filamentous pelvic fins located under the head, on the throat; the fins extend to about the middle of the base of the anal fin. The first dorsal fin is a single long, thin ray located on the top of the head. The second dorsal and anal fins are long based and bilobed, with the anterior portions of both symmetrically higher than the posterior portions. Adults are very difficult to tell apart; in fact, adult stellate codlets are unknown, suggesting this may be a neotenic species (never leaves the larval morph). Keys codlets have been suggested as being the same species as one of the other codlets. Florida species (minus the Keys codlet) are readily distinguished by looking at details of larval pigmentation and meristics—that is, the number of body parts, such as fin rays, vertebrae, or myomeres (muscle groups). Unless you are a patient larval-fish biologist, you will not see or care / be able to identify these to species.

Distribution and Habitat

All but Keys codlets (found in the Florida Keys) occur over outer continental-shelf and slope waters in the eastern Gulf off Florida; antenna codlets have a statewide distribution. Codlets are midwater, vertical migrators that probably contribute to the deep scattering layer (a biological zone in the ocean). They are most common in the upper 1000 ft and occur at least as deep as 1800 ft. Codlet larvae are found abundantly, often ranking among the most numerous fish larvae represented in plankton collections.

Natural History

Spawning by antenna codlets probably occurs year round in the Straits of Florida. The sheer abundance of codlet larvae of all species doesn't quite match up with catches of adults, suggesting that mortality at the larval stage is high. Codlets are planktivores; antenna codlets feed primarily on copepods and, as they get larger, on sea skaters (siphonophores), arrow worms (chaetognaths), and perhaps fish eggs and larvae.

Codlings
Family Moridae

- **Metallic codling** (*Physiculus fulvus*)

Background

The Moridae is one of three cod-like families found in Florida. Most species in this family are found on the continental slope, in depths too deep to be considered in this book. The depth range of one species, however, does include the deepest of Florida's continental-shelf waters. Metallic codlings look very much like phycid cods, but the former differ in having the second dorsal and anal fins equal

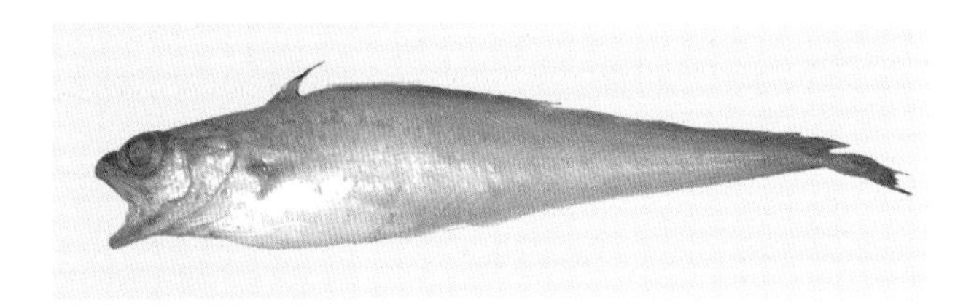

Metallic codling (*Physiculus fulvus*), Straits of Florida, Florida. *Photo by George H. Burgess.*

in length, with origins directly over each other (in phycids the anal fin is shorter and originates farther back). Those fins and the tiny tail are edged in black; otherwise, metallic codlings are brownish-grey.

Distribution and Habitat

Metallic codlings are known from offshore of Jupiter in southeast Florida, the Straits of Florida, and in the Gulf of Mexico, over mud- or sandbottom, in depths of 600–1200 ft.

Natural History

Nothing is known about the life history of metallic codlings.

Merlucciid Hakes
Family Merlucciidae

- **Offshore hake** (*Merluccius albidus*)
- **Luminous hake** (*Steindachneria argentea*)

Background

Merlucciid hakes, once included in the cod family (Gadidae), differ from the related phycid hakes and morid cods in having elongated, laterally compressed, silvery bodies; moderately elongated snouts, with a V-shaped ridge on the top of the somewhat flattened snout; and large, sharp teeth. They also lack a chin barbel. The two species in our waters are very different in appearance and easy to identify. Offshore hakes, which reach lengths of about 2.5 ft, have a well-developed, straight-edged caudal fin, while luminous hakes (1 ft) have a posteriorly tapering body, with the dorsal, anal, and caudal fins merging as a terminal filament. Luminous hakes have widely separated anal and genital openings, quite unlike most fishes, which have a single urogenital opening, the cloaca.

Distribution and Habitat

Offshore and luminous hakes are primarily continental-slope species, but they also appear in the deep waters of the outer continental shelf. Luminous hakes occur off Florida's Gulf coast, the Straits of Florida, and northeast Florida, to depths of at least 1900 ft and (rarely) as shallow as 120 ft. Offshore hakes are found statewide, in depths of less than 500 ft; in the Gulf they are most common from 1100 to 3300 ft. Offshore hakes segregate by size, with juveniles, subadult females, and males found in upper-slope waters of less than 1800 ft, while adult females inhabit the lower slope, at greater depths (to at least 3839 ft).

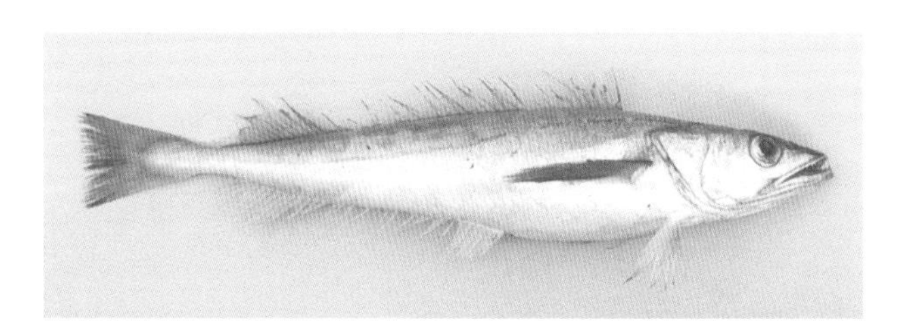

Offshore hake (*Merluccius albidus*), Straits of Florida. *Photo by George H. Burgess.*

Natural History

Offshore hakes eat fishes (including many of its own species, as well as lanternfishes), crustaceans, and squids. Males are smaller than females. Spawning occurs from late spring to early fall, on or near bottom, in depths of 1000–1800 ft. Each female produces 340,000 eggs, and these ova are buoyant when released. Luminescent hakes get their vernacular name because they produce biological light. Symbiotic (partner) bacteria living within the fish give off a faint blue glow on the underside of the body and head, including the mandible and the entire operculum. A donut-shaped, luminescent organ lies under the skin surrounding the anus. An elongate cavity, containing gelatinous tissue, is found between the anus and the genital opening, located posteriorly in front of the base of the anal fin; at the anal fin this cavity splits, forming a Y posteriorly around each side of the base of the anal fin. Muscle tissue on the lower part of the body posteriorly is translucent, with an underlying sheath that serves as a reflector. Gelatinous material is also found under the striated skin, where luminescence occurs.

Phycid Hakes
Family Phycidae

- **Gulf hake** (*Urophycis cirrata*)
- **Carolina hake** (*Urophycis earllii*)
- **Southern hake** (*Urophycis floridana*)
- **Spotted hake** (*Urophycis regia*)

Background

Phycid hakes are cod-like in appearance and historically were included in the cod family (Gadidae), although the cold-water fishes in this family are not found in Florida. Hakes are soft fleshed and round in cross-section, with very long, thin, white pelvic fins. The first dorsal fin is short and higher than the second, which, like the shorter anal fin, is long and low; the anal fin originates well behind that of the second dorsal fin. The tail is small and rounded. The mouth is large, extending past the eye; the snout is moderately long and rounded; and there is a small, median chin barbel.

Southern and spotted hakes are distinctive in having a dozen or more small white spots on the lateral line, diffuse dark spots on the second dorsal fin, and a dark margin on the anal fin. Their heads are dotted with several small, well-defined dark spots; they also have a diffuse dark blotch on the upper operculum and one or more vague to discrete dark lines radiating backward from the eye. Their base body color is tan to light-brown above and laterally white below. Spotted hakes differ in having the first dorsal fin marked by a white-bordered dark spot; a broad, dusky marginal tail band (lacking in southern hakes); and longer pelvic fins that reach or extend past the anal-fin origin (short of the origin in southern hakes).

Carolina hakes are almost uniformly dark–chocolate-brown, the only variations in this coloration being slightly darker margins that are thin on the dorsal fin and wider on the anal fin, and the lateral line, which can be paler. The first dorsal

Top left: Southern hake (*Urophycis floridana*). *Top right*: Gulf hake (*Urophycis cirrata*). *Bottom left*: Carolina hake (*Urophycis earllii*). *Bottom right*: Spotted hake (*Urophycis regia*). *Photos by George H. Burgess.*

fin is narrower and more pointed (falcate) than those of other hakes; the pelvic fins fail to reach the anal-fin origin; and the chin barbel is large. Gulf hakes have the longest pelvic fins, going well beyond the anal-fin origin. This species is greyish-tan, with a silvery belly; the dorsal and anal fins have thin, dark margins. The head has a slightly more-convex forehead profile than other hakes, and the chin barbel is tiny or absent.

Distribution and Habitat

Phycid hakes generally prefer cool water temperatures and usually are found over softbottom. Southern hakes occur in the Gulf of Mexico, from Cedar Key (in the Big Bend ecoregion) northwestward across the panhandle, in depths of 3–600 ft (but these hakes get to depths 1300 ft elsewhere). Gulf hakes are a deepwater species (records of 262–928 ft in Florida, to >2000 ft elsewhere) that are found in the Gulf of Mexico, from the panhandle south to the Straits of Florida. They occasionally occur off extreme northeastern Florida, where they are replaced, albeit uncommonly, in shallower (<300 ft) inner-shelf waters by Carolina hakes; the latter species surprising prefers hardbottom habitats (rocky outcrops and ledges). Spotted hakes occur statewide, in a wide range of depths (3–1335 ft). Young spotted and southern hakes are found in shallow waters (spotted hakes commonly enter estuaries), while adults inhabit deeper waters.

Natural History

Regardless of adult habitat choice, reproduction appears to occur well away from the coastline. Phycid hakes are obligate bottom dwellers that consume benthic invertebrates (primarily crustaceans and squids) and small fishes.

Pearlfishes
Family Carapidae

- **Pearlfish** (*Carapus bermudensis*)
- **Chain pearlfish** (*Echiodon dawsoni*)

Background

The Carapidae are odd little relatives of the cusk-eels. Pearlfishes lack caudal and pelvic fins, and the anus is forward on the body, located under the head. They have thin, elongate bodies that taper to a point; dorsal and anal fins that are continuous around this point; and anal-fin rays that are longer than the dorsal-fin rays. Should you happen to be handy with a microscope, the teeth on the upper jaw are small and heart shaped. Their bodies are translucent, with sliver blotches and black spots. Chain pearlfishes are similarly elongated and tapered, with translucent bodies, upper jaws that overhang lower jaws, and—don't put away that 'scope—canine teeth that are present in the upper jaws.

Distribution and Habitat

Both pearlfish species are found around the state. The pearlfish lives wherever its host (the sea cucumber) occurs, in water depths ranging from 3 to 700 ft. Chain pearlfish are free living and found around the state, over rubblebottom in deeper waters (200–600 ft).

Natural History

The pearlfish leads a bizarre existence, residing in the body of sea cucumbers. A pearlfish enters the host tail-first, through the anus of the sea cucumber. Usually only one individual is found per sea cucumber, but as many as four have been taken from a single host. Several sea cucumber species live in southeast Florida and the Florida Keys, but the most commonly inhabited species is the five-toothed sea cucumber (*Actinopyga agasszi*), which lives in seagrass beds and sandy areas near reefs. Individual

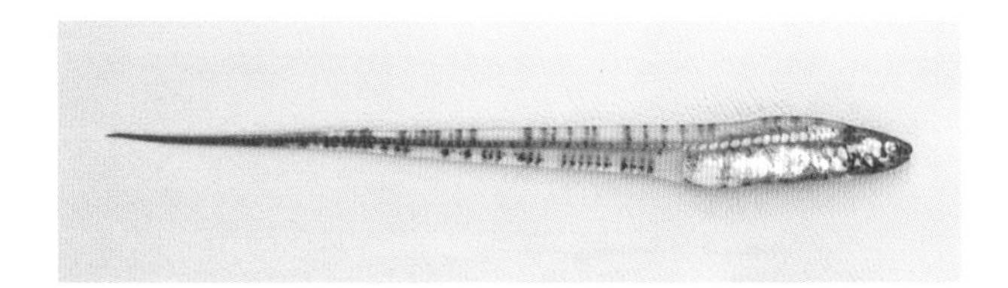

Pearlfish (*Carapus bermudensis*), Florida. *Photo by George H. Burgess.*

pearlfishes leave the host at night to forage in the surrounding waters. Many sea cucumbers, when frightened, spew out their guts, readily offering them—and any pearlfish crashers—in exchange for the opportunity to regenerate their innards. Chain pearlfishes apparently live a more-traditional, no-cucumber existence.

Pearlfishes spawn by releasing egg mats. Larval development includes two distinct larval phases, known as vexillifer and tenuis. These larval stages were originally named as distinct new species by early naturalists; they look so strange that it took a while for ichthyologists to correctly connect them with their adults. Vexillifer is the earlier stage, which has an elongate dorsal projection that looks like a stem with leaves. Larvae at the tenuis stage are very long (reaching 8 in) and thin. During this stage the pearlfish larva enters the host sea cucumber and then shrinks as it transforms into a juvenile.

Cusk-Eels
Family Ophidiidae

- **Atlantic bearded brotula** (*Brotula barbata*)
- **Blackedge cusk-eel** (*Lepophidium brevibarbe*)
- **Mottled cusk-eel** (*Lepophidium jeannae*)
- **Fawn cusk-eel** (*Lepophidium profundorum*)
- **Twospot brotula** (*Neobythites gilli*)
- **Stripefin brotula** (*Neobythites marginatus*)
- **Longnose cusk-eel** (*Ophidion antipholus*)
- **Shorthead cusk-eel** (*Ophidion dromio*)
- **Blotched cusk-eel** (*Ophidion grayi*)
- **Bank cusk-eel** (*Ophidion holbrookii*)
- **Crested cusk-eel** (*Ophidion josephi*)
- **Striped cusk-eel** (*Ophidion marginatum*)
- **Colonial cusk-eel** (*Ophidion robinsi*)
- **Mooneye cusk-eel** (*Ophidion selenops*)
- **Sleeper cusk-eel** (*Otophidium dormitator*)
- **Polka-dot cusk-eel** (*Otophidium omostigmum*)
- **Dusky cusk-eel** (*Parophidion schmidti*)
- **Redfin brotula** (*Petrotyx sanguineus*)

Background

Cusk-eels are little-known and secretive bottom dwellers. They have elongate, tapered bodies but are not true eels; rather, they are related to pearlfishes and live-bearing brotulas. The dorsal and anal fins merge to form a continuous pointed tail, and in most species the pelvic fins (consisting of only 2 rays) are placed far forward, under the eye or gill opening; some species have a forward-projecting spine under the skin of the snout.

The largest (>3 ft) and most distinctive member of the family is the Atlantic bearded brotula. This species is easily separated from the others by its chocolate-brown body and fins, the latter with a narrow white margin. Atlantic bearded brotulas have 6 barbels on the chin and snout. The smaller (5 in) redfin brotulas are dark-red (their young are bright-red), with a relatively small eye and pelvic fins located under the operculum. Stripefin brotulas are moderately sized (9 in), yellow-brown cusk-eels, with relatively long pelvic rays placed below the rear portion of the operculum. They have 2 discontinuous brown stripes: one stripe located from the eye backward along the body and fading out as it moves posteriorly, and a dark-brown stripe that bisects the dorsal fin. The related twospot brotulas are tan, with irregular dorsal blotches, the 2 largest blotches extending onto the dorsal fin.

The three similar-sized (ca. 10 in maximum) species of the genus *Lepophidium* are distinguished from other cusk-eels by having a small spine projecting forward from the snout, scales on the head, and body scales in straight rows. The dorsal-fin margin of mottled cusk-eels has alternating black blotches overlaying white and tan lateral splotches; the anal fin has a black margin. Blackedge cusk-eels are tan to gray, with a continuous black edge on the dorsal fin (and sometimes on the anal fin). Fawn cusk-eels are gray-brown, with a row of white spots just above the midline; the posterior portions of the dorsal and anal fins are edged with black.

The smaller sleeper (3 in) and mooneye (4 in) cusk-eels also have nose spines. Sleepers are pallid (sometimes with a orangish or golden hue), and mooneyes are silvery, with large eyes. Other cusk-eel species lack the nose spine, and their body scales form a basketweave pattern. Polka-dot cusk-eels (4 in) are tan, with a black spot above the opercle and dark-brown blotches irregularly spread over the flanks; they also have a series of dark blotches on the margin of the dorsal fin and a dark stripe along the base of the anal fin. Blotched cusk-eels have bold, dark-brown blotches and spots on the flanks and dorsal fin, and dark anal and pectoral fins. Colonial cusk-eels, small (6 in) and basally tan, also are distinctly marked with 2 rows of off-centered, brown, lateral blotches on the upper body, producing a checkerboard effect. Male and female crested and striped cusk-eels are sexually dimorphic, with adult males growing humped foreheads. Striped cusk-eels (9 in) are light-yellow to gold laterally, with 4 lines on the flank, the lower 1–2 often fading out posteriorly; the dorsal fin and the rear portion of the anal fin are dark edged. Crested cusk-eels (11 in) are tan, with a series of small blotches and spots that coalesce to form 3–4 discontinuous stripes along the flanks; the entire margins of the dorsal and anal fins are dark.

Bank cusk-eels (12 in) are uniformly brown, with a dark margin on the dorsal (and occasionally anal) fin; the pelvic fins are relatively long; and the head profile is bilaterally symmetric. Longnose cusk-eels (10 in) also are brownish, with a dark margin on the dorsal fin, but they have short pelvic rays and an asymmetric profile, the forehead being somewhat elevated. Shorthead cusk-eels basically can be identified only by microscope. They are a smaller (7.5 in) version of longnose cusk-eels, differing in having 4 gill rakers on the first gill arch. Dusky cusk-eels, another small (4 in) species, are dusky brown, have large scales on the head, and are the only cusk-eel species with 2 equally long pelvic rays.

Distribution and Habitat

Most cusk-eels live on softbottom. Their depth preferences serve to sort out the group over the continental shelf and the slope edge. From shallow to deep, these are as follows: dusky cusk-eels (1–35 ft), redfin brotulas (5–107 ft), polka-dot cusk-eels (2–164 ft), crested cusk-eels (2–180 ft), sleeper cusk-eels (3–180 ft), striped suck-eels (24–79 ft), longnose cusk-eels (36–237 ft), blackedge cusk-eels (18–300 ft), bank cusk-eels (18–305 ft), blotched cusk-eels (2–370 ft), mottled cusk-eels (60–510 ft, to 918 ft), mooneye cusk-eels (20–212 ft, to 1061 ft), colonial (73–273 ft), Atlantic bearded brotulas (2–570 ft, to 1500 ft), twospot brotulas (197–755 ft), shorthead cusk-eels (101–819 ft), barred cusk-eels (600–1230 ft), fawn cusk-eels (79–1811 ft), and stripefin brotulas (243–4744 ft). Virtually all species occur state-

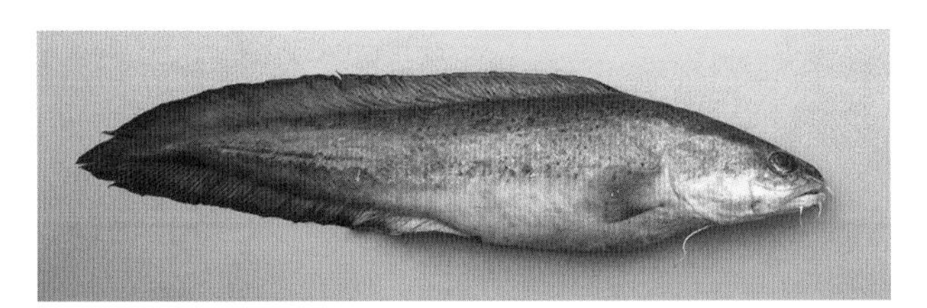

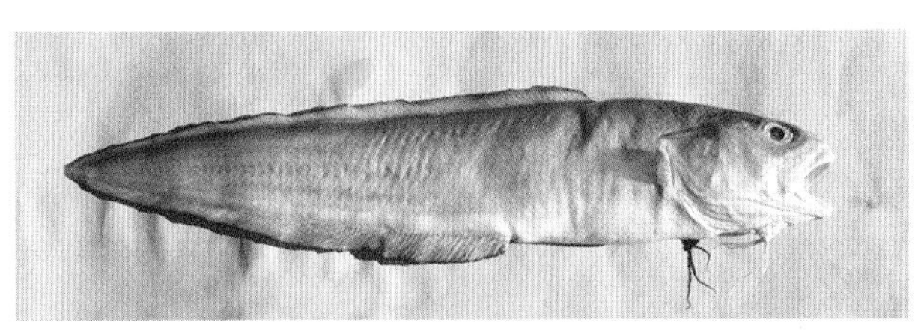

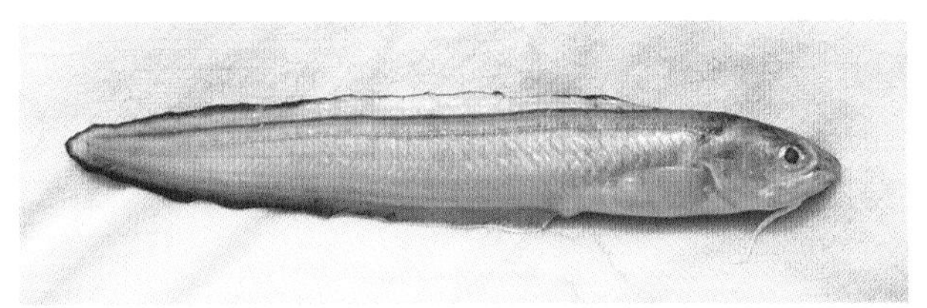

Top left: Atlantic bearded brotula (*Brotula barbata*), Dry Tortugas, Florida. *Top right*: Bank cusk-eel (*Ophidion holbrookii*). *Bottom left*: Polka-dot cusk-eel (*Otophidium omostigmum*). *Bottom right*: Stripefin brotula (*Neobythites marginatus*). *Photos by George H. Burgess.*

wide; exceptions include redfin brotulas (Florida Keys), striped cusk-eels (northeast Florida only), mooneye cusk-eels (east coast and Keys), and dusky and sleeper cusk-eels (southern Florida). Polka-dot cusk-eels are not found in southern Florida, the Florida Keys, or southwest Florida. Among the cusk-eels, redfin brotulas are the only reef dwellers, and dusky cusk-eels are the only seagrass inhabitants in south Florida's tropical waters.

Natural History

Little is known about cusk-eel reproduction. The best-studied species is the striped cusk-eel. The bump on the head of male striped cusk-eels and crested cusk-eels houses special muscles, used to produce sounds that attract prospective mates. Males and females pair-spawn, and females release a mass of eggs. Cusk-eels are most active at night. Individuals bury themselves in the sediment (tail-first), where they remain during daylight hours, emerging at night to feed and mate. In some coastal or inshore areas where light penetration is limited by poor water clarity, cusk-eels may be more active during daytime. Cusk-eels feed on bottom-dwelling shrimps, crabs, amphipods, worms, and (occasionally) fishes. Atlantic bearded brotulas are occasionally caught by deep-dropping anglers and commercial fishers seeking groupers and snappers; most of the other cusk-eels are taken in shrimp trawls.

Viviparous Brotulas
Family Bythitidae

- **Reef-cave brotula** (*Grammonus claudei*)
- **Gold brotula** (*Gunterichthys longipenis*)
- **Key brotula** (*Ogilbia cayorum*)
- **Curator brotula** (*Ogilbia sabaji*)
- **Shy brotula** (*Ogilbia suarezae*)
- **Black brotula** (*Stygnobrotula latebricola*)

Background

As their family name implies, viviparous brotulas give birth to live young. They are uncommonly secretive and seldom seen. All Florida species are small (<6 in), elongate, have blunt snouts, and swim by undulating their median fins. Two species—reef-cave brotulas and black brotulas (also known as "black widows")—have dark-brown to black bodies, with long black dorsal and anal fins that join to form a united and pointed tail. The snout of black brotulas is short and truncated, blunt and rounded, while that of reef-cave brotulas is more elongated and pointed. Gold, Key, curator, and shy brotulas have separate caudal fins, and adult males have pairs of stiff claspers. In gold brotulas, the body and the diagonal halves of the dorsal and anal fins are pale-gold–yellow. Most of the caudal fin is dark-brown, and the rear portions of the dorsal and anal fins have similarly colored submarginal stripes that gradually and markedly grow in width posteriorly. The darkened fin areas have gold-yellow fin bases and margins.

Members of the genus *Ogilbia* are very difficult to identify, owing to their small sizes and similar

Black brotula (*Stygnobrotula latebricola*), Juno, Florida

Black brotula (*Stygnobrotula latebricola*) associating with a spotted moray (*Gymnothorax moringa*), Hobe Sound, Florida

morphology (one of us historically referred to the group as a taxonomic morass!), and they have only recently been sorted out. The body and basal portions of the median fins of Key brotulas are pale-orange to brown, with tiny, dark, peppering; the margins of those fins are clear. Life colorations for curator and shy brotulas are unknown (they are brownish in preservative). Differentiation between the three *Ogilbia* species in Florida requires expert examination of specimens, involving microscopic morphological differences.

Distribution and Habitat

All but one species in this family are restricted to the southern part of Florida. Reef-cave brotulas and black brotulas are known only from cavities and caves on reefs and ledges in southeast Florida and the Florida Keys. Curator and shy brotulas have only recently received scientific names, and all known specimens are from the Keys and the Dry Tortugas. The Key brotula similarly has been definitively identified only from those areas, but *Ogilbia* records from shallow seagrass and algae-rich habitats in Florida Bay and the southwestern and southeastern coasts most likely will prove to be that species. Gold brotulas occur along the west coast of Florida and through the Florida Keys, where they burrow into the mudbottom in water depths ranging from shoreline to about 30 ft.

Natural History

Little is known about the lives of viviparous brotulas. Key brotulas live under rubble and clumps of the calcareous green alga *Halimeda*. Habits are not well known, but apparently Key brotulas and gold brotulas are more active at night. All species reproduce by internal fertilization; males have a penis that delivers sperm into the genital pore of the female. Only Key brotulas have been studied in any detail. Insemination in this species involves the delicate joining of male and female copulatory apparatuses; fertilization takes place in follicle, and development occurs in the lumen of the ovary, where the embryos remain until the yolk reserves are depleted. Ovulation and hatching apparently occur simultaneously, shortly after fertilization. Brood size is 14 young. The transparent young remain with the parent for some length of time, until they are about 0.5 in long. Black brotulas are the species most likely to be seen by divers. These brotulas are a remarkable sight when observed swimming, using continuous undulations of the median fins and looking very much like dark, swimming nudibranchs (gastropod marine mollusks). Black brotulas associate with large green morays (and occasionally spotted morays), apparently cleaning the eels; they can even be seen entering the open mouths of gaping eels.

Toadfishes
Family Batrachoididae

- **Gulf toadfish** (*Opsanus beta*)
- **Leopard toadfish** (*Opsanus pardus*)
- **Oyster toadfish** (*Opsanus tau*)
- **Atlantic midshipman** (*Porichthys plectrodon*)

Background

Toadfishes are among the homeliest of fishes. The three toadfish species in the genus *Opsanus* are nearly identical in morphology, differing chiefly in subtleties of their coloration and their distribution. Gulf and oyster toadfishes are shallow, inshore species, while leopard toadfishes occur in deeper waters. The overall appearance of all three of these species is certainly toad-like, having a large head, with eyes set on top. The wide mouth has a row of fleshy tabs projecting from below the lower jaw and around the eyes. The pectoral fins are wide and fan-like; the body is slender, with 3 short, thick dorsal spines. The color is usually brown to gray, with small dark spots on the head; rosette-shaped blotches on the body; and dark, wavy bars on the median fins. A wavy diagonal line crosses the eye. All of these toadfishes reach lengths of about 1 ft. The fourth species in this family, Atlantic midshipmen, are smaller (maximum size about 9 in) and more slender than the three *Opsanus* species and easily distinguished by the presence of photophores on the ventral portion of the body.

Distribution and Habitat

Gulf toadfishes inhabit hardbottom areas, oyster reefs, and seagrass meadows around the entire west coast, from southeast Florida (the upper Florida Keys and Biscayne Bay) to the panhandle, in inshore waters. Leopard toadfishes are also found along the Gulf coast, from southwest Florida through the panhandle, on hardbottom habitats in water depths from 30 to 800 ft; a toadfish resembling this species similarly occurs off the east coast. Oyster toadfishes distributionally replace Gulf toadfishes along Florida's east coast, from Jupiter northward, on oyster reefs, in inshore seagrass meadows and estuaries, and on structured hardbottom. Atlantic midshipmen are found around the state, on sedimentary bottoms, in water depths from nearshore to 120 ft.

Natural History

Probably no non-game marine fish is as well studied as the oyster toadfish. Commonly encountered and captured, moderately sized and easy to maintain in captivity, chock full of unusual behavior, and the possessor of large eggs, the

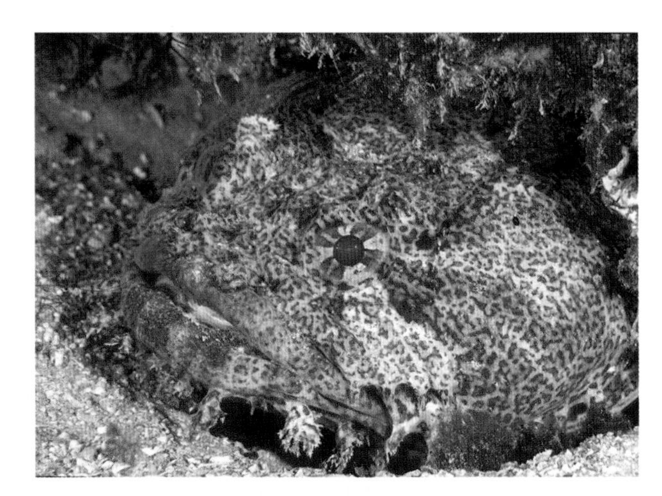

Top: Oyster toadfish (*Opsanus tau*), Stuart, Florida. *Bottom*: Gulf toadfish (*Opsanus beta*), Venice, Florida.

Leopard toadfish (*Opsanus pardus*), Florida Middle Grounds

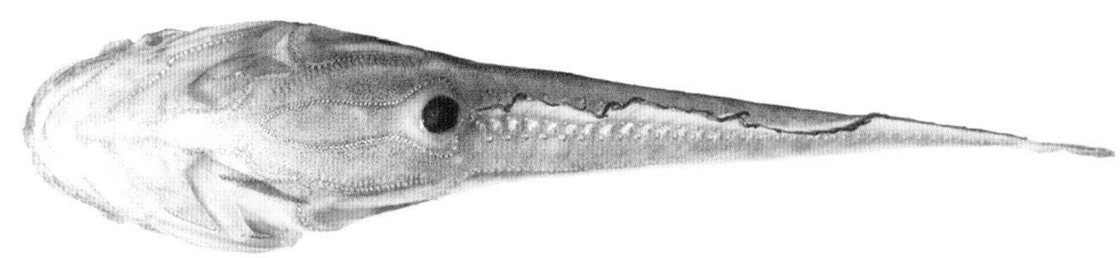

Top: Atlantic midshipman (*Porichthys plectrodon*), southwest Florida shelf. *Bottom*: Underside, showing rows of photophores. *Photos by George H. Burgess.*

toadie is a study waiting to happen. And researchers have obliged with vigor, as this species has become the marine guinea pig of physiological and anatomical studies. More than one scientific career has been intimately tied to oyster or Gulf toadfishes. Our understanding of osmotic salt-balance, liver and swim-bladder function, sound production and communication, and neural physiology have benefited enormously from this unattractive and crabby critter.

The reproductive process in Gulf and oyster (and probably leopard) toadfishes follows a similar sequence, which begins with males establishing nest sites. These sites include rocks, clam or pen shells, cans, boards, or burrows (e.g., those constructed by stone crabs). Individual males attract prospective mates to nest sites by producing boat-whistle sounds using specialized muscles attached to the swim bladder. If you are snorkeling or scuba diving in shallow waters and hear something that sounds a bit like an owl hooting underwater, it is probably a toadfish talking. A receptive female deposits relatively large (about 0.2 in) adhesive eggs on hard substrate, often on the upper surface of the nest site; then the male uses double-tailed sperm to fertilize them and guards the eggs for up to 40 days. Males produce sounds when guarding eggs and young, and they vigorously attack intruders if the eggs are threatened. The eggs hatch into free-swimming, tadpole-like hatchlings that remain with the male for a short period, during which time they will even cling to him.

Toadfishes have the classic morphology of lie-in-wait predators—eyes close set on top of the head, strong jaws, a broad mouth gape, and numerous teeth—characteristics that enable them to capture and consume mud and swimming crabs, shrimps, and small fishes. Experiments show that toadfishes don't eat enough of these predators to increase oyster production, but they may aid clam production. Gulf and oyster toadfishes live about 6 years, and Gulfs grow faster than their oyster toadfish cousins.

Toadfishes produce a toxic substance in their mucous that is thought to fend off predatory fishes, and they have the peculiar habit of excreting urea instead of ammonia with their urine. This is odd, because most fishes do not excrete urea. Female toadfishes and nonbrooding males frequently bury themselves in the sediment—an environment high in toxic ammonium. Originally, researchers thought this habit of excreting urea was a way of not fouling the nest, but some suggest that it also may be a antipredator strategy. If so, the latter strategy is not entirely successful, because a variety of fishes and birds, as well as bottlenose dolphins, prey on toadfishes. Bonefishes in the Keys hold Gulf toadfishes in particularly high regard.

Little is known about the natural history of Atlantic midshipmen. In the Gulf of Mexico, spawning occurs during spring and summer. Females produce an average of 140 eggs. Like the three toadfish species, Atlantic midshipmen make sounds that probably are communicative in nature. They are the only Florida shallow-water fishes that have photophores, light-producing organs that also may be used in communication and species recognition. Individuals bury themselves in the sand by day, emerging at night to feed on small crustaceans.

Goosefishes
Family Lophiidae

- **Reticulate goosefish** (*Lophiodes reticulatus*)
- **Blackfin goosefish** (*Lophius gastrophysus*)

Background

Goosefishes look like frogfishes that have been stepped on, with their huge heads—more than half of the total length of the fish—flattened dorsoventrally (on the back and sides). The tapered body (ovoid in cross-section) and caudal fin look normal sized. The mouth is large and abundantly armed with long, depressible, needle-like teeth—what goes in isn't coming out! Oval gill openings are located behind the broad, horizontally oriented pectoral fins. The first pair of dorsal spines are separated from the rest of the fin and located on the head, with this first pair of spines highly modified into what looks like a fishing pole and lure. Blackfin goosefishes, which reach 2 ft long, have black edges on the inner pectoral fins and a black band on the lower underside of the fins. The pectoral fins of the smaller (10 in) reticulate goosefishes, appropriately enough, resemble a network of thread-like veins.

Distribution and Habitat

Goosefishes are bottom-dwelling, deepwater species that occur on the outer continental shelf and upper continental slope. Blackfin goosefishes have been taken in waters from off of Cape Canaveral south through the Keys and the Dry Tortugas and westward in the Gulf, in depths of 305–1837 ft. Reticulate goosefishes occur in the Gulf and off the Dry Tortugas, in somewhat shallower (413–740 ft) waters.

Natural History

Well-developed dentition and huge mouths, which have led fishers to call the related goosefish

Blackfin goosefish (*Lophius gastrophysus*). *Photo by George H. Burgess.*

(*Lophius americanus*)—a cold-water species known only from northeast Florida, in very deep slope waters—the "all-mouth," signal that members of the family are very efficient predators. They consume fishes, squids, and other invertebrates. The larger (4 ft; 50 lb) goosefish, which occupies shallower waters in the northern part of its range, also picks off waterfowl at the water surface!

Frogfishes
Family Antennariidae

- **Longlure frogfish** (*Antennarius multiocellatus*)
- **Dwarf frogfish** (*Antennarius pauciradiatus*)
- **Striated frogfish** (*Antennarius striatus*)
- **Ocellated frogfish** (*Fowlerichthys ocellatus*)
- **Singlespot frogfish** (*Fowlerichthys radiosus*)
- **Sargassumfish** (*Histrio histrio*)

Background
Frogfishes are shallow-water anglers, wielding a pole and a lure. The pole (technically the illicium) is actually a modified dorsal spine, and the lure (esca) is a fleshy extension on the end of the spine. Frogfishes appear as extremely well-camouflaged, amorphous blobs that are difficult to distinguish from their surroundings. Individual species can be separated out using spot and pigment patterns, and ichthyologists examine the details of the esca structure to fully distinguish the various species. Two species formerly included in the genus *Antennarius* recently were moved into the genus *Fowlerichthys*.

Dwarf frogfish (*Antennarius pauciradiatus*), Lake Worth Lagoon, Florida

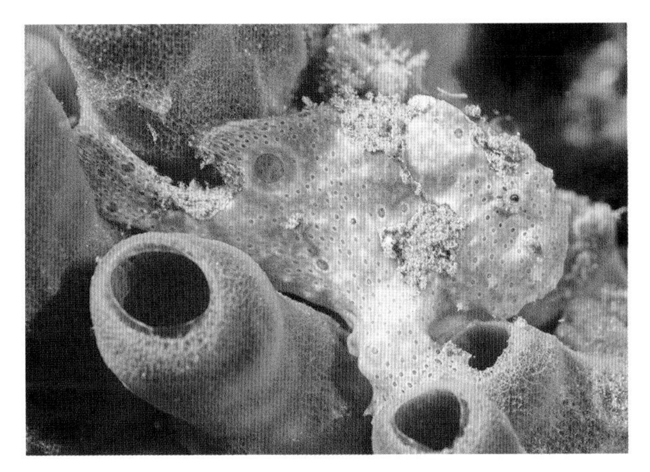

Left: Longlure frogfish (*Antennarius multiocellatus*), Palm Beach, Florida. *Right*: Longlure frogfish, Jupiter, Florida.

Top: Striated frogfish (*Antennarius striatus*), Palm Beach, Florida. *Bottom*: Striated frogfish, black form, Jupiter, Florida.

Longlure frogfishes are highly variable in color—red, yellow, orange, pale-green, or dark-brown—with 3 ocellated spots on the tail. Likewise, ocellated frogfishes can be brown, orange, white, or pale-pink; they have a single ocellated spot on the dorsal fin and another ocellated spot below, on the flank. Singlespot frogfishes are pale-yellow to brown, with a single spot on the flank. Dwarf frogfishes range from yellowish to brown, with a small, faint spot on the base of the second dorsal fin. Striated frogfishes typically have wavy lines or dark bars on the body, but some individuals are jet-black, with bright-white spots inside the mouth cavity. Sargassumfishes are colored exactly like their floating *Sargassum* seaweed home.

Distribution and Habitat

Frogfishes are associated with sand, rubble, shells, reefs, and other structured habitats in shelf and inshore waters. Longlure and dwarf frogfishes are found along the east coast and Florida Keys, in depths ranging from 1 to 300 ft. Ocellated, striated, and singlespot frogfishes occur throughout state waters, in depths of 1–800 ft. Ocellated frogfishes are common on the continental shelves of west Florida and northeast Florida (north of Jupiter). Sargassumfishes are found worldwide, in tropical waters wherever drifting sargassum occurs.

Natural History

Frogfishes have a unique reproductive style. Gravid females literally swell like balloons with maturing eggs. Like many other fish species, the relatively smaller males join the much larger females in a rapid ascent off the bottom, where they release sperm and eggs. The unique feature of frogfishes is the gelatinous egg sheet produced by a ripe female. The egg sheet is released so that the male—or in some cases, several males—can fertilize the eggs, which stay together as a drifting mass. These egg masses contain almost 300,000 eggs. Observations of striated and longlure frogfishes indicate that spawning occurs at night. Fertilized eggs of sargassumfishes float pelagically for about 60 days. The planktonic larval stages look like miniature adults, prior to the larvae settling to the bottom or, in the case of sargassumfishes, to floating sargassum.

When it comes to feeding, frogfishes are masters of the lie-in-wait strategy also employed by other species, such as lizardfishes, scorpionfishes, and stargazers. Characteristics of this feeding mode include a background-matching color and shape, a large mouth capable of a blindingly rapid response to fishes entering the strike zone, and a sacculate (stretchy) stomach capable of ingesting and digesting items much larger than the individual capturing them. Many an unwary aquarist has made the mistake of introducing the weird little fish

Sargassumfish (*Histrio histrio*), Jupiter, Florida

found in the seaweed into their populated home aquarium, only to later find but a single fat, frumpy fish and no tankmates. Individual frogfishes establish a feeding area, remaining there for varying lengths of time. They are masters at disappearing in plain sight, and although they often are found nestled in the sponges they so closely resemble, individuals also sit on open bottoms, representing the only structural feature within the immediate area. Unsuspecting fishes seeking cover are drawn into the deadly strike zone, a circle of death for the dupes that enter it.

As members of the anglerfish clan, frogfishes are deft anglers that use their poles and lures with utmost efficiency. The lures resemble—sometimes to a remarkable extent popular food items, such as worms or crustaceans. Lures are presented and displayed by twitching the pole. The lures, and even the fishing patterns, vary with individual species. For example, striated frogfishes have a worm-lure that mimics a polychaete worm in appearance and size, and a longlure frogfish's lure resembles a small shrimp. If a hungry prey item bites the esca off the end of the illicium, the frogfish can regenerate it over time.

Locomotion in frogfishes is often cited as an analogy of an early land-dwelling animal's style. The pelvic fins are used to walk along the bottom. The pectoral fins appear to have elbows, further enhancing the look of a walking animal. Frogfishes also employ jet propulsion, funneling streams of water outside the rear-facing gill openings.

Batfishes
Family Ogcocephalidae

- **Atlantic batfish** (*Dibranchus atlanticus*)
- **Pancake batfish** (*Halieutichthys aculeatus*)
- **Spiny batfish** (*Halieutichthys bispinosus*)
- **Gulf batfish** (*Halieutichthys intermedius*)
- **Longnose batfish** (*Ogcocephalus corniger*)
- **Polka-dot batfish** (*Ogcocephalus cubifrons*)
- **Slantbrow batfish** (*Ogcocephalus declivirostris*)
- **Shortnose batfish** (*Ogcocephalus nasutus*)
- **Spotted batfish** (*Ogcocephalus pantostictus*)
- **Roughback batfish** (*Ogcocephalus parvus*)
- **Palefin batfish** (*Ogcocephalus rostellum*)
- **Tricorn batfish** (*Zalieutes mcgintyi*)

Background

Batfishes are among the most peculiar of shallow-water fishes. Their dorsoventrally flattened, rounded- or triangular-shaped bodies are covered with rough scales (bucklers), making them look like plucked chickens. All have a pointed and upturned rostrum (snout), giving some species a Pinocchio look. Tucked into a small pocket under the snout is an esca,

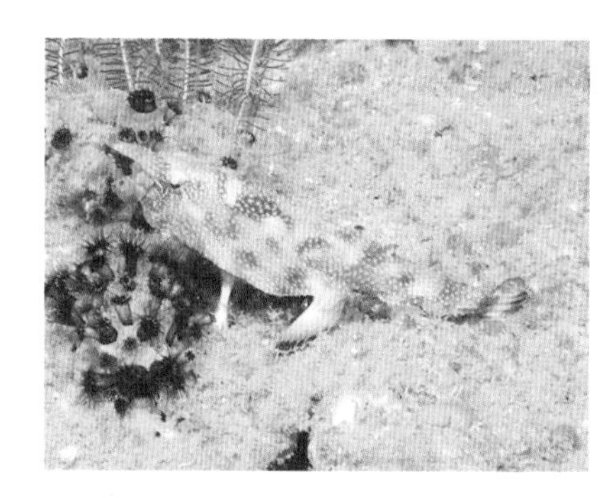

Longnose batfish (*Ogcocephalus corniger*)

a small, bulb-like lure like those found in the related frogfishes and goosefishes. The gill openings are small, round orifices located behind the pectoral-fin bases. When viewed from above, the rounded body shape of Atlantic, pancake, Gulf, and spiny batfishes separates these species from other batfishes that possess more-angular, triangular bodies. These four species are smaller and more flattened than other batfishes. The rear margin of each side of an Atlantic batfish's body has a prominent spine, with little spinelets; the other three species lack this feature and instead are distinguished from each other by small differences in buckler sizes and coloration.

The angular-bodied species can be equally confusing to identify, especially since coloration is variable and the rostrum length of individual species changes with growth. Tricorn batfishes are unique in having obvious recurved spines on each side of the rostrum, and longnose batfishes are readily identified by having the longest rostrum of any batfish. The body of a longnose batfish is covered with small, pale-white spots; the margins of the pectoral and caudal fins look like they've been dipped in reddish-brown ink, a characteristic shared by roughback batfishes, which have a short, thick snout. The body color of the medium- to short-snouted polka-dot batfishes ranges from pale- to dark-gray, with patches of orange or red. This species has white-ringed dark spots on the sides, just above the pectoral fins, and wavy spots on the pectoral fins and body; the lips often are red. Palefin batfishes also share dark spots on the front of the body, and they (appropriately) lack spotting on the pectoral fins, except on the bases. Despite their vernacular name,

Shortnose batfish (*Ogcocephalus nasutus*)

Polka-dot batfish (*Ogcocephalus cubifrons*), Lake Worth Lagoon, Florida

Roughback batfish (*Ogcocephalus parvus*), Lake Worth Lagoon, Florida

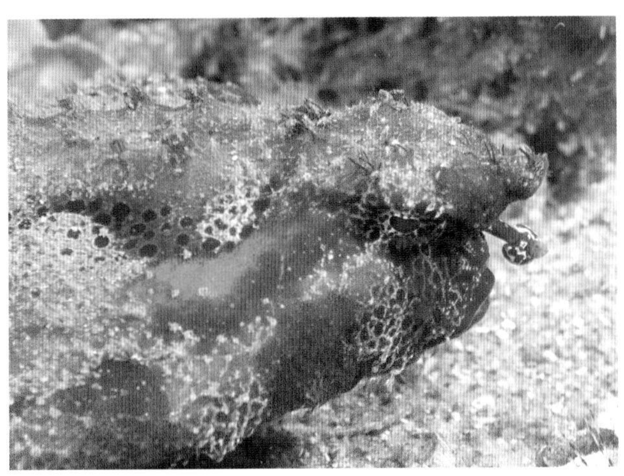

Polka-dot batfish (*Ogcocephalus cubifrons*), with lure extended, Lake Worth Lagoon, Florida

shortnose batfishes usually have a moderately long and somewhat upturned rostrum. They are uniformly gray or light-brown in color, with a banded caudal fin. Adult spotted batfish bodies are relatively smooth when compared with those of other batfishes, and they have small, pepper-like spots over a tan or brown base color. The caudal fin has a dark margin. Slantbrow batfishes have relatively narrow bodies; are brown to gray, with small pale spots around the eyes and mouth; and have a short rostrum that tends to point downward.

Distribution and Habitat

Batfishes are bottom dwellers that prefer sedimentary bottoms, often near hardbottom features. Atlantic, pancake, Gulf, and spiny batfishes generally inhabit waters deeper than 150 ft, with pancake batfishes reaching depths in excess of 1000 ft, and Atlantic batfishes even deeper (to 1000 ft). Tricorn batfishes also are a deepwater form; they frequent 300–600 ft depths. Most other species have rather broad continental-shelf depth distributions, although polka-dot and spotted batfishes are most common in depths less than 200 ft. All species are widespread in Florida waters, except for slantbrow batfishes, which are known from deep water (1200 ft) in the Straits of Florida and the northwestern Gulf of Mexico,

Natural History

Little is known about the life history of batfishes. Reproduction has not been observed, but some batfish species have ovaries similar to those found in frogfishes, suggesting that an egg sheet is produced. Species reaching the largest sizes are shortnose batfishes (15 in), spotted batfishes (12 in), and longnose and polka-dot batfishes (both 9 in); the smallest species (all ≤6 in long) are the four round-bodied batfishes and roughback batfishes. Polka-dot batfishes live at least 16 years, but the smaller *Halieutichthys* species only live for 2 years.

Batfishes feed mostly on bottom-dwelling invertebrates, such as snails, worms, shrimps, and crabs. A batfish's esca is thought to be used in attracting prey, but not in the manner employed by anglerfishes. Its structure is glandular, and a chemical is secreted that apparently attracts snails and other invertebrates close to the quiescent batfish. The guts of polka-dot batfishes are loaded with parasitic nematodes, so many that they often can be seen squirming out of the fish's cloaca!

Batfishes rest on the seafloor, propped up by their pelvic fins, which they use to manoeuver on the bottom. They also swim in short bursts, using typical tail propulsion; water taken in through the mouth and expelled through rear-facing gill openings may provide a bit of a boost.

Mullets
Family Mugilidae

- **Mountain mullet** (*Agonostomus monticola*)
- **Striped mullet** (*Mugil cephalus*)
- **White mullet** (*Mugil curema*)
- **Liza** (*Mugil liza*)
- **Redeye mullet** (*Mugil rubrioculus*)
- **Fantail mullet** (*Mugil trichodon*)

Background

Mullets are inshore species that are revered throughout Florida. The reasons for this adoration differ, depending on which coast you are on. To uppity southeast-coast residents, mullets are prized as bait for larger, more-prestigious game fishes; along the down-to-earth west coast, mullets are esteemed for their flavor, either when smoked or when split open and deep-fried whole. In fact, the panhandle town of Niceville even sponsors an annual festival honoring mullets. Mullets have silvery sides, with blunt heads, large scales, adipose (transparent) eyelids in all but one species, widely separated dorsal fins, and pectoral fins set above the midline. Of the six mullet species in Florida, mountain mullets are the smallest (4 in) and most distinctive: they are golden-brown dorsally; the median fins and tail are dull yellow, with a large dusky spot on the base of the tail; and they have no adipose eyelids.

The remaining species fall into two groups: one with stripes on the body and with scales only on the lower part of the second dorsal and anal fins (striped mullets and lizas), and second group without stripes but with those fins almost completely scaled (white, redeye, and fantail mullets). Striped mullets—most commonly known as "black mullets" in Florida—are tan to gray dorsally, with several dark, longitudinal stripes on the flanks and a blue-black spot above the gill opening. They are the largest mullet species in Florida, reaching a length of 3.5 ft. The nearly equal-sized lizas resemble striped mullets but have a narrower, more-pointed

Mountain mullet (*Agonostomus monticola*), Loxahatchee River, Florida

Striped mullet (*Mugil cephalus*), Crystal River, Florida

White mullets (*Mugil curema*), Lake Worth Lagoon, Florida

head; less-pronounced lateral stripes; and a more-forked caudal fin. The origin of the second dorsal fin is notably behind a vertical line drawn from the anal-fin origin (the two origins are over the top of each other in striped mullets).

White mullets (3 ft), commonly known as "silver mullets" in Florida, are dark-gray to tan dorsally. They have a dark spot at the base of the pectoral fin, a black margin on the tail fin, and a distinctive gold spot on the gill cover. Redeye mullets look very much like white mullets in having a golden spot on the gill cover, but the pectoral fins are large (the tips reach the origin of the dorsal fin) and in life they have a distinctive red-orange iris. Fantail mullets also look like white mullets, but they are smaller (10 in); have an olivaceous dorsum; and have duskier dorsal and caudal fins, with a darkened dusky margin to the caudal fin, which is more forked than the tail of white mullets. Fantail mullets have a narrow, crescentic marking at the base of the pectoral fin.

Distribution and Habitat

Adult and subadult mountain mullets reside in the freshwater reaches of coastal rivers. We have encountered this species mainly in southeast and southwest Florida, in coastal rivers and in water-control canals. Striped and white mullets are widespread around the state, found in estuaries, lagoons, rivers, and freshwater springs. Lizas and redeye mullets are restricted to the clear, shallow waters of southeast Florida and the Florida Keys. Fantail mullets are most common in the Florida Keys but range from Cape Canaveral on the east coast to about Tampa on the west coast. All species reside in shallow water, usually over softbottom until spawning season, when adults move offshore.

Natural History

Mullets spawn up in the water column, and most species spawn in offshore waters. Striped mullet adults aggregate and move offshore to spawn from October to January. When they are well offshore, we have observed striped mullets in funnel-like schools, spinning into a maelstrom while presumably spawning. Each fall off the east coast, prior to their offshore spawning movement, striped mullets emerge in great numbers from estuaries and lagoons and spill into the coastal ocean in the legendary mullet run. Large schools of these fishes move methodically southward, occasionally ducking into inlets, only to be pummeled by snooks, tarpons, crevalle jacks, bluefishes, sharks, and other predators. Walking on an east-coast beach while keeping pace with a football field–sized school of striped mullets can be like a watching a fireworks show, as these predators periodically shower the mullets in every direction, sometimes running the mullets (and occasionally themselves) right onto the beach. The motivation for the mullet run is not known, but it seems to be a prelude to spawning—falling water temperatures and the lunar cycle no doubt are environmental initiators. Young, immature striped mullets (known widely as "finger mullets") also move during the run—these are not spawning individuals, but maybe they're just practicing!

White mullets spawn in April and May on the east coast, but their aggregations are much smaller than those observed in striped mullets. Eggs and larvae drift offshore

Redeye mullets (*Mugil rubrioculus*) and white mullets (*Mugil curema*), Lake Worth Lagoon, Florida

for some time and then transform into tiny, silvery juveniles. Very young white mullets looking like tiny white dots; they travel in small groups, right on the surface but well away from shore, where they jump in synchrony. In the spring, these young make their way from offshore into shallow inshore areas, where they spend most of their lives.

Mullets feed on detritus (decaying plant material) embedded in the sand or even as a film on the water surface. You can watch mullets in shallow water methodically vacuuming sand and ejecting the excess through their gill openings. They digest the detritus in an alimentary tract that includes a gizzard-like structure. Striped mullets are well known for their habit of haphazardly jumping out of the water like stiff planks. The uninitiated might think the backwaters of Florida are absolutely teeming with fishes. Any mullet will jump when being chased by a crevalle jack or a similar predator. Even the tiny juveniles observed in blue water are quick to jump in unison when disturbed. But the leap of a large striped mullet is a long-standing mystery. We think that for adult striped mullets—like eagle rays, blacktip sharks, and sailfish that jump from the water for no apparent reason—free-jumping may at least partially involve some form of communication.

The other mullet species do not seem to jump without just cause, but when you watch white or striped mullets school in shallow water, inevitably an individual will break from the school to snap its tail on the surface. With time, you can learn to recognize the species (striped or white) by the sound of their water-surface tail snap. White mullets produce a single sharp snap, and striped mullets make a sort of double snap (the white mullet's tail is more truncated than the striped mullet's forked tail). The consistently distinctive nature of the tail snaps suggests some form of communication—a subject for future investigation. One researcher has alternatively suggested that mullets break the surface to quickly re-aerate after putting their heads in low-oxygenated bottom water or anaerobic sediment when feeding.

New World Silversides
Family Atherinopsidae

- **Rough silverside** (*Membras martinica*)
- **Inland silverside** (*Menidia beryllina*)
- **Key silverside** (*Menidia conchorum*)
- **Atlantic silverside** (*Menidia menidia*)
- **Tidewater silverside** (*Menidia peninsulae*)

Background

New World silversides are small, slender, silvery fishes that fall easily into the "baitfish" category. The five species look very similar superficially: a slender, upturned mouth;

a mylar-looking lateral stripe; pectoral fins inserted high on the flanks; and 2 separate dorsal fins. Rough silversides are so named because their scales have rough edges that don't require a microscope to see—a finger is all you need. The pelvic fins are located closer to the anal fin than in the other silverside species; the basic body color is yellowish to green; and the back is bluish. Rough and inland silversides reach about 4 in long, while Atlantic (5 in) and tidewater silversides (6 in) are a bit larger. Inland, Atlantic, tidewater, and Key silversides look a lot alike, but their distributions—especially Atlantics and Keys—and ecological preferences help in identification. Atlantic silversides have a much longer anal fin than the others, and its margin is almost straight. This species is a translucent blue-green to gray on the back, with white undersides. In contrast, the shorter anal fin of similarly colored inland silversides has a concave margin. Inland and tidewater silversides are readily confused. The short anal fin of tidewater silversides is a bit less concave than that of inland silversides (but not as straight or long as that of Atlantic silversides), but its swim bladder does not extend beyond the anal-fin origin (it does in inland silversides; this can be seen through their semitranslucent abdominal wall). Biochemical genetic studies confirm the distinction between these two species. Key silversides are a tidewater look-alike that differs from the latter species in having fewer anal-fin rays. Its validity as a good (distinct) species is debatable, as genetic evidence is not convincing; more study is required.

Distribution and Habitat

Rough silversides are found around the state, in shallow estuarine and marine waters. Inland silversides are distributed throughout Florida; while they have been recorded in salinities ranging from 1 to 33 psu, inland silversides are predominantly a freshwater, low-salinity species. On the other side of the spectrum are Atlantic silversides, a marine and estuarine warm-temperate species confined to northeastern Florida, reaching its southern limit just south of Cape Canaveral. Tidewater silversides, which occur from Cape Canaveral around the peninsula, are found in estuarine and marine situations and thus can co-occur with rough or inland silversides. Key silversides are only known from the lower Florida Keys (especially Big Pine Key), in varying salinities.

Natural History

New World silversides spawn in shallow water, often in groups. Eggs are shed over vegetation; they have sticky tendrils that attach to emergent vegetation or other substrates. Spawning takes place in shallow water, just as the tide begins to recede. Multiple males compete for mates in a mass reproductive frenzy. Tidewater silversides spawn from March to June, in Gulf marshes with estuarine salinities (23 psu), when water temperatures are between 62°F and 86°F. Inland silversides spawn at water temperatures of 58°F–80F°, in much lower salinities (averaging 1.6 psu). There are two reproductive peaks in many areas. Older fishes spawn from March to May, and their young settle in May.

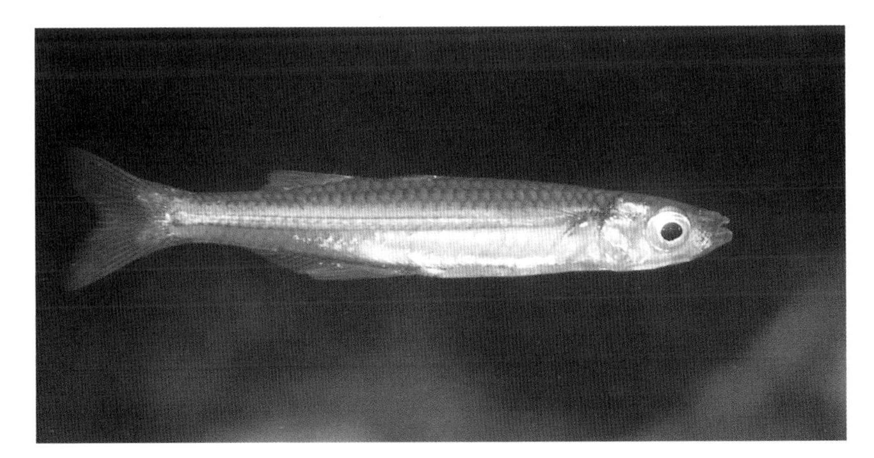

Tidewater silverside (*Menidia peninsulae*), Crystal River, Florida

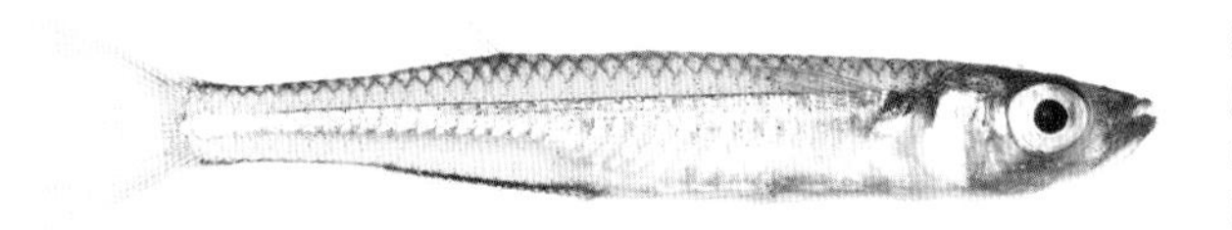

Inland silverside (*Menidia beryllina*), Loxahatchee River, Florida

By fall these young are ready to spawn themselves.

Depending on their size, New World silversides feed on detritus, zooplankton, bottom-dwelling invertebrates, and larval fishes. Tidewater silversides studied in detail in the Big Bend ecoregion undergo three feeding stages as they grow. In April, newly settled young (0.4–1.0 in) ingest detritus and tychoplankton (small, bottom-dwelling organisms swept into the water column by wave action or currents) at the water's edge. From May to November, nonreproductive adults and juveniles (1.0–2.0 in) move into deeper water and feed on zooplankton (selectively eating certain copepods and barnacle larvae). Reproductively mature individuals (2.7–3.5 in) shift from zooplankton to larger, bottom-dwelling crustaceans and larval fishes in February and March. Other silverside species may follow a similar feeding progression with growth, but they have not been studied in detail. Research suggests that inland silversides consume more bottom-dwelling organisms than zooplankton. Rough silversides feed at night, mostly on zooplankton.

Old World Silversides
Family Atherinidae

- **Hardhead silverside** (*Atherinomorus stipes*)
- **Reef silverside** (*Hypoatherina harringtonensis*)

Background

Old World silversides closely resemble New World silversides (family Atherinopsidae) in being small (4 in) and having a prominent midbody silver stripe. Both species in our waters have blunter heads, bigger eyes, and shorter anal fins than members of the related New World silversides, and their silver stripes are bordered dorsally by dark stripes. The head of hardhead silversides is wider than the body (more slender in reef silversides). They have tiny black spot on the scales of the body (only above the stripe in reef silversides), and the body is chunkier than that of the slimmer reef silversides. The tips of the caudal fins in both species are dusky to black in adults. Juveniles are best identified by counting the spines in the first dorsal fin (4–6 in hardheads, 5–7 in reefs).

Distribution and Habitat

Reef silversides are known from southeastern Florida, Florida Bay, and the Florida Keys, on shallow (<30 ft) seagrass meadows, patch reefs, and fringing reefs. Hardhead silversides also are restricted to that area, where they school near sandy beaches, patch reefs, seagrass meadows, mangrove shorelines, and artificial structures, usually near the surface in waters less than 30 ft deep.

Natural History

Both species are found in small to large schools and often are seen in immense shoals near reefs or docks. Such schools may be mixed with anchovies and/or dwarf herrings. Not much is known about reproduction in Old World silversides. Reports indicate that hard-

Hardhead silverside (*Atherinomorus stipes*), Islamorada, Florida

head silversides spawn at night, between midnight and daybreak. Based on descriptions of other members of the genus, reef silverside eggs should also be negatively buoyant and have tendrils. Both species are planktivorous, and they themselves are important forage items for needlefishes, barracudas, snappers, jacks, snooks, and many other species in shallow-water environments.

Flyingfishes
Family Exocoetidae

- **Margined flyingfish** (*Cheilopogon cyanopterus*)
- **Bandwing flyingfish** (*Cheilopogon exsiliens*)
- **Spotfin flyingfish** (*Cheilopogon furcatus*)
- **Atlantic flyingfish** (*Cheilopogon melanurus*)
- **Clearwing flyingfish** (*Cypselurus comatus*)
- **Oceanic two-wing flyingfish** (*Exocoetus obtusirostris*)
- **Fourwing flyingfish** (*Hirundichthys affinis*)
- **Sailfin flyingfish** (*Parexocoetus brachypterus*)
- **Bluntnose flyingfish** (*Prognichthys occidentalis*)

Background

Flyingfishes are the ultimate open-water prey fishes. If you're a small, delectable fish and your backyard—the open blue sea—lacks hiding places, you need some means of survival, and wings will serve you well. Flyingfish wings are the paired pectoral and, in some species (the four-winged flyingfishes), pelvic fins used to glide (not fly) over the water surface. All species exhibit classic countershading (dark-blue above and white below), but juveniles are different in coloration and, often, in morphology (form and structure). For example, on calm days flying juveniles of some of the four-winged species look like tiny moths, the two pairs of fins being larger than the body.

Color patterns on the fins help separate the four-wing species. Both margined and bandwing flyingfishes have distinctive black blotches on their dorsal fins. In margined flyingfishes (16 in), the caudal fin is entirely black and the pectoral fins are uniformly dark-blue, with thin white edges. The similar but smaller (12 in) bandwing flyingfishes differ in having just the lower lobe of the tail be black (the upper lobe is clear) and a broad, white central band on the pectoral fins. Pelvic-fin margins are dusky in bandwings and clear in margineds. Juvenile bandwing flyingfishes have long barbels under the chin, which are about 40 percent of a juvenile's total length. Like bandwings, spotfin flyingfishes (14 in) have black pectoral fins, with a white central band; they differ in having dark-edged pelvics (clear in bandwings) and clear dorsal and caudal fins. Atlantic flyingfishes (13 in) have clear fins, except for the pectorals, which are black, with a central white area that is widest at the rear edge of the fin. Juveniles have very short barbels under their chins. All fins of clearwing flyingfishes

Margined flyingfish (*Cheilopogon cyanopterus*), Jupiter, Florida

(10 in) are clear, except for the dusky pectorals. Clearwings have a single long, flattened chin barbel and reduced pelvic fins that are placed forward on the body. Juvenile clearwings do not have barbels under the chin. Fourwing flyingfishes (11 in) have dusky pectoral fins, with a clear central area. The other fins are largely transparent, and the dorsal-fin origin is located over that of the anal fin. Bluntnose flyingfishes (10 in) indeed have a blunt snout; the outer halves of their short pectoral fins are dusky. Two two-winged species, with small-sized pelvic fins, are found in our waters. Sailfin flyingfishes (6 in) have a high dorsal fin, with a large, dipped-in-ink black blotch. Oceanic two-wing flyingfishes (9.7 in) have very long, clear pectoral fins and a quite blunt head.

Distribution and Habitat

All the flyingfishes found off Florida are widespread in distribution and live in the upper layer of the open ocean, generally offshore of the shelf.

Natural History

Fourwing flyingfishes are the best-studied of the various species, because of their importance as a fishery species in the Caribbean. They spawn from November to June, near the water surface, releasing nonbuoyant eggs with sticky threads (tendrils) that attach to sargassum or flotsam (palm leaves, coconut husks, logs, etc.). Bluntnose and oceanic two-wing flyingfishes have buoyant eggs, with smooth surfaces and no tendrils. Other species have been observed spawning in mass aggregations in the absence of any flotsam. Larvae develop in more than a week and settle up toward the surface.

Flyingfishes feed on zooplankton and neuston (organisms that live on top of or just under the water surface), and several species are most active at night. Flyingfishes are highly attracted to lights on vessels of all sizes in the open ocean. Dead flyingfishes are commonly found on decks in the morning after a night spent running with lights or having them turned on while at anchorage. On calm nights in blue water, flyingfishes are predictably seen gliding through the lighted water and are readily dipnetted. Bottlenose and other dolphins chase flyingfishes into the sides of large vessels, presumably to stun them, making for an easier catch.

Halfbeaks
Family Hemiramphidae

- **Hardhead halfbeak** (*Chriodorus atherinoides*)
- **Flying halfbeak** (*Euleptorhamphus velox*)
- **Balao** (*Hemiramphus balao*)
- **Ballyhoo** (*Hemiramphus brasiliensis*)
- **False silverstripe halfbeak** (*Hyporhamphus meeki*)
- **Atlantic silverstripe halfbeak** (*Hyporhamphus unifasciatus*)
- **Smallwing flyingfish** (*Oxyporhamphus micropterus*)

Background

All Florida halfbeaks (except hardhead halfbeaks and smallwing flyingfishes) have very long lower jaws and very small, triangular upper jaws. In all species, the dorsal fin is situated almost symmetrically over the anal fin, near the posterior part of the body. Hardhead halfbeaks are easily confused with New World (family Atherinopsidae) and Old World (Atherinidae) silversides in having a mylar-like, silver lateral stripe and a dusky back, but they differ in having the lower lobe of the tail slightly longer than the upper lobe, a characteristic of halfbeaks (Hemiramphidae) and flyingfishes (Exocoetidae). Smallwing flyingfishes look like members of the flyingfish family, but they have much shorter pectoral fins and actually are members of the halfbeak family. Balaos have a red tip on the end of the extended lower jaw, and the very-similar-looking ballyhoos have a yellow upper tail lobe. Even more confusing are Atlantic silverstripe halfbeaks and false silverstripe halfbeaks, both of which are silvery, with blue-green backs, dusky-black margins to the caudal fin, and red tips on the lower jaw. Small differences in shape and gill-raker counts separate these two species.

Distribution and Habitat

Like their flyingfish relatives, halfbeaks are surface dwellers. Hard-

Ballyhoo (*Hemiramphus brasiliensis*), Palm Beach, Florida

head halfbeaks occur in nearshore waters of southeastern and southwestern (south of Tampa Bay) Florida and the Keys, in depths of less than 20 ft. False silverstripe and Atlantic silverstripe halfbeaks are found inshore and in estuaries, the former statewide (except in Florida Bay and the Keys) and the latter restricted to southern Florida, from Tampa and St. Lucie Inlet through the Keys (and southward throughout the Caribbean). Ballyhoos also share inshore waters statewide, but their close relative, balaos, are most abundant in somewhat deeper waters offshore of reef tracts in southeast Florida (but not in Florida Bay) and the Florida Keys. Ballyhoos and balaos tend to occur over seagrass meadows, reefs, or hardbottom. Atlantic silverstripe halfbeaks occupy blue water over the outer shelf in the Keys, and smallwing flyingfishes prefer oceanic waters even farther offshore.

Natural History

Halfbeaks are broadcast spawners. Balaos and ballyhoos spawn daily during spring and summer months in southeastern Florida. Female ballyhoos produce an average of 1100 eggs per spawning event, and balao females produce three to five times more eggs per spawning event. Their relatively large eggs are demersal and are sometimes found attached to vegetation. These two *Hemiramphus* species differ, however, in a peculiar way—ballyhoos are parasitized by an isopod crustacean that lodges itself on the tongue of young adults, but balaos are not afflicted this way. Surprisingly, the stomach contents of parasitized ballyhoos did not differ from those without the parasite, although it would seem impossible to feed with this creature attached to the tongue. Balaos eat planktonic fish larvae, copepods, crustaceans, polychaete worms, and siphonophores; ballyhoos feed mostly on floating pieces of seagrasses and the epiphytic algae that live on (by being harmlessly attached to) them. Young false silverstripe halfbeaks eat planktonic zooplankton, and larger individuals consume seagrasses and epiphytes.

Halfbeaks are preyed on by bottlenose dolphins and a Who's Who of predators. Not surprisingly, the four *Hemiramphus* and *Hyporhamphus* species are the gold standard of trolling bait for offshore anglers seeking dolphinfishes, sailfishes, wahoos, tunas, and king mackerels. Ballyhoos are the primary bait species, but the other species may be used when available. A small-bait fishery occurs in the upper Florida Keys. In southeastern Florida, anglers use oatmeal and ground-up fishes to attract ballyhoos to their boats.

Needlefishes
Family Belonidae

- **Flat needlefish** (*Ablennes hians*)
- **Keeltail needlefish** (*Platybelone argalus*)
- **Atlantic needlefish** (*Strongylura marina*)
- **Redfin needlefish** (*Strongylura notata*)
- **Timucú** (*Strongylura timucu*)
- **Atlantic agujón** (*Tylosurus acus*)
- **Houndfish** (*Tylosurus crocodilus*)

Background

At first glance, needlefishes all look like silvery soda straws, with elongate bodies and long jaws lined with numerous, needle-like teeth. The species are distinguished on the basis of intricacies of body and fin shapes. Flat needlefishes, which grow to 4 ft, are the easiest to identify, as they are the only needlefish species possessing a laterally compressed body. Their coloration is unique (a series of 10–12 dark bars on the flank), and they have the largest soft dorsal fin of any needlefish. Keeltail needlefishes are smaller (reaching a maximum of about 20 in) and have a keel (a fleshy, flattened lateral projection) on each side of the somewhat flattened caudal peduncle. The jaws are particularly long, and the body is more-or-less round in cross-section. All other needlefishes (except flat needlefishes) are ovoid, including the similar-looking Atlantic needlefishes, redfin needlefishes, and timucús (which grow to a maximum length of about 18 in). All three of these species have bluish-green backs, with silvery flanks, but only redfin needlefishes have dorsal, anal, and caudal fins with a distinctive red cast to them, as well as a thin black bar on the mid-operculum. Atlantic needlefishes and timucús are the toughest to separate, but timucús have a dusky stripe on the operculum (behind the eye) and the area from in front of the eye to the base of the lower jaw is dusky as well (silvery on Atlantic needlefishes). Timucús differ from other needlefishes by having almost equal-sized caudal-fin lobes and by lacking caudal peduncle keels. Atlantic agujóns and houndfishes differ from all other needlefish species (except keeltail needlefishes) in having a lateral keel on each side of the caudal peduncle, and they have much shorter jaws than keeltails. Atlantic agujóns and houndfishes appear very similar to each other, but the beak of Atlantic agujóns is noticeably longer than that of houndfishes, which are also larger (to 5 ft) than Atlantic agujóns (3 ft). In addition to the subtle difference in their coloration—Atlantic agujóns usually are cobalt blue, and houndfishes are greenish-blue—the dorsal-fin lobe is a bit higher and blackish in agujóns of all sizes, while it is a bit lower and black only in juvenile houndfishes.

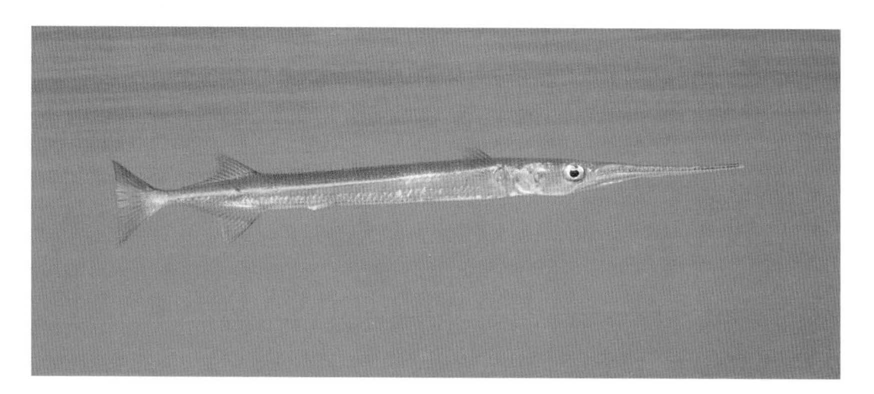

Redfin needlefish (*Strongylura notata*), Crystal River, Florida

Distribution and Habitat

Flat needlefishes are found throughout Florida, usually in oceanic waters. Keeltail needlefishes also occur around the state, usually offshore, but occasionally in schools in nearshore waters. Both of these species are known to associate with drifting sargassum. Atlantic needlefishes occur in shallow, primarily estuarine waters around the state and routinely ascend coastal rivers into freshwater. Redfin needlefishes and timucús generally are found in higher-salinity waters in nearshore seagrass and mangrove habitats along the west and eastcentral coasts of Florida, down to the Keys. Atlantic agujóns and houndfishes occur around the state, from inshore to oceanic waters, and are often associated with sargassum;

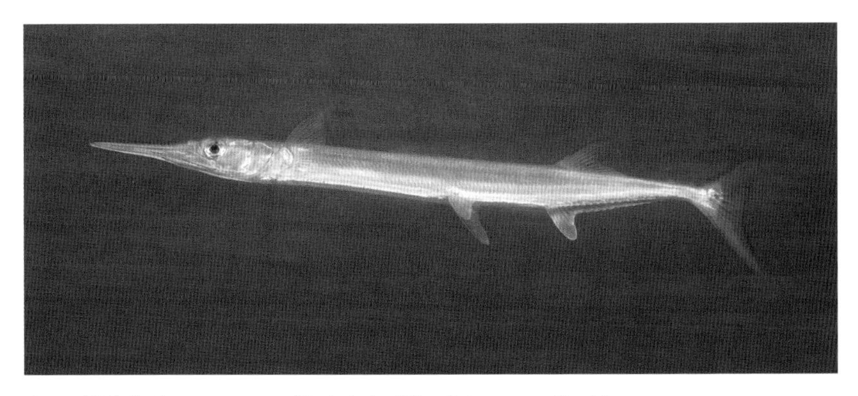

Houndfish (*Tylosurus crocodilus*), Lake Worth Lagoon, Florida

houndfishes tend to be the more coastal of these two species.

Natural History

Needlefishes are inhabitants of the upper layer of the water column, regardless of water depth. On calm days it is not unusual to see needlefishes skipping (greyhounding) across the water surface when startled by the approach of a boat. They feed mostly on small fishes (herrings, anchovies, mullets, jacks) near the water surface and often are seen skyrocketing into schooling fishes in nearshore waters. Needlefishes, in turn, are eaten by such predators as bottlenose dolphins, spotted seatrouts, great barracudas, snooks, and little tunnies, as well as members of their own family. Spawning has not been well studied, but needlefish eggs are known to be relatively large, with thread-like tendrils that attach to the bottom, to floating plant debris, or to other eggs. Some needlefishes are parasitized by a tiny crustacean that lodges itself in the lateral line's canal system in the head. Abnormal swelling occurs around the parasite, often on the lower jaw in these individuals.

New World Rivulines
Family Rivulidae

- **Mangrove rivulus** (*Kryptolebias marmoratus*)

Background

Rivulines are small topminnow-like fishes that, as group, are more often seen in tropical peripheral freshwater environments than in marine habitats. The only species of the group in Florida is the mangrove rivulus, which is brownish or tan, with dark mottling, and has large dorsal and anal fins located far back on the body, just in front of the caudal fin. Females have a prominent eye-spot on the caudal peduncle. Their unique habitat choice facilitates identification.

Distribution and Habitat

Mangrove rivulus are restricted to extremely shallow inshore waters, from just north of Cape Canaveral (Volusia County) on the east coast to southwest Florida, in direct correlation with the distribution of red mangroves.

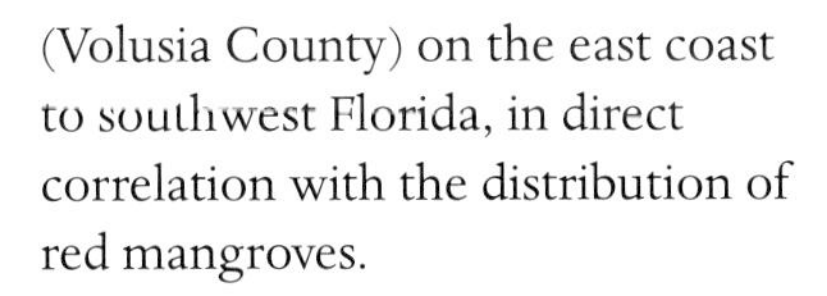

Mangrove rivulus (*Kryptolebias marmoratus*), Vero Beach, Florida

Natural History

Mangrove rivulus have an interesting life history, as the fish that mates with itself. Most individuals are hermaphroditic, with an ovotestis capable of producing both eggs and sperm. This species is the only vertebrate known to self-fertilize; up to 75 eggs are deposited at night in the water or on damp mud, where embryos develop in about 16 days. Mangrove rivulus inhabit stagnant mangrove swamps, where dissolved oxygen is low, salinity is high, and toxic hydrogen sulfide is prevalent. To cope with these conditions, individuals even rest out of the water. In Indian River Lagoon, for example, their preferred habitat is located at and above the upper reach of the intertidal zone. When water levels recede in these areas, they take refuge in small pools and (especially) in the burrows of land crabs. If no crab burrows are available, they crawl inside termite-riddled trunks of mangrove trees, hollow logs, coconuts, and even beer cans, in search of moisture during dry periods. In the Everglades and southwest Florida, they take refuge in standing water above the tideline. Mangrove rivulus feed on a variety of terrestrial and aquatic invertebrates, including snails, worms, insects, mosquito larvae, and even members of their own species.

Pupfishes
Family Cyprinodontidae

- **Sheepshead minnow** (*Cyprinodon variegatus*)
- **Goldspotted killifish** (*Floridichthys carpio*)
- **Flagfish** (*Jordanella floridae*)

Background

Pupfishes, formerly lumped with topminnows, are short-bodied, dumpy little (2–3 in) fishes that can be distinguished by their color patterns. Sheepshead minnows have 5–7 dark, wavy bars on the flanks and a dark, thin band at the base of the caudal fin. Females have 2 dusky blotches on the base of the dorsal fin and 1 ocellated spot at the rear of the fin. Males have a dark marginal band on the caudal fin, a dusky margin on the dorsal fin, and spots on the dorsal, anal, and caudal fins that tend to form bands. When in breeding coloration, males exhibit iridescent blue flecks on the back, and the color on their median and pelvic fins ranges from brown to dark-orange. Goldspotted killifishes have gold or pale-orange spots, which often coalesce to form vague bars. The intensity of the orange spots increases in breeding males. Flagfishes have alternating orange-red and black stripes; small gold specks; and a large black mid-body spot.

Distribution and Habitat

Sheepshead minnows and goldspotted killifishes live in very shallow water (<6 ft) around the state. Brackish to fully marine seagrass meadows, marshes, mangrove shorelines, and detritus-laden tidal ditches attract sheepshead

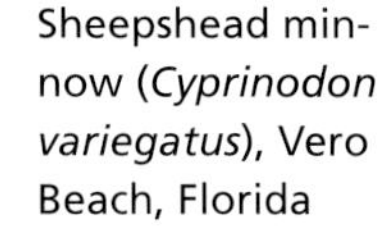

Sheepshead minnow (*Cyprinodon variegatus*), Vero Beach, Florida

Goldspotted killifish (*Floridichthys carpio*), Melbourne, Florida

minnows. Goldspotted killifishes occur in shallow marsh, seagrass, and mangrove habitats, where salinity can range from freshwater to hypersaline. Flagfishes are found from the St. Johns River system to around to the Ochlocknee River in the panhandle, in shallow freshwater and in low-salinity marsh and ditch habitats.

Natural History

Sheepshead minnows and goldspotted killifishes mature and spawn during their first year of life. Sheepshead minnows spawn from February to June in the St. Johns River. Females lay demersal eggs, and spawning may be drawn out over several days, as females might only release a few eggs at a time. Eggs (which sink) have sticky filaments or tendrils that attach them to submerged plants or debris. The brightly colored males compete vigorously for females while defending a mating territory against rival males. Once a receptive female enters a male's territory, the male holds her with his dorsal and median fins and spawning occurs. Goldspotted killifishes spawn over sandy bottoms, in very shallow water (<2 ft). Male and female pairs clasp their bodies closely when spawning; the female usually releases a single egg that sinks to the bottom. Flagfishes, which can live for at least 4 years, spawn in freshwater. Males reportedly build and defend nests that are located in heavy vegetation.

Sheepshead minnows feed on detritus, insect larvae, and small crustaceans. Goldspotted killifishes are opportunistic feeders that shift their diets, depending on habitat and salinity—in low salinity, filamentous algae is most important, although in higher salinities they consume a greater percentage of benthic crustaceans.

Topminnows
Family Fundulidae

- **Marsh killifish** (*Fundulus confluentus*)
- **Gulf killifish** (*Fundulus grandis*)
- **Mummichog** (*Fundulus heteroclitus*)
- **Saltmarsh topminnow** (*Fundulus jenkinsi*)
- **Striped killifish** (*Fundulus majalis*)
- **Seminole killifish** (*Fundulus seminolis*)
- **Longnose killifish** (*Fundulus similis*)
- **Diamond killifish** (*Fundulus xenicus*)
- **Rainwater killifish** (*Lucania parva*)

Background

Topminnows are a hardy group of soft-rayed fishes found in shallow marine waters and freshwater throughout eastern North America. Males and females usually show different color patterns and, during the reproductive season, males brighten up considerably. Eight species reside in Florida's estuarine and marine inshore habitats; many others are found exclusively in our freshwater locales. Male and female Seminole killifishes, the largest (6 in) of the freshwater species, are greenish, with linear rows of spots on the sides; females also have 15–20 darkgreen bars. Small (<2 in) diamond killifishes are the most distinctive—

Seminole killifish *(Fundulus seminolis)*, Rainbow River, Florida

Marsh killifish (*Fundulus confluentus*), Vero Beach, Florida

Rainwater killifish (*Lucania parva*), Crystal River, Florida

stubby and diamond shaped, with a concave forehead and pointed snout. Their background color is olive-green to tan, with 5–7 distinctive, thin white bars on the flanks; the bars on males are narrower than those on females; and males have a spotted dorsal fin. Marsh killifishes, saltmarsh topminnows, and rainwater killifishes also are small (2.5–3 in). Marsh killifishes have 14–18 dark bars on the flanks; these are thinner in males, which also exhibit a dark ocellated spot at the posterior margin of the dorsal fin. Similarly colored male and female saltmarsh topminnows are gray-green, with a row of small spots along the flanks. Rainwater killifishes are tan to greenish, with a pale midlateral stripe; the scales are cross-hatched, giving them an XXX-patterned appearance. The dorsal fins on males have red and orange edges.

Longnose killifishes (4.5 in) and striped killifishes (7 in) are closely related, and they have been lumped as a single species in the recent past. Male and female longnose killifishes are similarly colored: olivaceous dorsally, with 8–13 dark lateral bars and a spot on the caudal peduncle. Males may have black spot on the posterior base of the dorsal fin. Striped killifishes are dark-green to tan on the back; females have 2–4 dark, horizontal body stripes and 3 bars on the caudal peduncle; males have 10–15 dark, vertical bars. Males sometimes have a longer anal fin than females and often have a dark spot near the posterior base of the dorsal fin. Reproductively active males become darker, and their median fins become yellow, with black margins. Fatter-bodied mummichogs (5 in) and Gulf killifishes (7 in) are tougher to separate (but their distributional differences help), and they also were once considered a single species. Both have olivaceous backs and pale-yellow to silvery sides. Mummichog males have up to 15 pale bars; very small, whitish spots on the flanks and vertical fins; and (often) a dusky spot on the rear of the dorsal fin. Coloration is more pronounced in breeding adults; young usually are silvery-grey. Females look like young males and may have some dusky bars. Female Gulf killifishes have 12–15 narrow, dusky bars; males have pale spots, with faint or no obvious bars on the flanks. The caudal-fin margin of Gulf killifishes is nearly straight, while that of mummichogs is broadly rounded.

Distribution and Habitat

Topminnows tolerate a wide range of salinity and temperatures, which are characteristic of the shallow, tide-dominated marshes where they live. Seminole killifishes are a freshwater species that commonly move into the upper reaches of the St. Johns River estuary. Two species are restricted to the Gulf: diamond killifishes inhabit coastal marshes from the Florida Keys northward through the panhandle, and saltmarsh killifishes occur in upper marshes from the panhandle to Texas. Estuarine mummichogs are found only in northeast Florida, where the southern boundary of their range lies just south of Jacksonville. Marsh and rainwater killifishes prefer brackish water and live in coastal rivers and marshes around the state. Gulf killifishes are common in estuaries of the Big Bend ecoregion and panhandle area, but they also are found along the east coast, from Vero Beach northward. Mummichogs and Gulf killifishes co-occur in northeastern Florida and probably hybridize. The distribution of longnose killifishes and striped killifishes also overlaps along this same stretch of northeast Florida coastline and, yes, they hybridize to some degree as well, making identification a real chore in the region.

Longnose and striped killifishes prefer sandy bottoms in the littoral zone, often near marsh grasses, on sandbars, and near inlets, where they are known to occasionally bury themselves in the sand. Striped and longnose killifishes are found in higher-salinity waters than mummichogs and Gulf killifishes.

Natural History

The reproductive behavior of mummichogs has been well studied. Spawning begins in spring and lasts until fall, and it occurs when the tides are highest, during the new or full moon. The eggs are deposited in the shallows, at the high-water mark during high tide, so when the tide goes out, they will be exposed to the air in which they develop. Fertilized eggs incubate for about 2 weeks and are ready to hatch by the time the next month's high tides return and immerse the eggs, which then hatch. This process takes approximately 12 days to complete. The eggs have sticky fibrils of varying lengths that help hold eggs together with other eggs or help attach them to vegetation, shells, or sediment. Temperature is important in the development of the eggs—the optimal range is 59°F–86°F. Egg production by females follows a cycle that mimics the lunar-tidal cycle of the region. Spawning occurs during the 3- to 6-day period of biweekly spring tides. Individuals gather during these rising tides, usually within or near emergent marsh vegetation. Males bearing somewhat enhanced coloration (orange anal fin; dorsal fin with an orange margin; light, iridescent-blue spots; tail with a yellow margin) follow sometimes-circling ripe females. Multiple males may follow a single female. The female inspects the available substrate for a place to deposit the eggs. Once the female finds a suitable location, she adopts an S-shaped posture and extends her anal fin, which signals the male(s) to move in close. The two clasp in amorous bliss for a few seconds, during which a variable number of eggs (1–50, depending on the size of the female) and sperm are released. Topminnows feed on molluscs, ostracods, worms, shrimps, and insects found in the shallow marsh habitat.

Female mummichog (*Fundulus heteroclitus*), Merritt Island, Florida

Male mummichog (*Fundulus heteroclitus*), Indian River Lagoon, Florida

Livebearers
Family Poeciliidae

- **Eastern mosquitofish** (*Gambusia holbrooki*)
- **Mangrove gambusia** (*Gambusia rhizophorae*)
- **Sailfin molly** (*Poecilia latipinna*)

Background

Unlike the egg-laying killifishes (Rivulidae, Cyprinodontidae, and Fundulidae), livebearers are small killifishes that give birth to live young. The Poeciliidae is primarily a freshwater and brackish-water family, and eastern mosquitofishes are probably the most numerous freshwater fish in the state. They do wander into low-salinity estuarine waters, where they often share space with sailfin mollies, a species equally at home in freshwater and brackish water. Mangrove gambusias are pretty much the opposite of eastern mosquitofishes, replacing them in marine and brackish mangrove habitats.

Eastern mosquitofishes are small (<2 in), light-green to gray on the back, and lighter below. Males are smaller than females and have a pair modified pelvic-fin rays known as gonopodia (singular = gonopodium). Females exhibit a black spot on the abdomen, especially when they are carrying developing young. Mangrove gambusias reach 2 in long and are tan to pale-gray, with tiny black spots on the scales. The dorsal fin is golden-yellow, with black spots. As with eastern mosquitofishes, males are smaller than females and have a gonopodium. Sailfin mollies grow much larger (to 5 in); are pale-gray on the back; and are white below, with tiny, black, linearly directed spots on the upper sides of the body. The mouth is upturned and the head is flattened. Adult males grow a broad, flag-like dorsal fin (which they happily show off to admiring females) and develop a bright–blue-green breeding coloration. Females are pale, with rows of black spots along the flanks.

Distribution and Habitat

Eastern mosquitofishes and sailfin mollies are found in freshwater and estuarine waters around the state, usually where there is some emergent vegetation. Mangrove gambusias are only found from Ft. Lauderdale through the Florida Keys, where they live among the prop roots of red mangroves.

Natural History

Females brood their young inside their bodies, after being impregnated by males using their gonopodia. Mating usually involves multiple males attempting to mate with a single female. Not all matings are successful, but the female may end up with a brood where the eggs were fertilized by several different males. Mangrove gam-

Eastern mosquitofish (*Gambusia holbrooki*), Jupiter, Florida

Mangrove gambusia (*Gambusia rhizophorae*), Key West, Florida

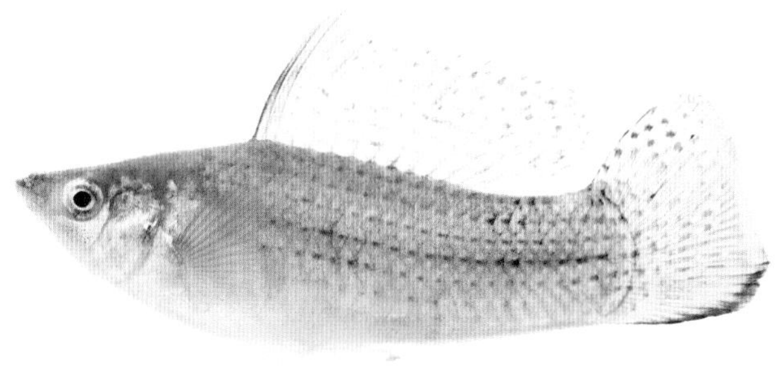

Top: Male sailfin molly (*Poecilia latipinna*), Indian River Lagoon. *Bottom*: Female, Merritt Island, Florida.

busias mate mostly in spring, with broods of 2–65 young. While male mosquitofishes and mangrove gambusias are smaller than females, the two sexes reach about the same size in sailfin mollies. When sailfin mollies mate, the color and intensity of the male's dorsal fin may determine if he gets lucky. Females produce between 10 and 40 young, the gestation time is 3–4 weeks. Broods have roughly equal sex ratios, but adult populations tend to be skewed toward females, possibly because the bright coloration of the males advertises them to predators as well as to females.

Livebearers feed on algae, organic matter, and invertebrates inhabiting the surface layer of the water. As their name implies, eastern mosquitofishes feed on the larvae of the state's most-famous biting insects and have been transplanted all over the world to feast on the critters only an entomologist can appreciate. Mangrove gambusias mostly consume terrestrial insects that float on the water surface. Sailfin mollies eat a lot of algae and detritus.

Squirrelfishes
Family Holocentridae

- **Spinycheek soldierfish** (*Corniger spinosus*)
- **Squirrelfish** (*Holocentrus adscensionis*)
- **Longspine squirrelfish** (*Holocentrus rufus*)
- **Blackbar soldierfish** (*Myripristis jacobus*)
- **Longjaw squirrelfish** (*Neoniphon marianus*)
- **Bigeye soldierfish** (*Ostichthys trachypoma*)
- **Cardinal soldierfish** (*Plectrypops retrospinis*)
- **Deepwater squirrelfish** (*Sargocentron bullisi*)
- **Reef squirrelfish** (*Sargocentron coruscum*)
- **Dusky squirrelfish** (*Sargocentron vexillarium*)

Background

Although squirrelfishes (including soldierfishes) are a group of fishes broadly characterized by texture (spines, serrations, and cutting edges) and color (red), individual species can be distinguished by paying careful attention to color pattern and body shape. Four species are totally red, short and heavyset, have large mouths that extend well past the eyes, and grow to lengths of about 5–8 in. Spinycheek soldierfishes have 3 long spines below each eye and large spines on the lower preopercle and upper opercle. The dorsal fin is completely red, in contrast to that of cardinal soldierfishes; the latter have pale membrane tips and a head that, while spiny, is not as well armed. Bigeye soldierfishes differ in having a body with 10–12 horizontal stripes. The more streamlined blackbar soldierfishes are uniformly red; the tail and median fins have thin white margins; and there is a distinctive black, diagonal slash running across the posterior edge of the gill cover.

Four somewhat smaller (4–7 in) squirrelfishes share striped bodies; have a reddish, diagonal band

Left: Spinycheek soldierfish (*Corniger spinosus*). *Right*: Blackbar soldierfish (*Myripristis jacobus*), Juno, Florida.

Top: Reef squirrelfish (*Sargocentron coruscum*). *Bottom*: Dusky squirrelfish (*Sargocentron vexillarium*).

extending from the eye to the corner of the preopercle; and have shorter jaws (not extending past the rear of the eyes). Longjaw squirrelfishes have alternating red and golden-yellow body stripes; a golden-yellow dorsal fin, with bright-white tips and a white line near the base; and a light-colored band extending from the jaw to the corner of the preopercle. Deep-water squirrelfishes have a golden-yellow dorsal fin and flanks, but their white body stripes are edged with black. Reef squirrelfishes are red, with thin white stripes on the flanks. The tips and base of the dorsal fin are bright-white, and there is a jet-black spot on the membranes between the first 3 or 4 spines. Dusky squirrelfishes have a similarly colored body but lack a black spot on the red dorsal fin; the dorsal spines are white, as are the adjacent membranes, giving the fin a candy-cane appearance.

The two largest (to 12–13 in) species—squirrelfishes and longspine squirrelfishes—have tall, pointed second dorsal fins; long, deeply forked tails, with pointed tips; a light-colored band extending from the jaw to the corner of preopercle; and bodies with alternating red and white-yellow stripes, usually overlaid with irregular white blotches or bars. The dorsal fin of squirrelfishes is yellow, with fleshy golden-yellow tips on the dorsal spines. The dorsal fin of the look-alike longspine squirrelfishes is reddish, with vertical, light-colored areas near the spines and fleshy white tips on the dorsal spines.

Left: Longjaw squirrelfish (*Neoniphon marianus*). *Right*: Deepwater squirrelfish (*Sargocentron bullisi*), Florida Middle Grounds.

Distribution and Habitat

All the Holocentridae are reef dwellers, and some prefer deeper reefs than others. Squirrelfishes are found on reefs and hardbottom around the state, usually in water depths ranging from 25 to 300 ft, while the easily confused longspine squirrelfishes occur in similar habitats and depths along the entire east coast, but rarely north of Tampa on the west coast. Blackbar soldierfishes primarily reside on reefs off the east coast and Keys, in water depths from 25 to 300 ft. Longjaw squirrelfishes are found mainly in the Keys, usually in caves by day, to depths of about 350 ft. Reef squirrelfishes most commonly occur on shallow patch reefs, in water depths less than 25 ft. Dusky squirrelfishes prefer very shallow, wave-swept hardbottom areas. Cardinal soldierfishes are rare in Florida waters, only reported from the Keys, in 60 ft of water. Spinycheek, bigeye, and deepwater squirrelfishes are sporadically found statewide, on deep reefs, at depths greater than 150 ft (bigeye squirrelfishes go as deep as 1200 ft).

Top: Squirrelfish (*Holocentrus adscensionis*), Jupiter, Florida. *Bottom*: Longspine squirrelfish (*Holocentrus rufus*), Florida Middle Grounds.

Natural History

Little is known about spawning in squirrelfishes or soldierfishes, other than that they are water-column spawners. The larvae are pelagic, and some species have an odd larval stage, known as rhyncichthys. This larval stage has a long rostral spine, reminiscent of a little billfish at first glance. Some species have yet another larval stage, known as meeki, that looks more like a small squirrelfish but is silvery. Planktonic duration for larvae is between 40 and 60 days, and the meeki stage adds another 10–70 days to their development.

All members of this family are nocturnal feeders, equipped with large, light-gathering eyes to make it easier to see their prey. The shallow-reef species—including squirrelfishes, reef squirrelfishes, longjaw squirrelfishes, longspine squirrelfishes, blackbar soldierfishes, and cardinal soldierfishes—consume benthic invertebrates, mostly shrimps and crabs. The deeper-reef species feed on larger plankters, as well as benthic invertebrates. Many (probably all) of the squirrelfish species are very vocal, creating sounds thought to be involved in courtship and territorial defense. The sounds are produced with special sonic muscles that drum on the swim bladder, with pulsing caused by moving the ribs.

Seahorses and Pipefishes
Family Syngnathidae

- **Pipehorse** (*Acentronura dendritica*)
- **Fringed pipefish** (*Anarchopterus criniger*)
- **Insular pipefish** (*Anarchopterus tectus*)
- **Pugnose pipefish** (*Bryx dunckeri*)
- **Whitenose pipefish** (*Cosmocampus albirostris*)
- **Crested pipefish** (*Cosmocampus brachycephalus*)
- **Shortfin pipefish** (*Cosmocampus elucens*)
- **Dwarf pipefish** (*Cosmocampus hildebrandi*)
- **Banded pipefish** (*Halicampus crinitus*)
- **Lined seahorse** (*Hippocampus erectus*)
- **Longsnout seahorse** (*Hippocampus reidi*)
- **Dwarf seahorse** (*Hippocampus zosterae*)
- **Opossum pipefish** (*Microphis brachyurus*)
- **Dusky pipefish** (*Syngnathus floridae*)
- **Northern pipefish** (*Syngnathus fuscus*)
- **Chain pipefish** (*Syngnathus louisianae*)
- **Sargassum pipefish** (*Syngnathus pelagicus*)
- **Gulf pipefish** (*Syngnathus scovelli*)
- **Bull pipefish** (*Syngnathus springeri*)

Background

Seahorses and pipefishes are small, mobility-challenged fishes that have tubular snouts and skin, without scales, over bony plates. This family is sexually dimorphic,

Left: Lined seahorse (*Hippocampus erectus*), Lake Worth Lagoon, Florida. *Right*: Lined seahorse, Lake Worth Lagoon, Florida.

Longsnout seahorse (*Hippocampus reidi*), Lake Worth Lagoon, Florida

with males bearing a pouch under the belly to carry fertilized eggs. Seahorses have horse-like heads that repose at a 90° angle to the body. These species move vertically and utilize prehensile tails. Lined seahorses vary widely in color. They can be bright-yellow but usually are tan, dark-brown, or reddish-brown. Males and females differ slightly in snout length, length of the trunk, and amount of ornamentation (in the form of skin filaments). The latter character is hugely variable—it can range from appearing relatively smooth to downright furry. Lined and longsnout seahorses grow to a height of about 7 in. Longsnout seahorses resemble lined seahorses but have a relatively longer snout, and the head and body are covered with tiny, dark spots. Longsnouts also come in a variety of colors. The miniscule (1.5 in) dwarf seahorses, in contrast, are drab-gray or light-brown, have a very stubby snout, and can be smooth to overtly hairy. Identifying juvenile lined seahorses and adult dwarf seahorses can be difficult, especially if done underwater.

Pipehorses are sort of a hybrid between seahorses and pipefishes—the head is directed slightly downward; the body is elongated; and they have a curled, prehensile tail. Pipehorses are small (3 in), and their bodies are adorned with feathery skin filaments.

Pipefishes can be a challenge to identify definitively, as all are elongated, with fairly drab color patterns. Regrettably, we have to rely on subtle or microscope-needed characters, such as the presence or absence of an anal fin; specific ridge placements; the number of body segments (trunk and tail rings); and fin ray counts. Color pattern is useful for differentiating some species. We hereby extend our apologies for what follows, but Mother Nature doesn't always make it easy for us!

Florida pipefishes can be separated into two artificial groups, based on an eyeballed snout length: one set being short-nosed, and the other long-nosed. The short-nosed contingent is composed of primarily tropical species: fringed, insular, pugnose, dwarf, crested, and banded pipefishes. Only the latter three species have anal fins. The largest (9 in) and most strikingly colored of these are banded pipefishes. They have pronounced reddish to purple bands over a white to yellow body, with the bands forming rings that circle around the entire body. The most common color phase (known as crinitus) is pale, with dark flecks and bands; the rarer ensenadae phase is pale-yellow, with dark-red to purplish-brown bands. Crested pipefishes (4 in) have a stouter body

Dwarf seahorse (*Hippocampus zosterae*), Indian River Lagoon, Florida

Shortfin pipefish (*Cosmocampus elucens*), Lake Worth Lagoon, Florida

Whitenose pipefish (*Cosmocampus albirostris*), Lake Worth Lagoon, Florida

Dusky pipefish (*Syngnathus floridae*), Lake Worth Lagoon, Florida

than most of their kin and bear a distinctive bony crest on the top of the head. The sides of the body have weak dark and light bars, which often alternate. Dwarf pipefishes are best described as "not as above"; while they are short-nosed and have an anal fin, their coloration is drab-brown.

Lack of an anal fin identifies the other three species in this group: pugnose, fringed, and insular pipefishes. Even by short-nosed standards, tiny (3 in) pugnose pipefishes have the shortest snout of all. Like most other pipefishes, the body has pale bars, or it may be mottled. Counting trunk rings (beginning with the ring bearing the pectoral fin and ending with the ring bearing the anus) is the best way to separate fringed and insular pipefishes: 14–16 in fringeds, and 17–18 in insulars. Both are tan-brown, with diffuse bars. Fringed pipefishes (3 in) usually have 3 backward-radiating, pale streaks behind each eye, as well as 3 black spots on the back, forward of the dorsal fin. Insular pipefishes (4.5 in) often have a dark streak on the snout, in front of each eye, and have thin, dark bars along the length of the body.

All long-nosed pipefishes, including whitenose and shortfin pipefishes and all members of the speciose genus *Syngnathus*, possess anal fins. Whitenose pipefishes (8 in) are the easiest to identify: the snout is whitish, and the head usually is darker than the body color. They have a series of pale bands that are equally spaced along the body. Chain pipefishes, one of the two largest (15 in) Florida species, also have the proportionately longest snout. They are distinctively colored, with a series of dark, diamond-shaped marks along the entire flank and a dark stripe running across the snout and through the eyes.

Three other species—opossum, dusky, and sargassum pipefishes—have very long snouts. Opossum pipefishes (8 in) are the only Florida pipefish species in which males brood the eggs in a pouch on the trunk (all other species have the pouch on the tail). They have a heavily ridged body, with a spine on each ring, and a 9-rayed caudal fin that (in adults) is reddish, with a central black stripe. Their coloration is light-brown overall, with red spots on the trunk rings, a silver line along the flank, and a sienna-brown upper snout. The very common dusky pipefishes (10 in) are a mottled brown, tan, or (more commonly) green, with barring being either distinct or totally absent. Sargassum pipefishes (8 in) are light-tan, with regularly spaced, dark bars composed of groups of 2–4 closely aligned, thin bars; pale rings around the tail; a dark stripe on the snout; and a pair of dark, radiating bands behind the eyes.

Shortfin, bull, northern, and Gulf pipefishes have moderately long snouts. Shortfin pipefishes (6 in) are highly variable in coloration but usually are gray to brown, with two-toned, pale and dark bands evenly spaced along the body. They have a short dorsal fin, with 21–25 rays. Bull pipefishes are large (15 in), and their pale body coloration is marked by a series of inverted-U, half-moon body saddles. Northern pipefishes (12 in) are dark-brown, with some lighter mottling; light bands may be present along the body; and the dorsal fin has 33–49 rays. Sexual dimorphism is the rule in the smaller (7 in) Gulf pipefishes: females are deep bodied, with V-shaped, keeled bellies, while males are more elongate and have more-traditional flat bellies (and therefore are box shaped in cross-section). Both sexes are dark-brown, with thin, silvery-white, Y-shaped marks on the sides, which are more prominently displayed by females.

Distribution and Habitat

Seahorses and pipefishes have an affinity for submerged or floating vegetation, but some live among reefs and hardbottom areas. Dwarf and longsnout seahorses are shallow-water species: longsnouts are found in depths of less than 50 ft, and dwarfs in even shallower waters, in seagrass beds. Surprisingly, lined seahorses, which are very common in nearshore seagrass beds, also occur offshore, in depths to 600 ft. Lined and

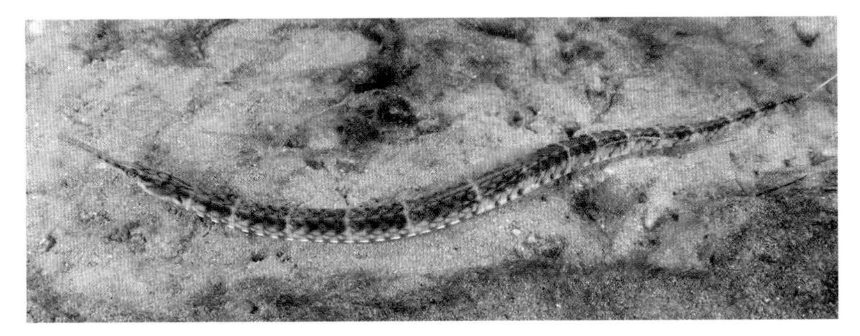

Chain pipefish (*Syngnathus louisianae*), Lake Worth Lagoon, Florida

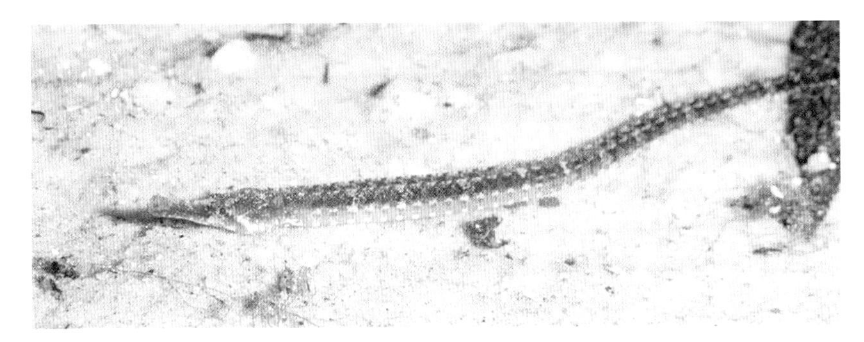

Gulf pipefish (*Syngnathus scovelli*), St. Johns River, Florida

Sargassum pipefish (*Syngnathus pelagicus*), Jupiter, Florida

longsnout seahorses associate with sparsely vegetated bottoms and calcareous algae, as well as with dense seagrass meadows. Both of these species can be found clinging to sea whips, sponges, and hydrozoans on reefs or hardbottom areas. Pipehorses occur around the state, in inshore and shelf waters. Individuals wrap their prehensile tails around seagrass blades, algae, and other bits of substrate.

Opossum pipefishes occur along the Florida Gulf coast and around the peninsula to just south of Cape Canaveral. What appears to be the only breeding population in the state resides in the freshwater reaches of the Loxahatchee, St. Lucie, and St. Sebastian Rivers. Adults associate with the emergent grasses *Panicum repens* (torpedo grass) and *Polygonum* spp. (knotweed) when they are in these rivers. Gulf pipefishes are found around the state, in shallow (<20 ft) waters, and are known from freshwater-spring runs. They are the only pipefish species, other than opossum pipefishes, that occur in freshwater. Fringed, insular, chain, dusky, crested, and whitenose pipefishes are also found statewide, in seagrass beds. Northern pipefishes occur only north of Cape Canaveral, in nearshore seagrass meadows. Banded, pugnose, and shortfin pipefishes frequent hardbottom and reef habitats, primarily in southern Florida. Dwarf pipefishes also prefer hardbottom habitats but occur around the state, in water depths from 10 to 200 ft. Bull pipefishes inhabit deeper waters (2–230 ft) than other pipefishes. They occasionally are found at the water surface, however, associated with flotsam, especially floating sargassum algae, the chosen habitat of sargassum pipefishes. Both species appear wherever sargassum is pushed by winds and currents.

Natural History

While males in some other species incubate already-fertilized eggs (e.g., sea catfishes and penguins) Seahorses and pipefishes are well known as the only members of the animal kingdom where the males become pregnant and bear the young. Females produce the eggs and males the sperm, but at mating time the female transfers her eggs into a brood pouch on the male's abdomen (or on the tail in most pipefishes). Mating details are best known for lined seahorses. After some courtship and romance, a female passes 200–1000 unfertilized eggs into her male partner's brood pouch. The male fertilizes the eggs and then incubates them for about 20 days. The pouch supplies oxygen and nutrients to the developing embryos through a network of capillaries, similar to a mammalian uterus. Developed young—miniature versions of the adults—are released into the water column to fend for themselves. Genetic studies have shown that seahorses mate for life. We suspect that longsnout and dwarf seahorses undergo a similar ritual.

Except for opossum pipefishes, all other pipefishes brood their eggs under the tail. Opossum pipefishes migrate upstream into freshwater rivers to spawn. Their eggs are incubated in an abdominal pouch, and the developed young apparently are released at dusk.

Seahorses and pipefishes use their tube-like snouts to feed on tiny invertebrates and larval fishes. Lined seahorses pluck their prey from the water column, near the bottom substrate. Snorkelers visiting seagrass beds can readily observe seahorses moving slowly over the bottom, pipetting little organisms from the surfaces of seagrass blades, fleshy algae, and sediments. Most pipefishes tend to be ambush feeders, hiding among seagrass blades as they seek prey. By virtue of their shape and coloration, pipefishes are masters at camouflage. Many live in seagrass meadows, where they blend in wonderfully. Even on bare sandbottom, a pipefish looks like a piece of grass, a pine needle, or other debris. As with most background-matching strategies, the adaptive significance of this camouflage involves getting food and avoiding becoming a meal.

Trumpetfishes
Family Aulostomidae

- **Atlantic trumpetfish** (*Aulostomus maculatus*)

Background

Atlantic trumpetfishes are an elongate, somewhat laterally compressed, tube-snouted species. Adults reach over 3 ft in total length and exhibit three base color patterns: either reddish-brown (most common); a grayish-blue to black body, with a bright-blue snout; or uniformly yellow or amber with a yellow snout. Regardless of the base color, a series of broken bars mark the flanks, and the median fins have 2 dark stripes. Individual color patterns seem to be controlled by the nervous system. The dorsal spines are separated and protrude in a single file along the dorsal midline. A barbel protrudes from the tip of the lower jaw.

Distribution and Habitat

Atlantic trumpetfishes are found throughout Florida but are most abundant in southeastern Florida and the Keys, where they resides on well-developed reefs and hard-bottom habitats (usually where sea whips are dense), from nearshore to 250 ft water depths.

Natural History

Little is known about spawning in Atlantic trumpetfishes, but apparently near dusk, pairs engage in a courtship dance over the reef. Eggs are cast into the water column, where fertilization occurs. After a lengthy 80 to 90 days spent as larvae in the plankton, transformed juveniles settle in seagrass meadows and other structured habitats, including their natal reefs. Atlantic trumpetfishes are very effective predators, often relying on stealth and deceit to dupe their unsuspecting prey. Adults occasionally swim closely alongside larger fishes (e.g., groupers) or schools of fishes (e.g., blue tang), striking prey that overlooked their presence or were spooked by their swim-mates. The most common predatory strategy involves individuals aligning themselves vertically and head down, gently swaying in sync with large sea whips, sponges, and even dangling ropes, to blend into the background. Young trumpetfishes also align themselves with stalked invertebrates, such as hydrozoans, to remain hidden from predators and prey. The mouth is situated at the end of an elongate tube and is capable of significant suction. Observations suggest that individuals of the three color morphs are more successful stalking and capturing fishes with similar color patterns. For example, yellow-morph individuals focus their hunting efforts on yellow, initial-phase blueheads, while the blue-color morphs are more successful pursuing blue reef fishes.

Atlantic trumpetfish (*Aulostomus maculatus*), Juno, Florida

Cornetfishes
Family Fistulariidae

- **Red cornetfish** (*Fistularia petimba*)
- **Bluespotted cornetfish** (*Fistularia tabacaria*)

Bluespotted cornetfish (*Fistularia tabacaria*), Lake Worth Lagoon, Florida

Background

The two species of cornetfishes have very elongated bodies and extremely long, tubed snouts. The body is somewhat compressed dorsoventrally (almost square in cross-section); the single dorsal and anal fins are symmetrical in size (short base, tall height), shape (falcate), and placement (far posteriorly); and the forked caudal fin has a very long central filament. They reach a large size (5–6 ft). The coloration of bluespotted cornetfishes, as their vernacular name suggests, consists of numerous electric-blue spots over a silvery-yellow base color. Red cornetfishes are uniformly salmon to gray in color. The related, but different, Atlantic trumpetfishes are laterally compressed and deeper bodied than the cornetfishes.

Distribution and Habitat

Bluespotted cornetfishes are found statewide, in high-structure habitats. While primarily a shallow-water species, they can occur in depths of up to 300 ft. Red cornetfishes also are found around the state, but generally in water depths greater than 80 ft. Young bluespotted cornetfishes frequent shallow-water seagrass beds.

Natural History

Limited information is available on reproduction in cornetfishes. Pelagic larvae are produced, which spend an unknown number of days in the plankton before settling on structured habitats, such as seagrass meadows or hardbottom. Feeding by these predators involves an inhalation slurp (akin to using a giant pipette) that is associated with a rapid, forward head movement, propelled by an S-shaped flexing of the midbody. Their prey includes fishes and shrimps. Fishes are eaten head-first, allowing the fin spines to fold down as they move through a cornetfish's straw-like mouth and onward into the gut. This allows them to consume lionfishes (members of the scorpionfish family) in their native Pacific waters. Cornetfishes are eaten by sharks.

Snipefishes
Family Macroramphosidae

- **Longspine snipefish** (*Macroramphosus scolopax*)

Background

Snipefishes are odd-shaped, small (up to 9 in, but usually half that size), deepwater fishes, with large eyes; a long, pipette-like snout; and a huge, slightly recurved and serrated second dorsal spine in the first dorsal fin. The 2 dorsal fins and 1 anal fin are placed far back on the moderately elongated body. Coloration is pale-red, pink, or orange. Two nominal species—the longspine snipefish and the slender snipefish (*Macroramphosus gracilis*)—have been recorded in

Longspine snipefish (*Macroramphosus scolopax*). Photo by George H. Burgess.

Florida and around the world, but the scientific jury is still out on whether they represent one or two species. The two differ subtly in body shape (slender snipefishes are more slender), as well as in dorsal-fin placement and the length of the second spine (longer in longspines!), but morphological intermediates are common. Larvae can also be separated into two groups. Importantly, independent genetic studies of the two morphotypes conducted in both the Atlantic and Pacific Oceans revealed no differences; the two kinds of fishes apparently represent ecophenotypes (variants based on where they live) of the longspine snipefish.

Distribution and Habitat

Longspine snipefishes are found over the outer continental shelf and upper slope; in Florida they have been taken in depths of 150–1800 ft. The two morphs occupy different habitats: the slender one up in the water column (nektonic lifestyle) and the deeper-bodied longspine near the bottom (benthic lifestyle). They swim slowly, in a head-down position.

Natural History

Snipefishes have short lives (by deepwater fish standards), living 5–6 years. The two morphs differ in their food habits, the nektonic one being essentially planktivorous, consuming pelagic copepods and ostracods, and the benthic model subsisting on crustaceans, worms, and other bottom-dwelling invertebrates. Snipefishes are often found in schools and, occasionally, in great abundance. They are eaten by a variety of predatory fishes, including dolphinfishes, and by seabirds, such as terns. Reproduction has been observed in a public aquarium, where activity began about 2 hours before dark. Males changed color and, while swimming parallel to each other, bumped bodies or charged at one other, leading with the large dorsal spine. A winning male swam parallel with a female near the bottom, eventually linking caudal peduncles, and they then slowly ascended upward in the water column while releasing eggs and sperm.

Flying Gurnards
Family Dactylopteridae

- **Flying gurnard** (*Dactylopterus volitans*)

Background

This family contains a single species—the flying gurnard. This species is conspicuous, with its wing-like pectoral fins and blunt snout. Adults are a mottled brownish (occasionally red) color, with electric-blue spots covering the back and the outer third of the pectoral fins. Flying gurnards can be distinguished from searobins, which also have wing-like pectoral fins, by the blunt head, larger eyes, and a prominent spine extending from each preopercle. The pectoral fins are separated into two sections: an anterior section, consisting of the first 6 pectoral rays, and one containing the remaining long rays, which form the wing. Adults reach a total length of about 18 in.

Distribution and Habitat

Flying gurnards usually are found throughout Florida, near structured habitats (in this case, reefs), but they also occur over unstructured bottom over the continental shelf and occasionally in lagoons, in water depths ranging from 10 to 300 ft. The pelagic young are found near floating sargassum and flotsam, over blue water.

Natural History

Flying gurnards spawn pelagic eggs, and the pelagic young, which can be mistaken for flyingfishes, do not glide out of the water like these associates. Adults and larger juveniles crawl across the seafloor on their extended pelvic-fin rays, as do searobins. The short, anterior pectoral rays are used like a rake to scratch through the sediment, dislodging the flying gurnards' favorite foods: crabs, mantis shrimps, clams, shrimps, and small fishes. This feeding action attracts followers (yellow jacks and coneys, and possibly other species) that trail a flying gurnard (a nuclear predator), ready to pounce on an easy meal spooked its foraging. Other nuclear predators include some parrotfishes, triggerfishes, surgeonfishes, and goatfishes. When threatened, flying gurnards fan out the long portions of their pectoral fins and swim in tight circles. Young flying gurnards fall prey to squids, seabirds (sooty terns, brown noddies), blackfin goosefishes, bigeyes, glasseye snappers, bluefin and yellowfin tunas, wahoos, blue and white marlins, sailfishes, and spearfishes.

Flying gurnard (*Dactylopterus volitans*), Lake Worth Lagoon, Florida

Scorpionfishes
Family Scorpaenidae

- **Blackbelly rosefish** (*Helicolenus dactylopterus*)
- **Spinycheek scorpionfish** (*Neomerinthe hemingwayi*)
- **Longsnout scorpionfish** (*Pontinus castor*)
- **Longspine scorpionfish** (*Pontinus longispinis*)
- **Spinythroat scorpionfish** (*Pontinus nematophthalmus*)
- **Highfin scorpionfish** (*Pontinus rathbuni*)
- **Devil firefish** (*Pterois miles*)
- **Red lionfish** (*Pterois volitans*)
- **Longfin scorpionfish** (*Scorpaena agassizi*)
- **Coral scorpionfish** (*Scorpaena albifimbria*)
- **Goosehead scorpionfish** (*Scorpaena bergii*)
- **Shortfin scorpionfish** (*Scorpaena brachyptera*)
- **Barbfish** (*Scorpaena brasiliensis*)
- **Smoothhead scorpionfish** (*Scorpaena calcarata*)
- **Hunchback scorpionfish** (*Scorpaena dispar*)
- **Dwarf scorpionfish** (*Scorpaena elachys*)
- **Plumed scorpionfish** (*Scorpaena grandicornis*)
- **Mushroom scorpionfish** (*Scorpaena inermis*)
- **Smoothcheek scorpionfish** (*Scorpaena isthmensis*)
- **Spotted scorpionfish** (*Scorpaena plumieri*)
- **Reef scorpionfish** (*Scorpaenodes caribbaeus*)
- **Deepreef scorpionfish** (*Scorpaenodes tredecimspinosus*)

Background

Scorpionfishes project a homely profile, yet most are beautifully colored (from the red portion of the spectrum). They have large, spiny heads; big mouths; fleshy head and body tabs; and, in some species, venomous spines. Body

Red lionfish (*Pterois volitans*), Marquesas, Florida

shape and details of coloration help differentiate many of the species, but more-intricate characters (fin-element counts, shape and size of the fleshy tab located near the eye, etc.) are required to identify others. The most easily identified pair of species, ironically, are the newest inhabitants of the area. The introduced devil firefishes and red lionfishes—two reddish-brown, Indo-Pacific species that get to about 15 in long—have huge, fan-like dorsal and pectoral fins that are broadly banded. The body has complex bars, and other fins have concentric rows of spots. Unfortunately, positively separating these two species requires counting fin spines.

Scorpionfishes of the genus *Pontinus* are red to red-orange, have unbranched pectoral-fin rays (the only four species with this character), and lack fleshy tabs on the head. These four species are identified by looking at the lengths of the head and dorsal-fin spines. Only longsnout scorpionfishes, which reach 1 ft long, have an elongated snout (snout longer than the eye diameter); the others have snouts about equal to the eye diameter. The similarly sized (10 in) highfin scorpionfishes and longspine scorpionfishes are appropriately named: highfins have a high dorsal fin, and longspines have a long third dorsal spine (twice as long as the first). Spinythroat scorpionfishes look a lot like longspines but are smaller (5 in) and have a shorter third dorsal spine.

The largest (16 in) of our scorpionfishes, spinycheek scorpionfishes, are reddish-orange, with a large head that lacks fleshy tabs. All fins (except the pelvic fins) contain well-defined, dark-brown spots; 3 large, diffuse spots are found on the rear half of the lateral line. Blackbelly rosefishes (12 in) are pale red, with white markings on the face and flanks; have 4 darker body bars (the last 2 uniting ventrally to form a V); and have 1 bar on the caudal peduncle. The roof of the mouth and the gill cavities are uniquely purplish black, and subadults usually have a dark blotch at the rear of the first dorsal fin; head tabs are missing in all sizes. Tiny reef (4 in) and deepreef (2 in) scorpionfishes also have dark spots on their dorsal fins. These two species are distinguished by the degree of spotting, the former bearing spots on the fins and head that are missing on the latter.

Scorpaena, the most diverse genus in Florida, includes variably red to brown species, with spiny heads that have fleshy cirri (eyebrows, if you will) above the eye (and often other fleshy tabs). This group includes some of the most common scorpionfishes, as well as others that are seldom seen. The three species that have a pair of broad, dark tail bands can be readily separated by looking at spot patterns near the pectoral fin. Barbfishes (9 in), one of the most common scorpionfish species, range from bright–red-orange to drab-brown. They have 2 large, squarish black spots above the operculum and the base of the pectoral fin, as well as small black spots on the base of the pectoral fin. The cirri are large, but not as big and ornate as those of plumed scorpionfishes (6 in), which are the brown to yellowish and have distinctive small white spots behind the pectoral fin. Spotted scorpionfishes (12 in) are even more striking; the armpit is a dark-purple patch, covered in large white spots. The body usually is drab-red and brown, often with a covering of algae on the skin, providing excellent

All images, barbfish (*Scorpaena brasiliensis*), Lake Worth Lagoon, Florida

camouflage. Reddish-brown and white mushroom scorpionfishes (3 in) are another species bearing a pair of tail bands; they have distinctive, inverted, mushroom-shaped skin tabs hanging over the upper third of the eye.

Other species are more difficult to separate. Coral (3 in) and longfin (6 in) scorpionfishes have relatively large eyes. Coral scorpionfishes have a dark, wedge-shaped saddle under the spiny dorsal fin, and the lower half of the anal fin is a mottled red. Elongated pectoral fins that reach to the end of the anal-fin base distinguish longfin scorpionfishes. Goosehead scorpionfishes (4 in) and smoothcheek scorpionfishes (6 in) have a black spot on the spiny dorsal fin and 3 bands on the caudal fin. Separation of these two species is best done by looking at details of their head spination, but smoothcheeks also have prominent bands on the pectoral and anal fins. Shortfin scorpionfishes, a totally pale-red species lacking in telltale markings, also require a microscopic examination of the head spines. Hunchback scorpionfishes (9 in) have a long eye cirrus and 2 dark bands on the tail. Dwarf scorpionfishes (2.5 in) are pale, with a web of dark and light lines located right under the dorsal fin. The upper portion of the pectoral fin is dark in smoothhead scorpionfishes (5 in), which have small fleshy cirri.

Distribution and Habitat

The blackbelly rosefish and spinycheek, highfin, longsnout, spinythroat, and longspine scorpionfishes reside on hardbottom habitats around the state, in water depths ranging from 150 to over 3000 ft (blackbelly rosefish). Most of the other species occur on hardbottom, reefs, seagrass meadows, and sandy bottoms, between 200 and 600 ft depths. Coral, goosehead, smoothcheek, mushroom, shortfin, dwarf, reef, and deepreef scorpionfishes are known only from Florida's east coast, while barbfishes and spotted, longfin, smoothhead, and hunchback scorpionfishes are distributed statewide. Plumed scorpionfish occur

Left: Plumed scorpionfish (*Scorpaena grandicornis*), Lake Worth Lagoon, Florida. *Right*: Mushroom scorpionfish (*Scorpaena inermis*), Lake Worth Lagoon, Florida.

Both images, spotted scorpionfish (*Scorpaena plumieri*), Palm Beach, Florida

along the southern portion of both coasts, where they inhabit seagrass meadows and algae stands. Devil firefish and red lionfish regrettably have successfully invaded all structured habitats, from shelf-edge reefs to the lower reaches of coastal estuaries.

Natural History

Little is known about age, growth, or reproduction in Atlantic scorpionfishes. Their better-studied Pacific relatives grow very slowly, live for as long as 25 years, and have externally fertilized eggs that are released in gelatinous masses. We have observed pairs of spotted scorpionfishes and barbfishes exhibiting what may have been courting behavior on east Florida reefs. One individual closely followed another of similar size and color. Invasive red lionfishes and devil firefishes are receiving considerable attention from scientists, and before too long we will know more about their ecology than all the native species combined.

Scorpionfishes are masters at blending into the background (disappearing in plain sight). Many bury themselves in the sediment during the day, with only their eyes protruding from the sand layer. From our years of diving and towing camera sleds over reefs and open sandbottom, it has become obvious that scorpionfishes emerge at night to feed, consuming fishes, shrimps, and crabs.

Their venomous dorsal spines can deliver potent pain. We have been stung by small and large scorpionfishes while sorting through trawl catches and working underwater around reefs. Pain from stings by the small species is at the level of a bee sting, and the immediate remedy is hot water. Stings from spotted scorpionfishes (a large species) produce a dull, long-lasting pain, with a puncture wound that is highly susceptible to infection. See a physician if there is any sign of swelling or prolonged pain.

The newest members of this family in Florida are red lionfishes and devil firefishes. Individuals from these two species have been reported by Florida scuba divers since the 1980s. We were aware of anecdotal accounts from wrecks offshore of Ft. Pierce well before the recent onslaught. The current invasion, which includes much of the western Atlantic, seems to have been prompted by more than just an occasional release of unwanted pets; more likely it is the result of a semicontinuous reseeding of fertilized eggs into the system at some time, probably from a public aquarium.

Searobins
Family Triglidae

- **Shortfin searobin** (*Bellator brachychir*)
- **Streamer searobin** (*Bellator egretta*)
- **Horned searobin** (*Bellator militaris*)
- **Spiny searobin** (*Prionotus alatus*)
- **Northern searobin** (*Prionotus carolinus*)
- **Striped searobin** (*Prionotus evolans*)
- **Bigeye searobin** (*Prionotus longispinosus*)
- **Barred searobin** (*Prionotus martis*)
- **Bandtail searobin** (*Prionotus ophryas*)
- **Mexican searobin** (*Prionotus paralatus*)
- **Bluewing searobin** (*Prionotus punctatus*)
- **Bluespotted searobin** (*Prionotus roseus*)
- **Blackwing searobin** (*Prionotus rubio*)
- **Leopard searobin** (*Prionotus scitulus*)
- **Shortwing searobin** (*Prionotus stearnsi*)
- **Bighead searobin** (*Prionotus tribulus*)

Background

Searobins are strict bottom dwellers, with wide, spiny, armored heads; narrow bodies; and winglike pectoral fins, the first 3 rays of which are free (not connected by a membrane) and can move independently. Their eyes are relatively large and face upward. Although similar in gestalt, the species can be distinguished by external characters, such as pectoral-fin length and coloration.

The three pink to red members of the genus *Bellator* are small (usually <4–5 in) and rarely seen (unless you are pulling trawl nets on the outer continental shelf). Horned searobins, the largest of the three (sometimes reaching 8 in), are pale-red to pink, and the snout has 2 hard, forward-pointing projections. Distinctive coloration includes yellow spots on the first dorsal fin, and yellow lines on the second dorsal and caudal fins. The first few fin rays of the long pectoral fin have black and white bands. Males have elongated filaments on the first 2 dorsal spines; only the first is elongated in males of the similar streamer searobins. Streamer searobins have a reddish-brown body, with yellow spots

Bandtail searobin (*Prionotus ophryas*), Lake Worth Lagoon, Florida

on the upper caudal fin and a red stripe below; the long pectoral fin has 2 dark bands and a yellow margin. Both sexes of shortfin searobins lack elongated dorsal spines and have short snout extensions. The body is brown to reddish dorsally; the short, light-edged pectoral fin has a dark blotch or spot; and the caudal fin has 2 vertical reddish bars.

Members of the genus *Prionotus* are generally larger (7–18 in) and are more commonly encountered in Florida waters. Shortwing searobins are the smallest (7 in) and most dully colored member of this genus. They have short, rounded pectoral fins (not extending beyond the origin of the anal fin when pressed against the body); a silvery-grey body; and a broad, dark-brown margin to the caudal fin. Bandtail searobins (8 in) are reddish-orange to brown, with 3 dark bands on the tail; the free pectoral rays, pelvic fins, and pectoral fins also are banded. These searobins have a short, fleshy tab over each eye and in front of each nostril. The lower pectoral fin is darker than the upper half and has reddish-brown spots. Bandtails often fan their long, broad pectoral fin perpendicular to the seafloor, rather than parallel to it, as is common with the other searobin species. Striped and bighead searobins are two of the three largest searobins, possessing larger heads and stouter bodies than their congeners (fellow members of the genus). Striped searobins (18 in) have 2 thin, dark stripes, running from the opercular margin to the base of the tail, and a dusky blotch on the first dorsal fin. The long pectoral fin is tan to brown, with thin, dark, wavy lines basally. At small sizes, this species can be confused with bighead searobins (14 in), but bigheads have 3 dark, oblique body bars on the rear half of the body, as well as a well-defined, dark dorsal-fin spot, with a white border beneath it. The somewhat shorter pectoral fins are dark-brown, with lighter, wavy cross bands; the tail has a broad, dark, submarginal band, with a thin, light edge.

Three species of searobins have long pectoral fins with electric-blue markings. Bluespotted searobins (8 in) have a light-brown body, with darker spots or blotches; a well-defined dark spot on the first dorsal fin; and 2 wide, dark bands on the tail, with another at its base. The pectorals are rounded, and the base color is dark to dusky, overlaid with brilliant-blue spots. Similar-sized (9 in) blackwing searobins look a bit like bluespotted searobins, but the pectoral fins have a relatively straight-edged outer margin that

Bighead searobin (*Prionotus tribulus*), Indian River Lagoon, Florida

Blackwing searobin (*Prionotus rubio*), Lake Worth Lagoon, Florida

Bluespotted searobin (*Prionotus roseus*), Lake Worth Lagoon, Florida

Leopard searobin (*Prionotus scitulus*), Lake Worth Lagoon, Florida

is brown, with lighter spotting forming diffuse bands; the inner margin is bright-blue. They also have a dusky dorsal fin and 2 poorly defined bands on the tail. Bluewing searobins look like blackwing searobins, but the margin of the pectoral fin is rounded, not truncated.

Spiny searobins (8 in) and Mexican searobins (7 in) have dusky spots on their first dorsal fins and diffusely banded caudal fins; they are the only two species with concave edges to the pectoral fins. Spiny searobins are readily distinguished by having overtly extended lower pectoral-fin rays that reach past the base of the rear anal fin (the upper and lower rays are equal in length in Mexican searobins). Bigeye searobins have a distinctive, dark band running across the anal fin; large eyes; and an ocellated spot on the dorsal fin. The short pectoral fins are uniformly dark, with small light spots; the largely unmarked caudal fin is more forked than in most searobins.

The most confusing species are the northern, barred, and leopard searobins. All are elongated, relatively thin searobins, with brown to tan spotting or mottling and a dark, diagonal bar under each eye. The most distinctive are leopard searobins (10 in), which have well-defined, reddish-brown spots on the body, dorsal and pectoral fins, and upper tail. They have 2 black spots on the first dorsal fin, and a wide, dusky, submarginal band on the anal fin. The pectoral fins are light-brown to yellow, with brown spots and 2 dark bands. The toughest call is separating northern and barred searobins, and there is still some controversy over whether they represent just a single species. Distribution information helps. Northern searobins are the largest member of the group, growing to 18 in. They have a dark spot (often ocellated) on the first dorsal fin; 3 broad, dark bands on the pectoral fins; and a concave margin on the tail. Barred searobins have 2 black spots on the dorsal fin, 1 between the first and second spines and 1 between the fourth and fifth spines. Their branchiostegal rays (the elements that support the lower—i.e., belly—side of the gill cover) are dusky, versus black in northern searobins.

Distribution and Habitat

Searobins reside over sandy or muddy bottoms, from nearshore to the outer continental shelf. Shortfin, horned, streamer, and shortwing searobins occur statewide and prefer waters deeper than 100 ft. Spiny and Mexican searobins segregate geographically, with spiny searobins occurring along both coasts of Florida and extending to the Mississippi delta off Louisiana, where they reside in depths of 120–600 ft. Mexican searobins start appearing off the panhandle and continue west all the way to Mexico. These two species overlap off the Florida panhandle and may hybridize in that region, as well. Striped and northern searobins occur only along Florida's northeast coast, where they live in nearshore waters, out to 300 ft depths. Barred, leopard, bighead, bluespotted, bandtail, and blackwing searobins are distributed statewide, in depths less than about 150 ft. Bigeye searobins are known only from the northeastern Gulf of Mexico, in depths ranging from nearshore to 100 ft. Bluewing searobins are common in the southern Gulf of Mexico and Caribbean but occur only sporadically in Florida waters, from 10–300 ft deep.

Natural History

Little is known about reproduction in searobins, except that fertilization is external and eggs are released into the water column. Northern and striped searobins produce sounds with specialized muscles and the swim bladder, and these sounds may be used to signal to prospective mates. Several species (shortfin, bluespotted, blackwing, leopard, barred) spawn during summer months. Bighead searobins spawn in winter months (September–March). Some species, such as bandtail and horned searobins, spawn year round. Following a larval stage, young individuals settle to the seafloor, where they bury themselves up to their eyeballs in sand or seek shelter within shell-hash (broken shells) substrate.

When disturbed, most searobins extend their pectoral rays like fans, presumably to deter predators by suddenly looking much larger and less edible. The pectoral fins may also function in courtship displays or interspecific communication, but such behavior has not yet been described by researchers. Searobins walk along the seafloor by standing on the tips of their 3 free pectoral rays. These rays have sensory capabilities—similar to taste buds—that help searobins find invertebrates buried in the sand. Searobins feed mostly on shrimps, worms, and other sand-dwelling invertebrates but will eat small fishes if they can catch them. Larger bighead searobins are known to prefer swimming crabs. Some species bury themselves in the sand during the daytime, emerging at night to forage.

Snooks
Family Centropomidae

- **Swordspine snook** (*Centropomus ensiferus*)
- **Smallscale fat snook** (*Centropomus parallelus*)
- **Tarpon snook** (*Centropomus pectinatus*)
- **Common snook** (*Centropomus undecimalis*)

Background

Snooks are prestigious inshore sport fishes in the southern half of Florida's peninsula. You will hear long-time native Floridians pronounce snook as "snuke," but most people pronounce it like "hook." Although common snooks are a prime target of sports fishers, three additional, largely unrecognized species of *Centropomus* also reside in Florida's inshore and coastal waters. Snooks are euryhaline and regularly ascend freshwater rivers in the southern portion of the state.

Common snooks, smallscale fat snooks, tarpon snooks, swordspine snooks, and the occasional largescale fat snook (*Centropomus mexicanus*) reside in Florida waters. The different snook species are superficially very similar, with long, duck-like snouts; elongate bodies; golden-brown backs; silvery flanks; and a thin, jet-black lateral stripe that gives rise to the name "linesiders," often used by fishers.

The different species have very similar coloration—olive to golden-brown backs and silvery sides, with a thin, black, lateral stripe—and are distinguished by subtle differences in morphology: the length of the second anal-fin spine, the size of the scales, and maximum adult size. Telling the snooks apart is best done in reference to the ubiquitous common

Common snook (*Centropomus undecimalis*), Jupiter, Florida

Smallscale fat snook (*Centropomus parallelus*), Jupiter, Florida

snooks. Smallscale fat snooks, often called "flatsides" by local fishers, can be distinguished by their relatively deeper body; larger eye; long second anal spine, which reaches the beginning of the tail fin when depressed; and maximum size of 28 in, second only to common snooks (>4 ft). Tarpon snooks look very much like smallscale fat snooks but have a slightly upturned tip of the jaw, a relatively short pectoral fin, larger scales, and a second anal spine that does not reach the base of the tail fin when depressed. The pelvic and anal fins are tipped with dark pigment. Swordspine snooks (maximum size 15 in) have a very long second anal spine that reaches into the tail fin when depressed.

Distribution and Habitat

Within the continental United States, this family is largely restricted to Florida waters, but common snooks (mostly juveniles) stray as far north as North Carolina and move into Texas from Mexico in the summer. Common snooks are the most abundant and widespread species in state waters. Common snook are the most flexible ecologically and are found from freshwater reaches of coastal rivers to offshore reefs. On Florida's Gulf coast, common snooks are frequently found as far north as Crystal River, where they can be seen swimming around the many springs in that area. The other three snooks are tropical peripheral species, found primarily along the southeastern coast from Cape Canaveral to Biscayne Bay, with centers of abundance outside of state waters. Smallscale fat snook and tarpon snooks will co-occur with common snooks on the southwest coast. Largescale fat snooks were recently documented from the Sebastian and Loxahatchee Rivers on the east coast. Swordspine snooks occur along the southeastern coast of Florida, usually in brackish water and freshwater. Very little else is known about this species in Florida. Common snooks are encountered less often in the Florida Keys than in the peninsular portions of southeastern and southwestern Florida. The largest individuals generally hail from the Indian River Lagoon, between Jupiter and Ft. Pierce.

Natural History

Common snooks are the largest of the family, reaching more than 4 ft long, weighing over 40 lb, and sometimes living for over 20 years. The other three species do not exceed 30 in long. Snooks are hermaphrodites, starting their lives as males and becoming females at lengths of 16–20 in. They aggregate at ocean inlets or passes to spawn during summer months. Acoustic tagging studies have revealed that these aggregations are not fixed groups and that there is a constant turnover of individuals, but individuals tend to return to the same spawning sites in consecutive years. Although successful fertilization of eggs requires a full-seawater medium (35 psu), all size classes (above the egg stage) are capable of residing in pure freshwater. The life history of common snooks generally follows a pattern where, with increased growth, there is a shift in habitat choice along salinity gradients, from freshwater to the coastal ocean. Over this gradient, individuals favor areas with natural or artificial structures, such as mangrove prop roots, seagrasses, oyster reefs, rock outcrops, and bridge or dock pilings. Common snooks feed—sometimes voraciously—on fishes, shrimps, and crabs. Common snooks generally eat during tidal changes and are very active at night. They are frequently seen around dock and bridge lights at night, with their snouts oriented into the flowing tide. Despite being tantalizingly close, most anglers will attest to the frustrating indifference of common snooks to lures or bait on these and other occasions. All snook species are very sensitive to temperature, and mortality or morbidity are the norm when shallow-water temperatures fall to 60°F. During winter 2010, when coastal and inshore water temperatures dropped to some of the lowest levels in a century, thousands of snooks are thought to have died. Many of these were in the Merritt Island Wildlife Refuge, where the snooks had colonized the area only recently.

Smallscale fat snooks and tarpon snooks, like common snooks, migrate as they grow from freshwater to marine salinities. Adults of both species have been observed mingling with adult common snooks assembled in east-coast inlets, such as St. Lucie and Jupiter Inlets. During fall and winter, smallscale fat snooks aggregate in Jupiter Inlet, St. Lucie Inlet, and probably

Tarpon snook (*Centropomus pectinatus*), Jupiter, Florida

Schooling adult snooks (*Centropomus undecimalis*), Jupiter Inlet, Florida

other east-coast inlets used by common snooks in summer months. They routinely assemble under dock and bridge lights at night to feed on the constant stream of small invertebrates and fishes passing through with the flowing tides. You can watch them feeding under a light and be confident that the tiny objects of their attention are beyond the capabilities of human visual perception.

Wreckfishes
Family Polyprionidae

- **Wreckfish** (*Polyprion americanus*)

Wreckfish (*Polyprion americanus*). *Photo by Steve W. Ross.*

Background

Wreckfishes, which are bass-like in appearance, are large (to 6.5 ft; >200 lb) and have twin head bumps: 1 on the forehead and 1 between the eyes. The top of the operculum has a marked horizontal ridge, ending posteriorly in a spine. The lower jaw protrudes well beyond the upper jaw, creating a distinctive underbite. The teeth are small and the mouth is protrusible. Wreckfishes are dark-brown to greyish above and lighter below; young are black, mottled with white. The single dorsal fin consists of a spiny anterior portion and a soft-rayed rear portion. The caudal fin is broadly rounded.

Distribution and Habitat

While adults have been taken in water as shallow as 160 ft, they normally are found in much deeper water (they have been recorded at depths of more than 2400 ft), in temperatures of 43°F–61°F, on deep, rocky slopes. Young (to about 2 ft) are pelagic, often found near the surface associating with flotsam; hence their vernacular name. Along Florida's east coast, wreckfishes are found at the continental shelf–slope break (600 ft) and deeper.

Natural History

Wreckfishes are a classic life-in-the-slow-lane species and are not capable of withstanding sustained, intensive fishing. They are long lived (to 78 years), slow growing, and late to reach sexual maturity (10–15 years; 32–37 in). Spawning probably occurs in the autumn and winter. Young wreckfishes are in the pelagic stage for an amazing 2 years, and then settle to the bottom when they reach lengths of 22–26 in. Deepwater fishes, squids, and crabs form the bulk of the adult diet, and pelagic juveniles eat fishes and isopods associated with floating objects. There is some indication that wreckfishes undertake a limited (330–500 ft) vertical migration at night in the water column.

Groupers
Family Epinephelidae

- **Mutton hamlet** (*Alphestes afer*)
- **Graysby** (*Cephalopholis cruentata*)
- **Coney** (*Cephalopholis fulva*)
- **Atlantic creolefish** (*Cephalopholis furcifer*)
- **Marbled grouper** (*Dermatolepis inermis*)
- **Rock hind** (*Epinephelus adscensionis*)
- **Speckled hind** (*Epinephelus drummondhayi*)
- **Red hind** (*Epinephelus guttatus*)
- **Atlantic goliath grouper** (*Epinephelus itajara*)
- **Red grouper** (*Epinephelus morio*)
- **Nassau grouper** (*Epinephelus striatus*)
- **Spanish flag** (*Gonioplectrus hispanus*)
- **Yellowedge grouper** (*Hyporthodus flavolimbatus*)
- **Misty grouper** (*Hyporthodus mystacinus*)
- **Warsaw grouper** (*Hyporthodus nigritus*)
- **Snowy grouper** (*Hyporthodus niveatus*)
- **Western comb grouper** (*Mycteroperca acutirostris*)
- **Black grouper** (*Mycteroperca bonaci*)
- **Yellowmouth grouper** (*Mycteroperca interstitialis*)
- **Gag** (*Mycteroperca microlepis*)
- **Scamp** (*Mycteroperca phenax*)
- **Tiger grouper** (*Mycteroperca tigris*)
- **Yellowfin grouper** (*Mycteroperca venenosa*)

Background

Ichthyologists recently removed groupers from the seabass family (Serranidae) and placed them in their own family, the Epinephelidae. Groupers and hinds are robust, reef-dwelling predators sought by fishers of all stripes. Although these two classes of fishes differ in body shapes and color patterns, telling them apart can be challenging, because many of the species change color, either with growth or based on behavior. Mutton hamlets—not to be confused with true hamlets (*Hypoplectrus* spp.)—are small (13 in), with a deep body, a rounded caudal fin, and eyes that are relatively larger than those in other groupers. Their color varies from bright-red to light-brown, with dark splotches and smaller spots on the flanks; there often is a dark band extending from the eye to the anterior base of the dorsal fin. Spanish flags, also small (11 in) by grouper standards, are pinkish-orange, with bright-yellow stripes running across the body; 1 stripe extends from the snout through the eye and to the caudal peduncle. They have a distinctive bright-red spot on the anterior third of the anal fin, and an equal-sized white spot on the lower body, just in front of the anal-fin base.

Atlantic creolefish (15 in), historically placed in the genus *Paranthias*, recently ware transferred to the genus *Cephalopholis*. Atypically for groupers, they have a deeply forked tail and a low dorsal fin. The body is purplish-red, with 3 small, light or dark spots under the rear half of the dorsal-fin base; the pectoral fin bears a dark-red, triangular blotch.

Top: Atlantic creolefish (*Cephalopholis furcifer*), Florida Middle Grounds. *Bottom*: Mutton hamlet (*Alphestes afer*), Jupiter, Florida.

Top left: Coney (*Cephalopholis fulva*), yellow phase, Juno, Florida. *Top right*: Coney, bicolor phase. *Bottom left*: Coney, brown phase, Jupiter, Florida. *Bottom right*: Juvenile

Coneys—small (≤12 in), slender minigroupers with a rounded tail fin—are more typical of the genus. They have three color phases, with varying base body colorations; all, however, have an overlaying peppering of small electric-blue spots. The base colors are yellow, reddish-brown, or bicolored, with the upper half of the body reddish-brown and the lower half (including the entire caudal peduncle and tail) white. Regardless of color phase, all individuals have 2 black spots on the tip of the lower jaw, and another pair on the top of the caudal peduncle. Graysbies also are small (≤12 in) and are shaped like coneys. The entire fish is gray-brown and covered with small dark-red spots; there are 3 regularly spaced white or black spots under the front half of the dorsal-fin base.

All other groupers achieved much greater sizes. Atlantic goliath groupers are aptly named: they grow to about 8 ft and weigh well over 600 lb (the hook-and-line world record of 670 lb was taken at Fernandina, Florida). Still called "jewfishes" (so named because of their historic popularity in the Jewish community, where they were prepared as gefilte fish, a Passover favorite of older Florida anglers), Atlantic goliaths are pale- to dark-brown, with dull-yellowish undersides. The flanks have broad, dark bars, covered with small brown spots; the eye is tiny, relative to the body size, with brown stripes radiating posteriorly; the mouth is large; and the tail is rounded. Youngsters (<16 in) have pronounced bars against a light (sometimes white) background, and some anglers refer to them as "bumblebee groupers." The spinous dorsal fin of red groupers (3 ft; almost 50 lb) is high and triangular, and the tail is emarginate. The background color is reddish-brown, with ill-defined, wide bars on the body; a dark, diagonal eye stripe; and tiny black dots surrounding the eye. Nassau groupers (3 ft; 25 lb) are similar to red groupers in body shape and in having a diagonal eye bar, but they also have a second bar on the nape that, in combination with the eye bar, has the appearance of a tuning folk. Nassau groupers are generally brownish-green to reddish, and have more-defined bars and a distinct, black caudal saddle. Marbled groupers (3 ft) have deep bodies and long, slightly concave to straight foreheads. Their background body color is dark- or light-brown to gray, with marbling consisting of dark blotches and small white spots. Juvenile marbled groupers are uniformly dark-brown, covered with scattered white spots and blotches, and have a rounded tail.

Speckled hinds (3.5 ft; >50 lb)—a species familiarly called "Kitty Mitchell" (supposedly named after a "working girl" of that name)—have a deep body, similar to marbled groupers, with a long sloping forehead. The bodies and fins of adults are uniformly dark-brown, with closely spaced, small white spots; juveniles (<8 in) are bright-yellow, with small white spots. Red hinds and rock hinds are smaller in size (2.5 and 2 ft, respectively). Red hinds are olive-green to off-white, overlaid with small reddish spots and 4–5 faint, dark, oblique bars; the soft dorsal, anal, and caudal fins have broad black margins, narrowly edged in white; and the dorsal spines have bright-yellow tips. Rock hinds are golden to olive-brown, with 3–5 black saddles along the base of the dorsal fin and 1 on the top of the caudal peduncle. Their bodies are covered with small, dark-red–brown spots and larger white blotches.

Graysby (*Cephalopholis cruentata*), Juno, Florida

Atlantic goliath grouper (*Epinephelus itajara*), Jupiter, Florida

The genus *Hyporthodus* contains the deep-reef groupers. The flanks of snowy groupers (4 ft; 70 lb) are gray-brown, with regularly spaced white spots, and the caudal peduncle has a black saddle; both fade with growth and are missing on larger adults. The pelvic and soft dorsal, anal, and caudal fins often have fine blue edges. Juveniles have pale-yellow pectoral and caudal fins. Yellowedge groupers (3.5 ft; 45 lb) might be confused with snowy groupers, but they are distinguished by having bright-yellow margins on the pectoral and dorsal (and occasionally anal and caudal) fins. The body color is generally a lighter brown, and the white spots are less well defined than in snowy groupers; like the latter species, the spots and a dark caudal peduncle saddle disappear in adults. Warsaw groupers, second only to Atlantic goliath groupers in size (7.75 ft; >400 lb), are uniformly dark-brown, with a bit deeper body than the others in their genus. The second (of 10) dorsal-fin spine is noticeably longer than others; large males have black anal fins. Young are dark-brown, with irregularly positioned, variably sized white spots (looking a bit like they were spattered with white paint) and a yellow caudal fin. Misty (sometimes called "mystic") groupers (4 ft; >100 lb) have 8–9 wide, dark bars (the 2 on the caudal peduncle sometimes merge) and irregularly placed, diffuse white spots over a lighter brown background; the bars are more pronounced in young, and fade in adults. Dark bands radiate posteriorly from the eyes, and a dark mustache appears over the rear portion of the upper jaw.

Red grouper (*Epinephelus morio*), Ft. Lauderdale, Florida

Nassau grouper (*Epinephelus striatus*)

Members of the genus *Mycteroperca* are generally more slender and have smaller heads than other large groupers. Individuals can vary considerably in coloration, often making it difficult to distinguish species in the field. Scamp (3 ft; 30 lb) and yellowmouth groupers (2.5 ft; 22 lb) are extremely similar in body shape, size, and coloration (including having yellow mouth corners and pale pectoral-fin margins), and they are subtly different in having more-angular head profiles than the other members of the genus. Adult and juvenile scamps vary from a uniform gray-brown coloration to a lighter background covered with darker cat's-paw markings; transformation between the two phases can occur literally before your eyes. Courting adult males adopt a gray head or a bicolor phase, with pale-gray on the head to about the pectoral fins and a uniform black on the rest of the body. Mature males develop broomtails: irregular caudal-fin margins (exsertions) and pointed tips on the soft dorsal and anal fins. Adult yellowmouth groupers have somewhat more-rectangular lateral blotches, and adults do not develop pronounced broomtails. The tips of the first dorsal-fin spines are yellow. Bicolored yellowmouth juveniles are distinctive: dark-brown above, white underneath, with bright–lemon-yellow on the dorsal spines. We have observed this color pattern on juveniles from 2 to 10 in long. The body shape of western comb groupers resembles that of scamps and yellowmouth groupers, but it is grayish, with dark lines radiating from the eye socket, and the flanks are covered with small white blotches and spots.

Gags and black, yellowfin, and tiger groupers are morphological look-alikes. Gags (4 ft; 80 lb) are pale-gray, with dark variegated marks over the flanks, often giving the appearance of lipstick-kiss marks; individuals, however, can change their background color to an unmarked uniform gray or almost black. Large males develop dark (to black) undersides that may extend onto the tail, prompting the anglers' nickname, "blackbellies." Black groupers, the largest of the *Mycteroperca* species (to 4.5 ft; 100 lb), range in coloration from uniformly dark to light–yellow-brown with dark, regularly placed rectangular blotches and small coppery spots on the flanks. The soft dorsal, anal, and caudal fins have broad black margins, with fine white edging; the pectoral fins have yellow margins. Yellowfin groupers (3.5 ft) also exhibit a variety of color patterns, ranging from a uniform black to a mosaic of black, white, and red. Regardless of flank coloration, the pelvic and soft dorsal, anal, and caudal fins have narrow black margins, sandwiched by fine white lines; the pectoral fins have bright-yellow margins. Note that black, tiger, and yellowfin groupers all have some degree of

Juvenile speckled hind (*Epinephelus drummondhayi*), Stuart, Florida

Red hind (*Epinephelus guttatus*)

Rock hind (*Epinephelus adscensionis*), Palm Beach, Florida

Juvenile snowy grouper (*Hyporthodus niveatus*), Lake Worth Lagoon

Juvenile yellowedge grouper (*Hyporthodus flavolimbatus*)

yellow on the pectoral-fin margin, which can confuse identification. Tiger groupers (3.5 ft; 30 lb) have slightly more-rounded head profiles and are marked with about 10 diagonal bars, composed of small, reddish-brown to black, pale spots on the head and body; the underside of the body is vermiculated (having irregular fine or wavy lines). Tail fin rays in large adult tiger groupers grow irregularly (exsertions), creating a rough margin on the fin and prompting the name "broomtails" or "ragtails" among fishers. Juvenile tiger groupers are bright-yellow, with a broad, dark, lateral stripe.

Distribution and Habitat

Groupers are broadly distributed in continental/insular and deep (>150 ft) / shallow (<150 ft) reef patterns. Mutton hamlets are insular, known only along Florida's east coast, in water depths less than 100 ft. Spanish flags are deep-reef dwellers, occurring around the state in water depths of 200–500 ft. Red groupers and gags are continental species that are widespread throughout Florida, found from shallow nearshore waters to deep (400 ft) structured habitats. Unlike other groupers that frequent wrecks and high-profile reefs, red groupers prefer low-relief ledges, spongebottom, and (off the west coast of Florida) limestone solution holes. Juvenile reds reside in shallow seagrass meadows for several years before moving offshore.

Atlantic creolefish occur around the state and prefer high-relief features in water depths ranging from 80 to 300 ft. Speckled hinds, marbled groupers, and scamps are also found around the state, on reefs and hardbottom, in water depths of 50 to 600 ft. Young speckled hinds, marbled groupers, and scamp settle on offshore reefs and do not use inshore habitats. In contrast, juvenile Atlantic goliath groupers prefer mangrove shorelines, especially in the Keys, Florida Bay, and the Everglades. Atlantic goliath groupers are distributed around the state but are most common off the southeastern coast, from Sebastian Inlet through the Florida Keys and up to at least Tampa Bay, where they love sunken ships or other high-relief hardbottom having caves and ledges. Nassau, yellowfin, yellowmouth, tiger, and western comb groupers are insular in character and generally restricted to the Florida Keys, the Dry Tortugas, and (occasionally) the Florida Middle Grounds, where reefs are well developed. We have observed Nassau grouper individuals as far north as St. Lucie Inlet, on the east coast. Nassau juveniles settle in seagrass meadows and other shallow habitats, while juvenile yellowfin, yellowmouth, and tiger groupers settle directly on reefs. Gag and black grouper young settle in seagrass meadows and move to shallow structured bottom as subadults.

Top: Scamp (*Mycteroperca phenax*), Florida Middle Grounds. *Bottom*: Juvenile, Hobe Sound, Florida.

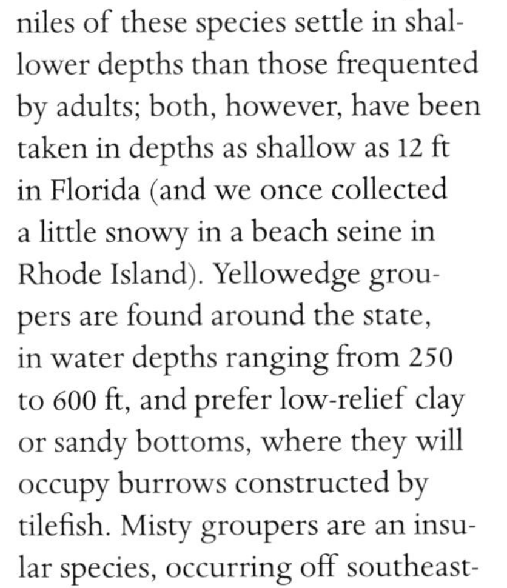

Adult snowy and Warsaw groupers are restricted to reefs and wrecks, in water depths ranging from 200 to 1200 ft statewide. Juveniles of these species settle in shallower depths than those frequented by adults; both, however, have been taken in depths as shallow as 12 ft in Florida (and we once collected a little snowy in a beach seine in Rhode Island). Yellowedge groupers are found around the state, in water depths ranging from 250 to 600 ft, and prefer low-relief clay or sandy bottoms, where they will occupy burrows constructed by tilefish. Misty groupers are an insular species, occurring off southeastern Florida and the Florida Keys, in water depths ranging from 250 to 1000 ft.

Graysbies, coneys, red hinds, and rock hinds occur around the state, on structured habitat, in water depths less than 200 ft. The young of these species settle to the adult habitat, rather than shallow inshore waters.

Natural History

Groupers and hinds are water-column spawners. Species exam-

Left: Yellowmouth grouper (*Mycteroperca interstitialis*). *Right*: Juvenile, Miami, Florida.

Gag (*Mycteroperca microlepis*), Venice, Florida

ined to date have proven to be protogynous hermaphrodites: all individuals are born as females but change to males later in life. Groupers and hinds are well known for their habit of aggregating en masse prior to spawning. Nassau groupers, yellowfin groupers, tiger groupers, and red hinds aggregate to spawn at specific sites, during particular moon phases, at numerous Caribbean and Bahamian locations. Such aggregations occur in Florida for black groupers, gags, Atlantic goliath groupers, and scamps, but these species are more difficult to observe than those in the southerly locations. Spawning aggregations might have been depleted or extirpated from historical sites, or grouper spawning in Florida's continental environment may be fundamentally different from that observed around Caribbean islands. Nassau groupers are well known to form spawning aggregations consisting of hundreds or thousands of individuals; no such aggregations have been reported for Florida waters. This suggests that Nassau groupers occurring in Florida are the result of haphazard (lucky) larval input from more southerly sources. Black groupers, gags, scamps, and Atlantic goliath groupers do aggregate to some degree in Florida waters. Riley's Hump, west of the Dry Tortugas, is one site that has been identified as a spawning site for black groupers (and mutton snappers). In Caribbean areas, multiple species may use the same spawning sites year after year. Gags from Florida to as far away as the Carolinas assemble to spawn off eastern Florida from December to May, with a peak in February. Groups of fishes spawn on mid- to outer shelf-edge reefs in 70–300 ft depths, apparently triggered by large, energy-producing groundswells generated by distant winter storms.

In the Gulf of Mexico, gags and scamps aggregate to spawn at reef sites, including Madison-Swanson and Steamboat Lumps Protected Areas (140–240 ft deep), offshore of the Big Bend ecoregion. Spawning fishes from these two species form social groups, composed of a few males and many females. Dominant males develop black undersides. Off eastern Florida, gags and scamps are known to spawn at the shelf-edge reefs, including *Oculina* thickets, at depths of about 300 ft. The gags seem to return to the same sites to reproduce, but not each year. Scamps also aggregate, and they interact in a haremic system, with a distinctively colored dominant male and several females. This male exhibits a bicolor phase, with a gray head and black posterior. Atlantic goliath groupers aggregate off southeastern and southwestern Florida, from August through October, on high-relief reefs or wrecks, in water depths between 60 and 150 ft.

Red groupers, the most common species in Florida waters, do not aggregate to spawn. Red groupers have a proclivity for low-relief habitats and limestone solution holes. In some areas, reds even excavate and maintain their own holes (depressions) in sandy bottoms. Depressions are built by a red grouper lying on the bottom and quivering the body and tail until a pit forms. Over time, algae and organic debris accumulate in the depression, forming the basis for a microecosystem that ultimately attracts small fishes and invertebrates the groupers can feed on. This phenomenon is referred to as habitat engineering. Atlantic goliath groupers also make room for themselves under ledges by fanning their formidable tails.

Left: Black grouper (*Mycteroperca bonaci*), Looe Key, Florida. *Right*: Juvenile, Lake Worth Lagoon.

Left: Tiger grouper (*Mycteroperca tigris*). *Right*: Juvenile.

Atlantic goliath groupers are one of the largest fishes found in Florida waters and represent one of the true natural treasures of the coastal ocean. They have become a charismatic symbol of Florida's marine environment. Reminiscent of sea-going buffalos and (until recently) going the way of their terrestrial counterpart, due in part to their trusting nature toward spearfishers, the harvest of Atlantic goliaths was prohibited in 1990 by the state of Florida, and more than two decades later they now are a common sight on reefs and wrecks along both coasts. The Florida spawning population may be the last within in its natural range that currently is not overexploited. These giants cut an impressive figure when seen underwater. When approached by divers, large adults often produce loud booming sounds, made by vibrating muscles on their swim bladders. Juveniles usually settle in inshore habitats with dense mangrove development. The extensive mangrove fringe off southwest Florida (in the Ten Thousand Islands) and throughout the Florida Keys is prime juvenile habitat for Atlantic goliath groupers. Young grow slowly and ply inshore habitats for at least 5 years. The southwest portion of the state supports a healthy population of Atlantic goliath groupers; the cold snap in 2010, however, killed large numbers of juveniles in southwest Florida. Adult Atlantic goliath groupers are often seen enveloped by schools of round scads.

In some grouper and hind species, the juvenile color pattern differs greatly from that of the adult. This phenomenon, known as ontogenetic color change, is most pronounced in yellowmouth and tiger groupers and speckled hinds. The adaptive significance of these color patterns is not well understood. In the case of tiger groupers,

Atlantic goliath grouper (*Epinephelus itajara*) spawning aggregation, Singer Island, Florida. *Photo by Wayne Shoemake.*

the juveniles, which bear a striking resemblance to initial-phase blueheads, have been observed using their wrasse-like color patterns, along with their pectoral swimming motion, to dupe unsuspecting prey—a phenomenon referred to as aggressive mimicry. Nothing is known about possible mimicry by the other two species. Young yellowmouth groupers resemble potentially toxic sharpnose puffers, so this is possibly a model for another mimetic association.

Groupers are important predators on reefs, at a range of depths. A simple rule of thumb on feeding habits, based on research done many years ago in the Caribbean, is that many of the *Epinephelus* and *Hyporthodus* groupers (rock and red hinds and red, Atlantic goliath, Nassau, snowy, Warsaw, and yellowedge groupers) feed on fishes and invertebrates, while several of the *Mycteroperca* groupers (gags, scamps, and black, yellowfin, yellowmouth, and western comb groupers) feed almost exclusively on fishes. The latter group—with their slender bodies, larger teeth, and more-active swimming behavior—are adapted for a fishy diet. The more sedentary *Epinephelus* species consume lobsters, crabs, shrimps, and mollusks, along with fishes. These dietary differences were thought to contribute to the higher incidence of ciguatera (a foodborne illness caused by certain algae-derived toxins) in the fish eaters. Much to the chagrin of fishers, adult Atlantic goliath groupers will readily engulf hooked fishes—even large fishes, such as sharks.

Sea Basses
Family Serranidae

- **Yellowfin bass** (*Anthias nicholsi*)
- **Swallowtail bass** (*Anthias woodsi*)
- **Streamer bass** (*Baldwinella aureorubens*)
- **Red barbier** (*Baldwinella vivanus*)
- **Yellowtail bass** (*Bathyanthias mexicanus*)
- **Twospot sea bass** (*Centropristis fuscula*)
- **Bank sea bass** (*Centropristis ocyurus*)
- **Rock sea bass** (*Centropristis philadelphica*)
- **Black sea bass** (*Centropristis striata*)
- **Threadnose bass** (*Choranthias tenuis*)
- **Dwarf sand perch** (*Diplectrum bivittatum*)
- **Sand perch** (*Diplectrum formosum*)
- **Longtail bass** (*Hemanthias leptus*)
- **Yellowbelly hamlet** (*Hypoplectrus aberrans*)
- **Florida hamlet** (*Hypoplectrus floridae*)
- **Blue hamlet** (*Hypoplectrus gemma*)
- **Golden hamlet** (*Hypoplectrus gummigutta*)
- **Shy hamlet** (*Hypoplectrus guttavarius*)
- **Indigo hamlet** (*Hypoplectrus indigo*)
- **Black hamlet** (*Hypoplectrus nigricans*)

- **Barred hamlet** (*Hypoplectrus puella*)
- **Tan hamlet** (*Hypoplectrus randallorum*)
- **Butter hamlet** (*Hypoplectrus unicolor*)
- **Undescribed tan hamlet** (*Hypoplectrus* sp.)
- **Eyestripe basslet** (*Liopropoma aberrans*)
- **Candy basslet** (*Liopropoma carmabi*)
- **Wrasse basslet** (*Liopropoma eukrines*)
- **Cave basslet** (*Liopropoma mowbrayi*)
- **Yellow-spotted basslet** (*Liopropoma olneyi*)
- **Peppermint basslet** (*Liopropoma rubre*)
- **Splitfin bass** (*Parasphyraenops incisus*)
- **Apricot bass** (*Plectranthias garrupellus*)
- **Roughtongue bass** (*Pronotogrammus martinicensis*)
- **Reef bass** (*Pseudogramma gregoryi*)
- **Freckled soapfish** (*Rypticus bistrispinus*)
- **Slope soapfish** (*Rypticus carpenteri*)
- **Whitespotted soapfish** (*Rypticus maculatus*)
- **Greater soapfish** (*Rypticus saponaceus*)
- **Spotted soapfish** (*Rypticus subbifrenatus*)
- **School bass** (*Schultzea beta*)
- **Pygmy sea bass** (*Serraniculus pumilio*)
- **Orangeback bass** (*Serranus annularis*)
- **Blackear bass** (*Serranus atrobranchus*)
- **Lantern bass** (*Serranus baldwini*)
- **Snow bass** (*Serranus chionaraia*)
- **Saddle bass** (*Serranus notospilus*)
- **Tattler** (*Serranus phoebe*)
- **Belted sandfish** (*Serranus subligarius*)
- **Tobaccofish** (*Serranus tabacarius*)
- **Harlequin bass** (*Serranus tigrinus*)
- **Chalk bass** (*Serranus tortugarum*)

Background

Several related groups of fishes historically have been amalgamated into or separated from the family Serranidae. Recent genetic studies have suggested that groupers and hinds warrant recognition in their own family (Epinephelidae), while four other lineages, some formerly recognized as families in their own right, are more closely related and properly should be allied within the Serranidae. These four evolutionary lines (sea basses, basslets, anthiine or streamer basses, and soapfishes) are a diverse lineage of small (most are <8 in) reef and bottom dwellers, many of which are hermaphroditic.

Black sea basses, the largest member of the family (2 ft long), have a dark-gray–black body (but the scales have white centers) and fins; the dorsal, anal, and caudal fins often have a bluish hue. A centrally located lateral stripe and diffuse body bars are most obvious in juveniles and subadults, but these markings can be turned on or off at will. Mature males develop humps on their heads and long extensions from the tips of their tails; fleshy tips on the dorsal spines also are more pronounced

Top: Black sea bass (*Centropristis striata*), Palm Beach, Florida. *Bottom*: Juvenile, Lake Worth Lagoon, Florida.

in adults. Black sea basses have a moderately trilobed caudal fin, but its tripartite nature is much more accented in the smaller (12 in) bank and rock sea basses. The central tail point of bank sea basses is longer than the upper and lower tail tips. They have a series of 6–7 black bars on the upper flank, which merge with rectangular blotches that straddle the lateral line; the head is yellowish-brown to orange, with blue lines on the face. The similar rock sea basses also have body bars, but they are distinguished by having a squarish spot under the mid-dorsal fin and 3 equally long tail-fin tips, and by being less deep bodied, with a more slender head profile. The truncate-tailed twospot sea bass, the smallest (6 in) member of the genus (although it probably doesn't belong in this genus), is light-brown, with a dark blotch on the midline and another at the base of the tail fin.

Top: Rock sea bass (*Centropristis philadelphica*). *Bottom*: Bank sea bass (*Centropristis ocyurus*), Lake Worth Lagoon, Florida.

The two species of sand perches—sometimes called "squirrelfishes" (erroneously, as they are not members of the Holocentridae) by Gulf anglers—are difficult to identify. Both have mildly forked tails, with the upper lobe slightly longer than the lower, and bright-blue lines on their faces. Dwarf sand perches (10 in) are light-brown, with 2 brown lateral stripes that are especially prominent in juveniles and become a series of symmetrical rectangles in adults. The upper lobe of the tail in adults has an elongated, thread-like ray. Their larger (12 in) relatives, sand perches, have a light-gray or brown background, with 6 thin blue stripes running from the snout to the tail. Juveniles are pale, with a dark-brown lateral stripe. Opercu-

Left: Sand perch (*Diplectrum formosum*), Lake Worth Lagoon, Florida. *Right*: Dwarf sand perch (*Diplectrum bivittatum*), Lake Worth Lagoon, Florida.

Left: Orangeback bass (*Serranus annularis*), Hobe Sound, Florida. *Right*: Lantern bass (*Serranus baldwini*), Lake Worth Lagoon, Florida.

lum spination patterns definitively identify these two species.

The dwarf sea basses (*Serranus* spp.) resemble one another in shape and size but differ in color pattern. Orangeback, lantern, and snow basses (2–2.5 in) can be confused with each other. The forward portion of the upper body of orangeback basses is reddish-orange, with a dark-red saddle under the front of the dorsal fin; the undersides are white, with yellow bars. Two distinctive, black-rimmed, yellow to orange squares lie just behind the eye. Lantern basses are gray to white, with reddish-brown to black blotches coalescing to form stripes on the upper body. Below these is a yellow midbody stripe, formed by black-edged, yellow scales; the lower body has 4–6 dark, elliptical blotches, bordered ventrally by yellow-orange bars. They have a vertical series of black spots on the base of the caudal fin that are absent in snow basses. The latter species has both wide and narrow, brown, radiating stripes behind the eye, and brown, rectangular, upper-body blotches that become bars on the lower body. Distinct white patches are found on the sides, behind the pectoral fin and on the caudal peduncle.

Perhaps the dwarf sea basses that are easiest to identify are harlequin basses (4 in) and belted sandfishes (3 in), thanks to their long straight foreheads, pointed snouts, and distinctive coloration. Harlequin basses are light-gray above and pale-yellow below, with 8 irregular black bars; the caudal fin is darkly spotted, with yellow tips. Belted sandfishes have a dark stripe through the eye and 7 reddish-brown lateral bars, the rear 4 being much darker than their predecessors; a large dark spot on the dorsal fin merges with the fourth bar. All fins are banded, and there is a white patch behind and below the pectoral fins.

Tobaccofishes (4 in) are slender, with burnt-orange mid- to lower flanks; the upper body is white, with brown and grey saddles. The

Snow bass (*Serranus chionaria*), Lake Worth Lagoon, Florida

Left: Harlequin bass (*Serranus tigrinus*), Juno, Florida. *Right*: Belted sandfish (*Serranus subligarius*), Jupiter, Florida.

Top: Tobaccofish (*Serranus tabacarius*). *Bottom*: Chalk bass (*Serranus tortugarum*), Lake Worth Lagoon, Florida.

caudal fin bears a brown angle-shaped (<) marking. The base coloration of chalk basses (2 in) is silvery-blue, with faint, reddish-brown, upper-body bars that barely extend onto the dorsal fin.

Blackear and saddle basses and tattlers are a bit more confusing. Blackear basses (3.5 in) have a dark spot on the edge of the operculum, and a light-brown to yellow body, with 6–8 brown bars; the particularly dark central bar extends diagonally from the tip of the dorsal fin to just below the belly. Tattlers (8 in) also have a dark-brown, oblique bar, but (unlike the blackear) it is bounded posteriorly on the lower part of the body by a silvery-white bar. Tattlers have a dark bar under the eye, a wide dark stripe that extend from midbody to the base of the caudal fin, and a thin stripe above it that terminates at the base of the rear dorsal fin. Saddle basses (4 in) are light-brown to white, with 4 dark-brown bars and 2 white patches, the first patch behind and under the operculum and pectoral fin, followed by a V-shaped dark bar, and then the second white patch. A dark spot on the rear of the dorsal fin extends onto the body as a dark saddle.

Although not in the genus *Serranus*, three slender species—pygmy sea basses (3 in) and school (4 in) and splitfin (2.5 in) basses—look like they could be. Pygmy sea basses are light-brown, with dark-brown bars; the 2 bars on the whitish caudal peduncle and immediately preceding it are especially dark and obliquely oriented. The belly under the operculum and the pectoral fin is white; the pelvic and anal fins are dark; and reddish lines radiate from the eye. School basses are golden- to orangish-

Top: Tattler (*Serranus phoebe*), Jupiter, Florida. *Bottom*: Pygmy sea bass (*Serraniculus plumilio*), Lake Worth Lagoon, Florida.

brown, with a series of small white blotches on the flanks, and have a yellow tail. Splitfin basses are rosy-red in color, with a dark, elliptical blotch on the first dorsal fin and yellow leading edges on the caudal fin.

The hamlets (genus *Hypoplectrus*) are favorites of evolutionary biologists, because they represent speciation in action. They have stumped ichthyologists for decades, because there are few (if any) distinguishing morphological characteristics. The only reliable way to separate them is by color pattern. For some time, the hamlets were considered a single species (ironically, under the species name *unicolor*, meaning one color!) exhibiting highly variable color patterns, but a closer examination of geography, genetics, mating behavior, and habitat preferences suggests that hamlet color morphs are good species in the traditional taxonomic sense. Ten (possibly eleven) species, all which reach lengths of about 5 in, are reported from Florida, including the recently discovered Florida hamlets, a light-brown to tan species with 6 dark-brown bars, the first one transecting the eye. The third (and widest) bar extends from under the spinous dorsal fin (this bar pattern is shared by several other species), and the fourth has a weak break at the lateral line. Pelvic, anal, and dorsal fins have thin blue margins, and 2 distinctive dark spots appear at the base of the tail. Barred hamlets (the Florida hamlet was formerly lumped within this species) share the 6 light- to dark-brown bars against a yellowish-white background, but they have a continuous fourth bar and lack the 2 caudal spots. Black hamlets are the easiest to identify—they are uniformly black.

Two hamlet species are blue in coloration. Blue hamlets are uniformly sky-blue, with dark-blue to black margins on the leading edges of the 2 tail lobes, while indigo hamlets have alternating white and dark-blue bars. Four other hamlet species are predominantly yellow. Butter hamlets have pale-yellow

School bass (*Schultzea beta*), Hobe Sound, Florida

Left: Florida hamlet (*Hypoplectrus floridae*), Florida Middle Grounds. *Right*: Barred hamlet (*Hypoplectrus puella*), Jupiter, Florida.

bodies (the upper portions are frequently light-grey) and fins, with a black saddle on the caudal peduncle; the head and operculum are crisscrossed by electric-blue lines, and a prominent black spot often is found in front of the eye. Golden hamlets are bright–golden-yellow, with blue-rimmed black spots on the snout and jaws. They often have blue lines radiating from the eye, including a broad one that ends at the corner (preopercle). In yellowbelly hamlets, the lower half of the body, the caudal peduncle, and all fins but the front of the dorsal fin are yellow; the remainder of the dorsal fin and the body are dully blue to brown. Shy hamlets have dark-brown flanks, with a golden-yellow back and fins, and a blue-rimmed black spot on the snout. The head, belly, and fins of yellowtail hamlets are yellow, and the flanks and caudal peduncle are dark. Tan hamlets are uniformly golden-brown and have spots on the snout and the base of the pectoral fins (and sometimes on the upper base of the caudal fin); the pelvic fins have light-blue edges. Another tan hamlet, most likely an undescribed species related to black hamlets, may be in our waters.

Black hamlet (*Hypoplectrus nigricans*), Lake Worth Lagoon, Florida

Liopropoma species are similar in size (≤5 in) and shape (slender, with tapered snouts), but they have distinctive color patterns. The largest (5 in) member of this genus, wrasse basslets, are pale-red to pink, with a wide, dark-brown lateral stripe, bordered by thin yellow lines. The similarly sized eyestripe basslets are orange above, with small yellow spots that become orange below the lateral line; the underside of the head and the belly are white. They have a yellow line that runs from the tip of the snout, through the eye, to the edge of the gill cover; the dorsal, anal, and caudal fins are yellow. Very similarly colored yellow-spotted basslets (3.5 in) have uniformly yellow spots over a yellow base coloration. Cave basslets (3.5 in) are red to pink-orange, with small black spots on the dorsal and anal fins and 2 spots that often merge, forming a bar near the rear of the tail. Candy

Left: Blue hamlet (*Hypoplectrus gemma*), Jupiter, Florida. *Right*: Indigo hamlet (*Hypoplectrus indigo*), Jupiter, Florida.

Butter hamlet (*Hypoplectrus unicolor*), Jupiter, Florida

basslets have 5 purple stripes, bordered by thin red lines, running the length of the yellow to light-brown body. The second dorsal fin has a blue-edged or a ringed black spot, and the tail lobes have 2 such spots. Peppermint basslets look like candy basslets, but with a different color scheme: they have 5 red stripes over a yellow base coloration. The posterior tips of the dorsal, anal, and 2 caudal-fin lobes have white-edged black spots.

Yellowtail basses do not have the same look of a basslet. They are slender and orangish, with yellow leading edges on the caudal fin and an orange spot on the base of the upper pectoral fin. Yellowfin basses (10 in) have a pinkish-lavender body, with yellow spots and blotches, and 3 yellow stripes on the head. The fins are yellow, and the deeply forked tail has a lavender posterior margin. Threadnose basses (4.5 in) are slender, rosy above and yellow below, with regularly spaced white and yellow spots along the flanks and the dorsal and anal fins. The pointed tips of the tail are bright-red. Their vernacular name is derived from the elongated filament that extends from the second nostril.

Among the anthiines, roughtongue (8 in) and apricot (4 in) basses are a bit deeper bodied than the others. Roughtongue basses are a pale- to bright–reddish-orange, with broad yellow markings radiating from the eye onto the operculum and upper body, including a yellow line under the eye. A distinctive, dark–orange-brown bar occurs on the flanks, under the spinous portion of the dorsal fin. All fins but the pectorals have pale-blue edges. The body and front half of the dorsal fin of apricot basses is uniformly reddish-orange; all other fins are pale-yellow.

Longtail basses, the largest of the anthiines (18 in), actually have a short, truncated tail, with a reddish base and a yellow margin, a pattern common to all fins except the pectorals. The dorsal, anal, and

Left: Shy hamlet (*Hypoplectrus guttavarius*). *Right*: Golden hamlet (*Hypoplectrus gummigutta*).

pelvic fins have trailing filaments, and the third dorsal spine becomes a long streamer in adults. The dorsum is reddish-orange, and the silvery flanks are flecked in orange and yellow. A yellow stripe runs from the tip of the snout, under the eye, to the edge of the operculum. They can be confused with streamer basses (12 in) and red barbiers (10 in), but these latter species have deeply forked tails. The body of streamer basses is deeper, the snout shorter, the mouth more oblique, and the eye larger than these features in longtail basses and red barbiers. Streamer basses are strikingly colored inhabitants of deep reefs and wrecks. They are reddish-pink above the midline, and silvery to white below. Their eyes and dorsal, anal, and caudal fins are yellow; the dorsal spines have trailing flags. Red barbiers are reddish-orange, with 2 parallel yellow stripes extending to the base of the pectoral fin, 1 from the eye backward and 1 running posteriorly from the snout under the eye. Males are a vibrant red on the flanks and have red pelvic, yellow anal, and light-blue dorsal fins, the latter with multiple, highly elongated extensions. Swallowtail basses (10 in) are reddish-orange, with a broad, purple-edged yellow band extending from the large eye posteriorly above the midline. The dorsal, anal, and pelvic fins are elongated; in adults, the tips of the highly forked tail fin are so elongated that they almost are equal to the length of the body. The dorsal fin is yellow, as are the leading edges of the caudal fin.

Undescribed species of hamlet (*Hypoplectrus* sp.), Lake Worth Lagoon, Florida

Soapfishes are members of the genus *Rypticus*, and these species are easy to distinguish by their color patterns. Freckled soapfishes (6 in) are bicolored: brown above, and yellow-brown below. Their color pattern is the result of small, densely distributed spots. Irregularly placed, small light blotches often overlay the dark upper half of the body, and a broad, light-colored stripe bisects the forehead (nape). Whitespotted soapfishes (8 in) are golden-brown, with irregular white spots covering the lateral part of

Top: Wrasse basslet (*Liopropoma eukrines*), Florida Middle Grounds. *Bottom*: Peppermint basslet (*Liopropoma rubre*).

the body. A white stripe runs down the nape in all individuals, but it is most pronounced in juveniles. Greater soapfishes, the largest in the genus (13 in), are easily separated from their congeners by their charcoal-gray coloration. Juveniles are covered in closely spaced, black, gray, and brown spots, with a prominent white stripe on the forehead. The most difficult pair to separate are the slope and spotted soapfishes. They are both light-brown, with black spots covering the flanks, but slope soapfishes can be distinguished by their fin coloration—pale-yellow pectorals and outer halves of the dorsal, anal, and caudal fins (vs. dark in spotted soapfishes)—and spotting pattern, usually with spots on the belly and lower portion of the tail (absent in spotted soapfishes).

Reef basses (3 in) don't look much like the other soapfishes. They are small (3 in), with a short snout and a reddish-brown body marked by diffuse light spots. They are most readily identified by a large black ocellated spot on the operculum, and white stripes radiating posteriorly from the eye.

Distribution and Habitat

Sea basses, as a group, are continental species that occur most commonly off northcentral Florida and avoid the Keys. Rock sea basses are misnamed, as they rarely associate with rocks or hardbottom, instead preferring shelly or sandy bottoms on the inner shelf. Black sea basses are found on reefs and structured bottom, from nearshore to depths of more than 250 ft. Their numbers have been expanding off southeastern Florida over the past decade. They often numerically dominate on hardbottom habitats and artificial reefs

Left: Roughtongue bass (*Pronotogrammus martinicensis*). *Right*: Red barbier (*Baldwinella vivanus*).

Longtail bass (*Hemanthias leptus*), Jupiter, Florida

ranging along the entire east coast around to at least Tampa on the west coast. Barred and Florida hamlets are known from southeast Florida around to Tampa on the west coast. Sightings of yellowtail hamlets have been reported, but in the absence of any museum specimens, the jury is still out on its presence in the state.

Basslets in the genus *Liopropoma* are secretive; they live under ledges and, yes, in caves. Eyestripe basslets occur off southeastern Florida, the Florida Keys, and off the west coast, on deep reefs (250–700 ft). Wrasse basslets are found on both sides of the peninsula, on shallower deep reefs (80–300 ft). Cave basslet are only known from the east coast (depths of 80–350 ft). Candy and peppermint basslets have been observed from southeastern Florida and the Florida Keys (depths of 10–200 ft).

Anthiines are found exclusively on deep reefs in the mesophotic zone, from 150 to 2500 ft deep, on

from Palm Beach northward. Bank sea basses are more widespread and have been found off the Dry Tortugas. Juveniles of all three species tend to live in depths shallower than those inhabited by the adults. Sand perches occur around the state, over sandy or low-relief hardbottom, in depths from 10 to 330 ft. The related dwarf sand perches are most common on sandy bottoms in the Keys and the Gulf of Mexico. Twospot sea basses live in deep water (200–900 ft) and are found only off Florida's southern coast.

Most dwarf sea basses, hamlets, and soapfishes are tropical reef dwellers with insular affinities. They are restricted to southern Florida, but some (e.g., tattlers and freckled and whitespotted soapfishes) have more fully embraced continental situations and are widespread around the state. While most dwarf sea basses are found directly on hardbottom, some (e.g., blackear and saddle basses) live on open rubble or sand near hardbottom outcrops, and others (including school basses and chalk basses) hover in the water column. Most of the hamlets exhibit an insular distribution pattern that covers southeast Florida and the Florida Keys. Blue hamlets are known only from the southeast coast and the Florida Keys; golden, yellowbelly, yellowtail, and shy hamlets occur only in the Florida Keys, on reefs in water depths less than 120 ft. Butter hamlets are the most widespread species,

Freckled soapfish (*Rypticus bistrispinus*), Lake Worth Lagoon, Florida

Top: Whitespotted soapfish (*Rypticus maculatus*), Hobe Sound, Florida. *Bottom*: Juvenile, Sarasota, Florida.

the outer continental shelf and upper slope around Florida. Most species hover above the bottom, picking plankton from the water column, but a few stay close to or on the bottom. Red barbiers are hoverers that usually are found in schools, at depths of 120–1300 ft. Roughtongue basses also hover in groups above the substrate, in water depths ranging from 180 to 700 ft, and threadnose basses school from 200 to 2500 ft deep. Streamer basses may school or occur solitarily, in depths ranging from 300 to 2000 ft. Longtail basses are found between 150 and 2000 ft deep and generally are solitary or gather in small groups, on or just above the bottom. Swallowtail basses are solitary hoverers over thickets of deep-water coral or other structure, in depths of 280–1500 ft. Apricot basses are known only from the east coast through the Florida Keys, from 200 to 1200 ft deep, where they associate with *Oculina* coral thickets and other structured bottom. They stay near the bottom, rather than hovering in the water column like the other anthiines.

Whitespotted soapfishes occur around the state, on structured habitat. We have observed this species hiding within the branches of *Oculina robusta* coral on the west Florida shelf, in areas where there is little relief. Freckled soapfishes are widespread, but they are more common in deeper waters of the west Florida shelf, where they shelter in sponges, empty mollusk shells, and other cavities on low-relief bottoms. Slope and spotted soapfishes only reside off the southeast coast and Florida Keys. Although these two species overlap in their known depth distributions, slope soapfishes occur most frequently in 30–120 ft depths, where relief is high, and spotted soapfishes are most common in shallow, flat areas that are 2–20 ft deep. Reef basses are only found off southeastern Florida and the Florida Keys, in depths of 3–70 ft. They are secretive, living in the recesses of reefs and structured bottom, including rubble piles constructed by sand tilefishes.

Natural History

Sea basses are water-column spawners, and most species are hermaphroditic, which complicates their mating systems. Black sea bass are sequential or protogynous (beginning life as females) hermaphrodites. Adults assemble into harems, with a dominant male presiding over a group ranging from a few to many females. In their better-studied Indo-Pacific relatives, a sex change by females is driven by the social order in the harem. Once the presiding male dies or leaves, the next-largest (or oldest) female will change into a male and take over the harem. Roughtongue bass and red barbier males are more ornate than the females, and a haremic social system is thought to operate with these species.

Longtail basses change sex, but not all individuals are born as females. Dwarf sea basses and hamlets are simultaneous hermaphrodites and take turns switching sex with their partners in real time, in a process known as egg trading.

Top: Greater soapfish (*Rypticus saponaceus*), Jupiter, Florida. *Bottom*: Juvenile, Palm Beach, Florida.

Their mating system varies with the density of fishes in an area. Where densities are high, there is a haremic system, consisting of a dominant male and several to many hermaphrodites. The hermaphrodites have smaller territories within the larger males' territories. When fish densities are low, monogamy (pairing) is the norm. In lantern basses, tobaccofishes, chalk basses, and belted sandfishes, both types of mating systems are known.

Harlequin basses have monogamous pairs and spawn reciprocally (trading eggs) over periods of days; hamlets reproduce in a similar fashion. Hamlets spawn by pairing and then curving their bodies into complementary S-shapes. These actions are easy to observe, as many of the hamlet species court and spawn near dusk in southeastern Florida. On rare occasions (and this is why hamlets are so engrossing), you might see two different species pairing up.

Members of the sea bass family are carnivores, with some focusing on motile invertebrates and fishes, and others (particularly the anthiines) on plankton. Black sea basses are voracious, as anyone who has ever put a baited hook or Sabiki rig overboard along much of east Florida will attest. We often see hamlets, sand perches, or serranines swimming around with the antennae, legs, or claws of a crustacean protruding from their jaws.

Color patterns in the hamlets bear a striking resemblance to other reef fishes, prompting some researchers to assert that mimicry—possibly aggressive mimicry—is driving their speciation. Blue hamlets resemble blue chromises; black hamlets are similar to dusky damselfishes; shy hamlets favor rock beauties; and yellowtail hamlets bear a resemblance to yellowtail damselfishes. This pattern of color commonality may just be coincidental, but where there's smoke . . .

The name "soapfish" comes from the observation that their mucous covering contains a mild toxin (grammistin). This toxin was discovered serendipitously by a famous ichthyologist, who placed a speared soapfish inside his swimming suit and later suffered a painful skin irritation from direct contact with the mucous. The mildly toxic mucous is thought to make soapfishes unpalatable to predators.

Basslets
Family Grammatidae

- **Fairy basslet** (*Gramma loreto*)
- **Royal basslet** (*Lipogramma regia*)
- **Threeline basslet** (*Lipogramma trilineatum*)

Background

Basslets are distinguished from other bass-like fishes by their lack of lateral lines. These small species are coral-reef dwellers. The beautiful fairy basslets (3 in) cannot be confused with any other species. The anterior half of the head, body, and fins are violet to purple; the rear half of the body and fins are bright-yellow to gold. A narrow yellow streak connects the eye to the midedge of the operculum, and a similarly sized one extends from the tip of the lower jaw, through the eye, to the upper edge of the operculum. The front of the dorsal fin has a small dark spot. The scales on the rear half of the body, while primarily yellow in color, are edged in blue. Threeline basslets have 3 brilliant blue lines on the head: the first bisecting the forehead (from the tip of the snout to the front of the dorsal fin); the second extending from the top of the eye to a spot under the front portion of the dorsal fin; and the short third appearing under the eye. The lower-posterior operculum (cheeks) are pinkish-purple, and the fins are spotted. Royal basslets are easily identified by the ocellated spot (a blue-ringed black dot) at the posterior base of the dorsal fin. They have 6 lateral orange bars and 3–4 postocular stripes.

Distribution and Habitat

Fairy basslets are a popular aquarium fish and are highly sought by underwater photographers and sightseeing divers. They have only recently (and very rarely) been seen in the Keys, and there has been a single sighting near Jupiter. Fairy basslets either are just now colonizing Florida (they are found in the Bahamas and throughout the Caribbean) or, more likely, they represent an intentional introductions by humans. We are aware of a number of seeding events in which folks with less-than-honorable intentions have released other species of attractive, nonnative fishes onto Florida reefs; we do not have any indication that there is an established Florida fairy basslet population. They frequent ledges and caves, in depths of 3–155 ft. Threeline basslets occupy the same type of habitat, in deeper waters (45–550 ft throughout their range). Royal basslets have been observed only once in the United States, in 335 ft of water at the Pinnacles region (southwest of Pensacola) in the northeastern Gulf of Mexico.

Natural History

Basslets are secretive species that are attracted to dark recesses in the reef. Both Florida species frequently are observed swimming upside down, orienting to the upper ceiling of caves or ledges. Male fairy basslets are mouth brooders that protect their mate's fertilized eggs.

Fairy basslet (*Gramma loreto*)

Jawfishes
Family Opistognathidae

- **Swordtail jawfish** (*Lonchopisthus micrognathus*)
- **Yellowhead jawfish** (*Opistognathus aurifrons*)
- **Moustache jawfish** (*Opistognathus lonchurus*)
- **Banded jawfish** (*Opistognathus macrognathus*)
- **Mottled jawfish** (*Opistognathus maxillosus*)
- **Yellowmouth jawfish** (*Opistognathus nothus*)
- **Spotfin jawfish** (*Opistognathus robinsi*)
- **Dusky jawfish** (*Opistognathus whitehursti*)

Background

Jawfishes have big heads, big mouths, and bulging eyes, but relatively small bodies (all <6 in). Swordtail jawfishes are the most distinctive, having a long, lanceolate tail (one-quarter or more of its total length) and a laterally compressed body. The pelvic fins are blue, and the body is tan to gray, with thin, pale-blue bars anteriorly that morph into blue spots posteriorly. All other species have rounded tails.

Yellowhead and moustache jawfishes have light-colored bodies. The head of yellowhead jawfishes is appropriately yellow, fading to white or pale-blue on the otherwise unmarked body. There are a pair of black spots in the chin, but these are not to be confused with the facial feature of moustache jawfishes, which is a dusky mark (mustache) found above the upper jaw. The latter species is light-brown to olive on the sides, with thin blue stripes along the sides; the dorsal and anal fins are marked with blue bands, including thin blue fin margins. Yellowmouth jawfishes have a mouth with a yellow interior, tan sides, and several rows of brown spots on the dorsal fin.

The other Florida species fall into a dusky-brown coloration category and can be difficult to properly identify, especially when only the head is seen protruding from a burrow. From this perspective, the various species can be differentiated by looking at the pigment patterns on the head and jaws, but it can be challenging. Banded jawfishes are brown, with 4–6 irregular bars along the back that terminate on the lower dorsal fin; most have a dark spot on the yellowish dorsal fin, between the sixth and ninth dorsal spines. The upper jaw of males extends well posterior of the eye and is marked with several dark bands. The easily confused mottled jawfishes differ in having a series of pale blotches along the bases of the dorsal and anal fins. The upper and lower jaws are more mottled than banded. Dusky jawfishes are brown, with a series of obscure dark body bars and pale triangular blotches along the dorsum. There often is a faint bluish-green blotch between the second and fourth dorsal spines. Spotfin jawfishes are brown, with rows of irregular

Yellowhead jawfish (*Opistognathus aurifrons*), Looe Key, Florida

Banded jawfish (*Opistognathus macrognathus*), Lake Worth Lagoon, Florida

spots coalescing into dashes on the sides; the median fins are dark-chocolate-brown, with rows of small white spots and 1 large, elongated, ocellated black spot between the third and seventh dorsal spines. The pigment on the jaws of this species form several paired bands, differing from all of the other jawfish species.

Distribution and Habitat

Yellowmouth jawfishes are the rarest and deepest-dwelling member of the group, occurring at depths of about 300 ft; the only Florida record of this species is from off of Pensacola. Swordtail jawfishes are found on silty bottoms, in water depths of 70–280 ft, off Florida's eastern Gulf to as far south as the Keys. Moustache jawfishes are distributed on the outer shelf, around the peninsula, in water depths ranging from 60 to 300 ft. Yellowhead jawfishes occur over sand- or rubblebottom, usually in the vicinity of structured hardbottom, in depths from nearshore to 200 ft, from about Ft. Pierce on the east coast, throughout the Florida Keys, and westward along the Florida shelf in the eastern Gulf. The dusky-colored species are most prevalent on sand- and rubblebottom, in depths shallower than 40 ft, including inshore lagoons, but dusky, banded, and spotfin jawfishes also are found in waters as deep as 150 ft. Dusky jawfishes seem to prefer nearshore areas around hardbottom. Banded and spotfin jawfishes are found off both the east and west coasts of Florida, while mottled and dusky jawfishes are known only from southeast Florida and the Keys.

Mottled jawfish (*Opistognathus maxillosus*), Lake Worth Lagoon, Florida

Spotfin jawfish (*Opistognathus robinsi*), Lake Worth Lagoon, Florida

Spotfin jawfish (*Opistognathus robinsi*) constructing its burrow, Lake Worth Lagoon, Florida

Natural History

Several aspects of their natural history mark jawfishes as unique among shore fishes: they reside and spawn in burrows of their own construction; occur in groups or colonies; feed on macroplankton and invertebrates drifting by their burrows; and are sexually dimorphic (males have markedly larger mouths that facilitate the oral brooding of fertilized eggs). The species vary individually on these themes, and details are known for only a few of them. The large mouth and buccal cavity serve jawfishes admirably, as both a front-end loader and an incubator. Jawfishes are master engineers, capable of constructing elaborate burrows lined with carefully selected shell and rubble material; some species even construct a front door during resting periods.

Swordtail jawfishes live in small groups. Their burrows are conical pits dug in silty or muddy bottoms. They pair-spawn, but oral incubation has not yet been confirmed. Swordtails hover well above the burrow, feeding on plankton; in some regions they share their burrows with a small crab. Yellowhead jawfishes live in groups or colonies of up to 70, in sand and rubble areas adjacent to reefs and structured hardbottom. Individuals construct burrows consisting of an entrance shaft and a small chamber. Often a large rock forms the ceiling of the burrow chamber. Following some courtship and pair-spawning, incubated eggs transform into larvae that settle over sandbottom and immediately dig in.

Banded, mottled, and spotfin jawfishes are found in shallow areas where coarse, shelly material is available for burrow construction. Dusky jawfishes may also be found in these habitats, but from our observations they are more common around reefs, and sometimes within or near yellowhead jawfish colonies. Banded, mottled, spotfin, and dusky jawfishes have similar borrows, with chambers

Dusky jawfish (*Opistognathus whitehursti*), Palm Beach, Florida

constructed of shells carefully placed and locked down to form the entrance shaft. These species rarely hover above the burrow, so typically only the head is seen protruding above the entrance. Some species are monogamous, while in others males have multiple female partners. Females produce a batch of eggs several times a month; spawning may take place with some kind of lunar rhythm, but this has not been well studied. After some interaction and displays, a receptive female and the chosen male enter a burrow, where fertilization takes place. The male incubates the eggs by rolling them around as a cohesive clump.

Bigeyes
Family Priacanthidae

- **Bulleye** (*Cookeolus japonicus*)
- **Glasseye snapper** (*Heteropriacanthus cruentatus*)
- **Bigeye** (*Priacanthus arenatus*)
- **Short bigeye** (*Pristigenys alta*)

Background

Bigeyes are large-eyed, red, peculiar-smelling fishes. Basic aspects of their appearance can be used to separate the four species known from Florida waters. Short bigeyes have deep, oval-shaped bodies that distinguish them from the others, which have more elongate bodies. The soft dorsal, soft anal, pelvic, and caudal fins are white to pale-blue, with marked, dark–reddish-brown margins. Bulleyes are easily recognized by their long pelvic fins, which are longer than the length of the head. The rear edges of the dorsal, anal, and caudal fins are black-edged, and the other parts of the pelvic-fin membranes are inky-black. Bigeyes and glasseye snappers (the latter is yet another mistaken common name, as this fish could hardly pass for a snapper) are the toughest to identify. Both are deep-red to pinkish, but in life glasseye snappers usually

Glasseye snapper (*Heteropriacanthus cruentatus*)

Bigeye (*Priacanthus arenatus*), Hobe Sound, Florida

Short bigeye (*Pristigenys alta*), Florida Middle Grounds

have silvery blotches and bars on the flanks; these generally fade in death. Sometimes, even when glasseye snappers are alive, the sides are uniformly colored; to maximize confusion, the normally unicolored bigeyes may exhibit silvery bars on occasion. The soft portion of the dorsal fin is a bit higher than the spinous dorsal fin in glasseye snappers, while in bigeyes the two parts of the fin are pretty much equal sized, reminiscent of a drill sergeant's flat-top haircut! The tip of the lower jaw in glasseye snappers is about in line with the midline of the body, while the tip of the lower jaw in bigeyes is above the midline, giving them the appearance of a more pronounced underbite.

Distribution and Habitat

Bigeyes, as a group, are hardbottom inhabitants that are more active by night than by day. Juveniles generally occur in shallower depths than adults and can be found on jetties and dock pilings. Short bigeyes are found throughout state waters but are most common on the west Florida shelf, and less so off southeastern Florida, where they are found in water depths of 80–300 ft. Solitary individuals hover above small holes or ledges, where they hide when threatened. Glasseye snappers and bigeyes also are found statewide. Glasseyes are most common on shallow (5–60 ft) reefs and hardbottom areas off southeast Florida and the Florida Keys, but they occur in water depths ranging from 3 to 1000 ft. Bigeyes are found at depths of 30–800 ft but are most common at 60–300 ft. Bigeyes may be solitary or form small groups during daytime hours. Bulleyes occur only on shelf-edge reefs off eastern Florida, in deeper waters (300–1200 ft).

Natural History

Based on observations of related Indo-Pacific species, bigeyes may aggregate at specific locations to spawn. Juveniles often associate with drifting sargassum. We have found juvenile short bigeyes nestled within the branches of ivory bush coral or foliose algae on the west Florida shelf. Bulleye, glasseye snapper, and bigeye juveniles are rarely observed, probably due to their secretive habits. Short bigeyes and bigeyes feed on large zooplankton, transported over reefs from the surrounding water column. Bigeyes give off a unique odor for a fish (we cannot offer an explanation for the adaptive significance, if any, of this smell).

Cardinalfishes
Family Apogonidae

- **Bigtooth cardinalfish** (*Apogon affinis*)
- **Bridle cardinalfish** (*Apogon aurolineatus*)
- **Barred cardinalfish** (*Apogon binotatus*)
- **Whitestar cardinalfish** (*Apogon lachneri*)
- **Slendertail cardinalfish** (*Apogon leptocaulus*)
- **Flamefish** (*Apogon maculatus*)
- **Mimic cardinalfish** (*Apogon phenax*)
- **Broadsaddle cardinalfish** (*Apogon pillionatus*)
- **Pale cardinalfish** (*Apogon planifrons*)
- **Twospot cardinalfish** (*Apogon pseudomaculatus*)
- **Sawcheek cardinalfish** (*Apogon quadrisquamatus*)
- **Belted cardinalfish** (*Apogon townsendi*)
- **Bronze cardinalfish** (*Astrapogon alutus*)
- **Blackfin cardinalfish** (*Astrapogon puncticulatus*)
- **Conchfish** (*Astrapogon stellatus*)
- **Freckled cardinalfish** (*Phaeoptyx conklini*)
- **Dusky cardinalfish** (*Phaeoptyx pigmentaria*)
- **Sponge cardinalfish** (*Phaeoptyx xenus*)

Background

Cardinalfishes are a diverse group of small (<5 in total length), stout, toothy reef fishes, with big eyes and mouths. They are reclusive by day but commonly seen while diving at night. Of the three cardinalfish genera in Florida, the genus *Apogon* is the most speciose, with 12 species. Members of this genus have very similar body shapes, including 2 separated dorsal fins, large eyes, and large mouths, but the various species can be distinguished by differences in coloration. The background color of all *Apogon* species is red, red-orange, or pale-red, and the distinguishing marks are black vertical bars, saddles, or spots.

Six species have bars or saddles only, usually in two places along the body: anteriorly, under the posterior base of the second dorsal fin; and posteriorly, on the caudal peduncle. Barred cardinalfishes have thin bars at the posterior base of the second dorsal fin and on the posterior margin of the caudal peduncle. In contrast, mimic cardinalfishes have a broad, V-shaped anterior bar and a broad peduncle saddle, and broadsaddle cardinalfishes have a thin anterior bar and a very wide peduncle saddle, bounded by thin white bars. Pale

Top: Barred cardinalfish (*Apogon binotatus*), Lake Worth Lagoon, Florida. *Bottom*: Belted cardinalfish (*Apogon townsendi*), Jupiter, Florida.

Left: Twospot cardinalfish (*Apogon pseudomaculatus*), Lake Worth Lagoon, Florida. *Right*: Flamefish (*Apogon maculatus*), Lake Worth Lagoon, Florida.

cardinalfishes have relatively wide saddles at each location, and belted cardinalfishes have a thin, V-shaped anterior bar and 2 thin black bars bounding a broad, dusky caudal bar. Slendertail cardinalfishes have a long caudal peduncle, with an anterior saddle about the width of the dorsal fin base that extends to the anal-fin base. Another, less distinct bar occurs forward of the dorsal-fin origin.

Pale cardinalfish (*Apogon planifrons*), Jupiter, Florida

Four *Apogon* species have 1 or 2 spots. Twospot cardinalfishes have a black spot at the same place where the anterior bar is located on other *Apogon* species, and another on the caudal peduncle. Flamefishes have a similar anterior spot, but they have a faint wide bar on the caudal peduncle. A black spot, bordered posteriorly by a white spot at the rear base of the dorsal fin, identifies whitestar cardinalfishes. A single obscure spot, really a blotch, at the base of the caudal fin characterizes sawcheek cardinalfishes.

Bigtooth cardinalfishes lack bars, saddles, and spots; their base salmon coloration is punctuated only by a dusky stripe crossing through the eye. The similarly colored bridle cardinalfishes have small bars behind and under the eye; faint, thin stripes along the body; and yellowish median and tail fins.

The three species in the genus *Astrapogon* have silvery flanks, with black pigment on the back and fins; the pelvic fins are fan shaped and noticeably larger than those on *Apogon* or *Phaeoptyx* species. Bronze cardinalfishes have a brown to bronze background color, with small black flecks of pigment across the flanks. The iris is yellow. Blackfin cardinalfishes and conchfishes differ in being silvery to charcoal-gray in base color and having black lines radiating from the eyes. The fins of conchfishes are uniformly black or dark-gray, while the dark fins of blackfin cardinalfishes have pale margins.

Cardinalfishes in the genus *Phaeoptyx* are a little more robust than *Apogon* species, with pinkish to gray, translucent bodies covered with small black-pepper spots. The

Bigtooth cardinalfish (*Apogon affinis*), Singer Island, Florida

Top: Bronze cardinalfish (*Astrapogon alutus*), Indian River Lagoon, Florida. *Bottom*: Conchfish (*Astrapogon stellatus*).

three species can be separated by subtle differences in color pattern. Freckled cardinalfishes have a stripe at the base of the dorsal fins. Dusky cardinalfishes tend to be grayish and have one black fleck per scale. Sponge cardinalfishes have variably sized black flecks and are generally pale gray.

Distribution and Habitat

Apogon species associate with structured habitats, usually reefs, in water depths less than 300 ft. Most *Apogon* occur offshore of southeastern Florida and the Florida Keys. Three species—flamefishes and twospot and barred cardinalfishes—are found north of Stuart on the east coast and off the west coast and panhandle. Bronze cardinalfishes usually inhabit seagrass meadows. Two *Astrapogon* species, conchfishes and blackfin cardinalfishes, are restricted to southeast Florida and the Florida Keys, where live queen conchs, which constitute the favored habitat for these two species, are most abundant.

Natural History

Spawning behavior has not been described for all species, but in general cardinalfishes spawn in pairs, near the bottom, probably at night. Apparently, females court the males. Once a male gives in to the advances, the female extrudes the eggs to be fertilized. Following fertilization, the male scoops up the eggs in his mouth, where he broods them for an undetermined period. While males typically do the brooding, there are reports of both sexes brooding in *Phaeoptyx*. Brooding males have distended jaws and cannot feed; eggs, how-

Sponge cardinalfish (*Phaeoptyx xenus*), Palm Beach, Florida

ever, occasionally have been found their stomachs.

Most cardinalfishes skulk around in recesses of caves or ledges during daylight hours. Once night falls, they emerge from their daytime lairs to feed on tiny invertebrates and fishes drifting in the water column. A few species—such as bigtooth and twospot cardinalfishes and flamefishes—are less reclusive and are regularly seen during the day. Bigtooth cardinalfishes form loose groups over ledges or other hardbottom breaks.

Other cardinalfish species symbiotically associate with invertebrates. The aptly named conchfishes live within the mantle cavity of live queen conchs (*Strombus gigas*), but they also have been reported to take refuge inside pen shells (*Atrina* sp.) in the Dry Tortugas. This relationship is not mutualistic, as the fish gains a hiding place but does not contribute anything beneficial to the conch. Similarly, bronze cardinalfishes have been reported living within the mantle cavity of West Indian fighting conchs (*Strombus pugilis*). Sponge cardinalfishes spend their daylight hours inside tube sponges (*Verongia* and *Calyspongia*). We often have discovered sponge cardinalfishes inside PVC tubes set on the seafloor as sediment collectors. Sawcheek and bridle cardinalfishes frequently associate with sea anemones (*Condylactis* and *Bartholomea*). Occasionally, early-juvenile twospot cardinalfishes and sawcheek cardinalfishes are seen swimming within the spines of longspine sea urchins (*Diadema* spp. and *Astropyga magnifica*).

Tilefishes
Family Malacanthidae

- **Goldface tilefish** (*Caulolatilus chrysops*)
- **Blackline tilefish** (*Caulolatilus cyanops*)
- **Anchor tilefish** (*Caulolatilus intermedius*)
- **Blueline tilefish** (*Caulolatilus microps*)
- **Tilefish** (*Lopholatilus chamaeleonticeps*)
- **Sand tilefish** (*Malacanthus plumieri*)

Background

Tilefishes are familiar mostly to deep-dropping anglers, who ply the depths for edible species. Generally, they lump all members of the genus *Caulolatilus* into group called "gray tilefishes" and refer to the larger, easily recognized tilefishes (*Lopholatilus chamaeleonticeps*) as "golden tilefishes." The latter species is the largest member of the family, reaching about 4 ft long and weighing 60 lb. They have a unique yellow flap of skin on the head, just ahead of the dorsal-fin origin, that tends to be larger and more brightly colored on larger males. This may be used for signaling females or rival males during courtship or territorial displays. The color of this species is blue-gray to blue-green above and white below, with small bright-yellow spots covering the back and tail; a thick yellow margin runs the length of the dorsal fin.

The four gray tilefish species do look very much alike (generally pale-gray dorsally, with white undersides) and all of them (except blueline tilefishes) have a well-defined dark spot above the base of the pectoral fin. Blueline tilefishes also are darker gray than the other species and have a diagonal blue streak, bordering a yellow one, under the eye. They are the largest species in the genus *Caulolatilus* and grow to around 2.5 ft long, larger (by about 10 in) than goldface tilefishes. Keys to goldface identification are a distinctive horizontal yellow bar under the eye, and yellow spots on the caudal fin and rear two-thirds of the dorsal fin.

The colorful blackline tilefish is the only species having a rounded caudal fin, with extended tips. The spiny dorsal fin and predorsal ridge, the leading edges of the caudal fin, and spots on the upper body are bright-yellow, and there often is a black line along the base of the dorsal fin. Blackline tilefishes and anchor tilefishes are small (about 15–16 in). Anchor tilefishes have a distinctive black line running from the upper jaw through the eyes, connecting to a black predorsal ridge. Many don't realize that the small, elongate sand tilefishes are related to these deep-dwelling tilefishes, and pejorative names like "Boston bananas" or "green weenies," used by disappointed anglers, do not help the identity crisis. Sand tilefishes are elongate and are cylindrical in cross-section. Coloration is pale–gray-green on the back, with golden-yellow edges on the dorsal, anal, and caudal fins; there is a distinctive dusky smudge on the upper caudal lobe.

Distribution and Habitat

Tilefishes (with one exception, sand tilefishes) are most abundant in water depths greater than 250 ft, usually on sand- or mudbottom where most (and probably all) build burrows. Species-specific depth ranges are approximately as follows: sand tilefishes (30–500 ft); blueline tilefishes (100–800 ft); anchor tilefishes (150–950 ft); goldface tilefishes (300–650 ft); blackline tilefishes (150–1600 ft); and tilefishes (250–1800 ft). All but anchor tilefishes, which in Florida are restricted to the Gulf of Mexico, probably occur around the state within their preferred water-depth and substrate bands. These conditions are most prevalent along the east coast, from Jupiter northward. Blueline and anchor tilefishes construct burrows in sand, mud, or shell-hash substrate, but they are often observed or caught near hardbottom features, where individuals burrow under rocks. Off the east coast of Florida, blueline tilefishes burrow into sandy sediments at 300–500 ft depths, where bottom-water temperatures range from 55°F to 60°F. In contrast, tilefishes tend to be found on level bottom, where sediments are muddier and water temperatures are cooler (46°F–57°F). Goldface tilefishes, which may prefer rubble-bottom, are known from the Dry Tortugas in Florida.

Sand tilefish (*Malacanthus plumieri*), Palm Beach, Florida

Natural History

Owing to their abundance and importance as fishery species, the greatest amount of natural-history information is available for blueline tilefishes and tilefishes. Blueline tilefishes live for more than 40 years, mature at 3–4 years of age, and spawn up to 4 million eggs during summer months. Blueline tilefishes are hermaphroditic: individuals begin life as females and change into males with age and growth. Like other studied members of the family, blueline tilefish eggs are buoyant and the larvae are planktonic. Tilefish females mature when they are between 5 and 9 years of age. Females can produce up to 10 million eggs during spring and summer spawning seasons. Individuals change sex with growth and can be over 25 years old. All tilefish species are opportunistic bottom feeders that consume a variety of bottom-dwelling invertebrates, including brittle stars, crabs, shrimps, mantis shrimps, mollusks, and worms, as well as small fishes.

All studied tilefishes live in colonies of varying sizes, with burrows spread over the seafloor (in relation to the extent of their preferred sediment type). They are ecosystem engineers—their burrowing activities create habitat that attracts and supports a host of other fish and invertebrate species. These mounds readily appear on sonar traces. Blueline tilefishes dig elliptical or trench-like burrows in what is otherwise featureless softbottom. These burrows—measuring up to 5 ft long, with a shaft at least 2 ft deep at one end—provide shelter for crabs, shrimps, and small fishes. Adult conger eels are known to occupy burrows with both blueline tilefishes and tilefishes. Individual blueline tilefishes and tilefishes have been observed filing into the same burrow. Yellowedge groupers also use tilefish burrows; the extent to which the resident tilefish tolerates these unrelated guests is unknown. Bluelines have been observed going head-first into the burrows and backing out tail-first, indicating that the burrows are not elaborate homes used to ambush prey, but are more like a refuge to dive into when resting or if a predator swims through the area looking for a meal. Tilefish burrows must be a little more roomy, as individuals of this species are sometimes seen coming out head-first. The burrows probably are occupied and reworked by generation after generation of their kin. Active burrows have sloping sides in which squat lobsters and crabs burrow, usually leaving sidecast material.

The engineering efforts of sand tilefishes result in a relative large, neatly stacked pile of rubble gathered from surrounding areas. The extent of the pile may signify a male's attractiveness to females, but the more immediate effect of is in generating habitat for small fishes, which gain shelter and a feeding advantage from it. Studies in Caribbean waters have documented over 30 species of fishes that associate with sand tilefish rubble mounds. The most common were bicolor damselfishes, sunshinefishes, lantern basses, goldspot gobies, and cherubfishes; these mounds also are one of the few places where elusive reef basses can be found with regularity. Sand tilefishes dig a 4–6 in wide, horizontal trench by lying on the sandy rubblebottom and rapidly undulating their bodies. The trench is then covered with rubble, shells, fragments of coral, and other debris gathered by the fishes. The completed mounds can be 5 ft wide and 3 ft high. Individuals maintain the mound and the entrance to it. The adaptive significance of the mounds is refuge from predators—a must when you live on an open plain.

Sand tilefishes are colonial animals, grouped into mating units that constitute a harem. Multiple females reproduce with the male that is found in their feeding areas. Males chase off rivals in their territory, but they don't stop females that visit an adjacent peripheral male. Reproduction involves paired synchronous swimming, prior to an ascent that leads to the release of gametes. Juvenile sand tilefishes are uncommonly encountered, and at one time they were thought to be a distinct species. Sand tilefishes fall prey to patrolling barracudas and mutton snappers.

Bluefishes
Family Pomatomidae

- **Bluefish** (*Pomatomus saltatrix*)

Background

Bluefishes, the only member of this family, look a lot like jacks, but they lack the 2 detached anal spines characteristic of the latter group. The elongate body is blue-green dorsally, silvery laterally, and white on the belly. The low first dorsal fin has 7 or 8 short spines. The jaws of the large mouth have single rows of large, laterally compressed, dagger-like teeth that are replaced when broken or eroded. Adults grow to 3.6 ft and weigh over 30 lb.

Bluefish (*Pomatomus saltatrix*), Hobe Sound, Florida

Distribution and Habitat

Bluefishes are found around the state (although they are not common in the Keys), in oceanic, shelf, and inshore waters. They are a pelagic continental species that prefers 66°F–72°F water temperatures.

Natural History

Bluefishes, a staple of fall–winter recreational surf fishery on Florida's east coast, are the piscine equivalents of Yankee tourists, escaping to Florida when waters in the Middle Atlantic Bight and New England are cold and then returning north as waters warm up. Bluefishes spawn in the water column, and the young mature in inshore habitats. Females mature in 2 years and can release as many as 1 million eggs. There are several spawning groups along the U.S. East Coast: some spawn in the spring off the mid-Atlantic states, and another group reproduces in the summer off the northeastern United States. In the spring, large bluefishes caught off the shelf edge in Florida are emaciated and appear to have spawned in the region. Young adult fishes migrate into Florida waters in fall and winter. Small juvenile bluefishes (called "snappers" in the northeastern United States) are rarely observed in Florida waters south of Jacksonville. Larger bluefishes, pushed southward by passing cold fronts, are seasonally abundant in nearshore waters of both coasts. Migrating bluefishes often enter the mouths of coastal rivers and lagoons for varying periods.

Schools usually segregate into like-sized classes. The largest adults make their springtime northward trek well offshore, near the shelf break, and smaller bluefishes migrate in shallower waters. Studies show that most tagged fishes move less than 3 mi per day, but some migrating individuals average as much as 34–69 mi per day during what can be a 1200 mi journey, lasting over several months. These high-end movement rates suggest that bluefishes are adept at riding currents (like the Gulf Stream) in their travels.

Bluefishes live about 13 years. Although primarily piscivores, they also consume squids, crabs, shrimps, and, surprisingly, even snails as large as whelks. They feed in marauding packs, slashing through schools of anchovies, mullets, or herrings with abandon. Unlike most predatory fishes, blues take bites out of their prey rather than swallowing them whole. When they are in a full feeding frenzy and schooling baitfishes are abundant, bluefishes often regurgitate what they've just eaten and continue to chase prey, leaving an oily, self-made chum slick at the water surface. When bluefishes are in this behavioral state, even novice anglers can cast any lure in their tackle boxes and be successful. Under these conditions, we have hooked bluefishes on rigged soda-can pop tops and on unbaited shiny hooks. The provincial nickname "snapper" alludes to their proclivity to bite at anything that moves, including human fingers (if these digits get too close to the head). Feeding bluefishes have even bitten humans that were swimming amid schools of forage fishes, and deaths were attributed to this species during the Allies' World War II invasion of northern Africa. Makos and other shark species happily enjoy a meal of fresh bluefishes.

Jacks
Family Carangidae

- **African pompano** (*Alectis ciliaris*)
- **Yellow jack** (*Caranx bartholomaei*)
- **Blue runner** (*Caranx crysos*)
- **Crevalle jack** (*Caranx hippos*)
- **Horse-eye jack** (*Caranx latus*)
- **Bar jack** (*Caranx ruber*)
- **Atlantic bumper** (*Chloroscombrus chrysurus*)
- **Mackerel scad** (*Decapterus macarellus*)
- **Round scad** (*Decapterus punctatus*)
- **Redtail scad** (*Decapterus tabl*)
- **Rainbow runner** (*Elagatis bipinnulata*)
- **Bluntnose jack** (*Hemicaranx amblyrhynchus*)
- **Pilotfish** (*Naucrates ductor*)
- **Leatherjack** (*Oligoplites saurus*)
- **Bigeye scad** (*Selar crumenophthalmus*)
- **Atlantic moonfish** (*Selene setapinnis*)
- **Lookdown** (*Selene vomer*)
- **Greater amberjack** (*Seriola dumerili*)
- **Lesser amberjack** (*Seriola fasciata*)
- **Almaco jack** (*Seriola rivoliana*)
- **Banded rudderfish** (*Seriola zonata*)
- **Florida pompano** (*Trachinotus carolinus*)
- **Permit** (*Trachinotus falcatus*)
- **Palometa** (*Trachinotus goodei*)
- **Rough scad** (*Trachurus lathami*)
- **Cottonmouth jack** (*Uraspis secunda*)

Background

Members of jack family have small scales; 2 anal spines; and deeply forked tails, with narrow caudal peduncles (anatomical signs that they are built for speed). Aside from these few general attributes, individual species display a wide range of body shapes, from bullet shaped to deep bodied and laterally compressed. Jacks fall into two broad groups, one with scutes (enlarged, rough, bony plates) along the rear to middle portion of the lateral lines, and the other lacking those distinctive accoutrements.

The smooth group, in turn, consists of two subgroups: one deep bodied and laterally flattened, and the other with elongated bodies that are more rounded in cross-section. The latter include the brown to purplish-gray amberjacks, which have sloping foreheads and uniquely bear a diagonal stripe connecting the spiny first dorsal fin to the eye (and often continuing downward to the mouth). These species are superficially similar and difficult to tell apart. The most distinctive are almaco jacks, which have a deeper body, have long dorsal- and anal-fin lobes, and reach 3 ft long. Juveniles have 5–6 dark bars on the flank, some of which are divided.

Banded rudderfishes and greater and lesser amberjacks are more slender and have shorter dorsal-

Top: Almaco jack (*Seriola rivoliana*), Jupiter, Florida. *Bottom*: Juvenile, Jupiter, Florida.

Left: Greater amberjack (*Seriola dumerili*), Jupiter, Florida. *Right*: Juveniles, Jupiter, Florida.

Top: Banded rudderfish (*Seriola zonata*), Palm Beach, Florida. *Bottom*: Juvenile, Lake Worth Lagoon, Florida.

and anal-fin lobes. The origin of the relatively short anal fin of banded rudderfishes lies under the halfway mark of the dorsal fin, farther back than in other amberjacks. Juvenile and subadults have 6 well-defined, dark body bars that extend onto the dorsal and anal fins, giving this species its vernacular name (scientists once thought these subadults were adults!). The real adults lose the banding and, on Florida's east coast, are called "amberines" (pronounced like "tangerines"). Adults reach 2 ft long and have a slightly more elongate head profile and longer jaw than greater and lesser amberjacks.

The small (1.5 ft) and rarely encountered lesser amberjacks are known to school with greater amberjacks and almaco jacks. Adult lesser amberjacks are purplish to pinkish, with an amber midbody stripe situated between the eye and tail. The dorsal- and anal-fin lobes are poorly developed, and there are almost none in the dorsal fin. Juveniles are distinctive, with 8–9 broken dark bars on the body. In contrast, juvenile greater amberjacks have 5 or 6 divided bars that do not extend into the median fins. Adults lack the lesser amberjacks' amber stripe and have well-developed fin lobes (but not as big as those of almaco jacks). They are the largest of the amberjacks, reaching a maximum size of 5 ft and weighing over 100 lb. Anglers familiar with greater amberjacks call them "reef donkeys," alluding to their strong and stubborn fight when hooked.

Pilotfishes look like overgrown (about 2 ft long) juvenile banded rudderfishes in color and shape, but they lack the amberjack eye stripe, and the caudal lobes have dark submarginal bands, with white tips. Rainbow runners are large (3.5 ft) and have an elongated, mackerel-like shape, with a pointed snout; blue upper body; and deeply forked, bright-yellow tail. They have a small mouth and a single, 2-rayed finlet directly behind the dorsal and anal fins.

Leatherjacks (12 in) have multiple, partially detached finlets behind the dorsal and anal fins. The first 5 dorsal spines are separate, and poison glands are associated with the dorsal and anal spines. If you catch one of these guys, be careful handling it or be prepared for bee sting–like pain.

The deep-bodied, laterally compressed, scuteless group includes some important gamefishes—Florida pompanos and permits—and several less–highly regarded but equally silvery species. The most distinctive of these are Atlantic moonfishes and lookdowns. Both species look as though they were cut from sheets of aluminum foil, and they have extremely long, flattened foreheads. The forehead of Atlantic moonfishes (15 in) has a slightly concave profile and is a bit shorter than that of lookdowns, which have a nearly straight profile. Lookdowns (12 in) at all sizes have longer extensions of the dorsal and anal fins than Atlantic moonfishes, but these fins

Lesser amberjack (*Seriola fasciata*) juvenile, Jupiter, Florida

Rainbow runner (*Elagatis bipinnulata*), Jupiter, Florida

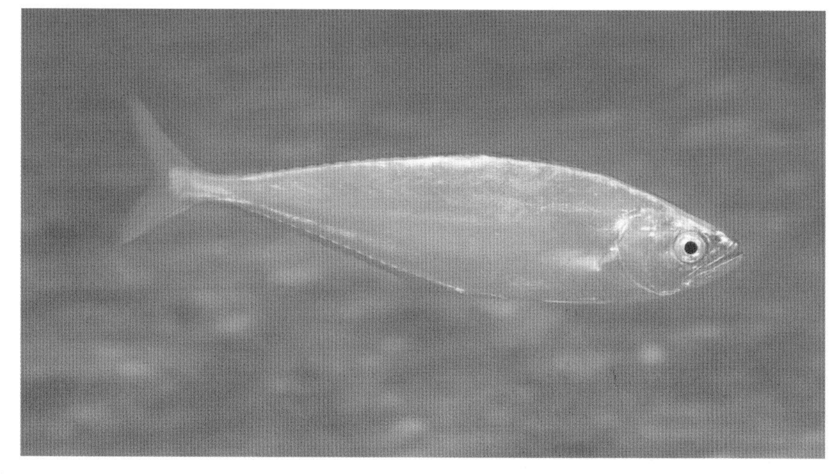

Leatherjack (*Oligoplites saurus*), Lake Worth Lagoon, Florida

Lookdown (*Selene vomer*), Lake Worth Lagoon, Florida

are enormously elongated in juveniles. Atlantic bumpers (12 in) have a symmetrically ovoid body; deeply forked yellow tail; and a diagnostic, prominent black spot on the upper caudal peduncle. Anglers on Florida's east coast call them "horn bellies," in reference to their prominent anal spines.

Florida pompanos, permits, and palometas have similar body shapes and head profiles. Their heads are small, with broadly rounded snouts; the first several dorsal spine are separate (no connecting membrane); the tail is deeply forked; and the dorsal and anal fins are elongated. In palometas those lobes are especially long and black in color. The leading edges of the caudal fin also are black, and the body has 4–5 thin black bars. Palometas grow to just under 2 ft long. Separating Florida pompanos (2 ft) and permits (2.5 ft) is more difficult, especially the large adults. Florida pompanos have a yellow to gold belly and anal fin; the tail is yellow tinged as well. Permits have only black on the median fins, and they often have a sooty midbody splotch.

All other jacks have lateral scutes on the caudal peduncle, often extending farther forward on the body. Scads in the genera *Decapterus*, *Selar*, and *Trachurus* all are small, cigar shaped, rounded in cross-section, and have a dark spot at the rear of the operculum, but they differ in their maximum sizes and coloration. Known throughout Florida as "goggle eyes," bigeye scads (12 in) differ from other scads in having very large eyes and a bit of a deeper and stouter body, and they often possess a moderately wide yellow stripe from the eye to the tail. The presence of prominent scutes along the entire lateral line separates the similarly shaped rough scads (16 in) from the others. Neither bigeye nor rough scads have the single finlet found behind the dorsal and anal fins of redtail, mackerel, and round scads. Redtail scads (16 in), called "firetails" by anglers who use them for live bait, are the easiest of the three to identify, with their steel-blue backs and reddish-orange tails. Round scads (known around the state as "cigar minnows") are distinguished by the series tiny black spots on the front part of the lateral line. They reach 9 in long, a bit shorter than mackerel scads (12 in), which are referred to as "speedos" by Florida's east-coast sport anglers. They have a silvery-blue stripe extending from the operculum to the tail.

The last group has well-developed scutes, deeper bodies, reach larger sizes (2–5 ft) and generally can be referred to as "true jacks." African pompanos are neither strictly African nor pompanos, but that is a discussion for another time. With its very steep forehead and extended dorsal filaments this species resembles a giant lookdown. Adults have a large eye and thick lips; they grow to over 3 ft long and weigh close to 50 lb. Anglers affectionately refer to African pompanos as "garbage-can lids." Young African pompanos are striking, with numerous, greatly extended dorsal and anal filaments. It has been suggested that these young are mimicking jellyfish.

Bluntnose jacks might, at first glance, be confused with Atlantic bumpers, inasmuch as they share the same ovoid shape, highly forked yellow tail, and are about the same size (18 in). Bluntnose jacks, however, lack the bumpers' unique black spot on the caudal peduncle and possess a sooty dorsal-fin margin; an upper caudal-fin lobe, with a sooty tip; and, of course, scutes.

While most true jacks are silvery or pale-yellow in coloration, two species normally are predominantly brown to black (some other species also get dark as large adults). Rarely seen cottonmouth jacks have 6–7 sooty black bars on their flanks that fade as they reach larger sizes, and the dorsal, anal, and caudal fins are dark. Their mouths (tongue and all) are white, hence the vernacular name. Horse-eye jacks are the second-largest (3 ft) of the speciose genus *Caranx*. They are dark-brown to black, with dark fins and a large, flat (actually, a bit concave) forehead-snout. Crevalle jacks (5 ft; >50 lb)—the largest member of the genus—and horse-eye jacks (2 ft) closely resemble one another in their blunt-headed profile, overall body shape, and color: the flanks are silvery, and the fins and tail are yellow. The two can be distinguished by eye diameter (larger in horse-eye jacks) and the position of black pigment: crevalle jacks have a conspicuous black spot on the pectoral fin and the upper gill cover; horse-eye jacks have a black tip on the dorsal fin.

The other three *Caranx* species (barjacks, yellow jacks, and blue runners) differ from their congeners in head and body shapes and in coloration. All are more fusiform (streamlined) and have more gradually sloping heads. Barjacks (2 ft), easily identified by the black stripe running along the upper back from near the origin of the

Top: Atlantic bumper (*Chloroscombrus chrysurus*), Jupiter, Florida. *Bottom*: Juvenile, associating with a jellyfish (*Chrysaora* sp.), Indian River Lagoon, Florida.

Florida pompano (*Trachinotus carolinus*), Juno, Florida

dorsal fin all the way to the edge of the lower caudal lobe, also have an electric-blue stripe that runs parallel to the black stripe but ends on the base of the tail. Yellow jacks appropriately have yellow fins and flanks, the latter often with wavy, oblique, steely-blue bars. Blue runners are blue-gray to blue-green dorsally; there is a small black spot on the operculum, and the tips of the caudal lobes are black. Juveniles less than about 5 in long have yellow tails. Early juveniles of all *Caranx* species are typically deeper-bodied and silvery, with dark or black bars or (in yellow jacks) greenish-brown mottling.

Permit (*Trachinotus falcatus*)

Palometa (*Trachinotus goodei*)

Bigeye scads (*Selar crumenophthalmus*), Palm Beach, Florida

Distribution and Habitat

Most jacks are distributed widely throughout oceanic, shelf, and coastal waters of Florida. Abundances of particular species vary, according to region and habitat. Some species are more prevalent in oceanic waters, making only occasional forays into coastal waters. These include greater amberjacks, almaco jacks, lesser amberjacks, banded rudderfishes, pilotfishes, rainbow runners, and cottonmouth jacks. Others—such as crevalle jacks, blue runners, Florida pompanos, Atlantic bumpers, lookdowns, Atlantic moonfishes, leatherjacks, and round scads—are most common in coastal waters. Crevalle jacks are ecologically flexible, ranging from as far as 100 mi upstream in freshwaters of the St. Johns River to the blue offshore waters of the continental-shelf edge. Coastal species favoring southeastern Florida waters include African pompanos, permits, horse-eye

Mackerel scad (*Decapterus macarellus*), Jupiter, Florida

Round scad (*Decapterus punctatus*), Jupiter, Florida

jacks, yellow jacks, bar jacks, redtail scads, and bigeye scads.

Florida pompanos are a migratory species, moving south along the east and west coasts in the fall and returning northward in the spring. As adults, permits are generally restricted to the southern half of the peninsula on both coasts, but juveniles are found by northern-peninsula beaches, alongside equally small Florida pompanos. Palometas occur primarily along the southeast coast and in the Florida Keys, and they are far less abundant than their bigger cousins. Scads are found in schools, often numbering in the hundreds or higher, usually around artificial or natural hardbottom structures in shelf waters. They are mainstays near sunken ships and artificial reefs and, in the absence of structures, are known to orient around large Atlantic goliath groupers. Bigeye scads are found in the Gulf but primarily occur offshore of the southeastern Florida coast. Round scads are the most common species and are found statewide, from the shoreline to about 300 ft. Mackerel scads are an east-coast–only species that occur in depths of about 30–200 ft. Rough scads inhabit deeper water (to about 300 ft) throughout the state and are replaced by redtail scads at even greater depths (500–650 ft) along the southeast coast.

Natural History

Spawning is incompletely known for most jack species, but it is safe to assume that all jacks are water-column spawners. The timing, location, and frequency of spawning vary among species. Evidence indicates that true jacks aggregate at traditional sites to spawn. Spawning may occur between pairs or among small groups of individuals that separate from larger aggregations. In some species, such as horse-eye jacks, blue runners, and barjacks, members of one sex (presumably the males) turn black during courtship. Greater amberjacks, almaco jacks, and banded rudderfishes spawn from March through May off southeastern Florida, usually over structured bottom, in 150–400 ft of water. Several of the amberjack species

African pompano (*Alectis ciliaris*), west Florida shelf

arrive seasonally to spawn offshore of the shelf edge, between Jupiter Inlet and Lake Worth Inlet. This spawning area is well known but not well documented in the scientific literature; it is very likely that spawning also occurs on the shelf in other areas of the state during this same time window.

The larvae and juveniles of all true jack species are found in open oceanic waters, often closely associated with objects such as flotsam, sargassum, and jellyfish. Blue runners, bar jacks, and yellow jacks are often numerically dominant species found under drifting rafts of sargassum. We have collected juvenile scads, amberjacks, true jacks, and rainbow runners around sargassum, boards, wooden shipping pallets, hawsers, ropes, lobster floats, buckets, fronds, trees, garbage, and other flotsam. Individuals of some species also associate with jellyfish, particularly Atlantic bumpers, which are often found with cannonball (*Stomolophius*), pink meanie (*Drymonema larsoni*), and moon (*Aurelia*) jellies. Several jack species associate with the Portuguese man-o'-war siphonophore (*Physalia physalia*). Once they leave the bobbing world of sargassum and flotsam, juveniles take up residence in coastal or shelf waters.

All jacks are active predators, and their body sizes and habitat choices influence their diets. Blue runners, barjacks, crevalle jacks, horse-eye jacks, and yellow jacks relentlessly herd schools of anchovies, mullets, pilchards, or silversides into shallow waters. Crevalles have mastered the art of smashing mullets against vertical seawalls. The blunt heads of foraging crevalles cutting across Florida's inland waters on calm mornings produce telltale surface wakes. All of these species break the surface while feeding, alerting seabirds to potential prey. Lookdowns and Atlantic moonfishes tend to be less boisterous in their feeding activities, consuming small fishes and crustaceans that drift in the water column; they are regulars at night in illuminated areas around dock and bridge lights, where such prey items concentrate.

Crevalle jack (*Caranx hippos*), Jupiter, Florida

Horse-eye jacks (*Caranx latus*), Jupiter Inlet, Florida

Juvenile jack (*Caranx* sp.), Lake Worth Lagoon, Florida

Scads feed on small planktonic organisms and accordingly form schools where currents transport their favorite bounty. Bigeye scads school during the day, usually around reefs or other structures, with individuals making occasional forays into the water column to feed on plankton. Adult round scads often school with Spanish sardines, and juveniles school with grunts; when doing so, they turn on a black lateral stripe that potentially helps them blend in with the juvenile grunts—a possible case of social mimicry.

A general characteristic of jacks is their propensity for associating with larger animals, particularly large sharks, turtles, and rays. One jack species that has carried this to the extreme is the pilotfish, a species that is rarely seen unless it is accompanying an oceanic whitetip shark. The relationship seems very tight between these two species. Others pairings are not so obligatory, but we have seen bar jacks, blue runners, yellow jacks, rainbow runners, and almaco jacks swimming closely with mantas, southern stingrays, yellow stingrays, reef sharks, tiger sharks,

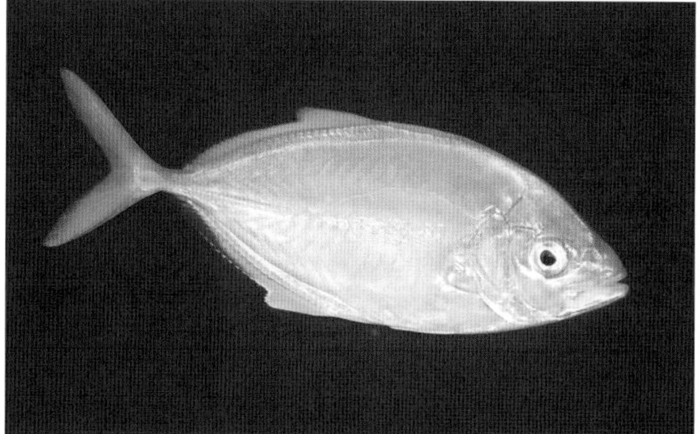

Left: Yellow jacks (*Caranx bartholomaei*), Palm Beach, Florida. *Right*: Juvenile, Lake Worth Lagoon, Florida.

Left: Bar jack (*Caranx ruber*), Jupiter, Florida. *Right*: Juveniles, Jupiter, Florida.

Blue runner (*Caranx crysos*), Jupiter, Florida

Juvenile round scads (*Decapterus punctatus*) exhibiting a black lateral stripe and schooling with tomtates (*Haemulon aurolineatum*), Jupiter, Florida

whale sharks, and bull sharks. On numerous occasions we have observed yellow jacks following spotted snake eels as they move over the reef; presumably these jacks benefit by grabbing any small fishes or shrimps spooked by the eels. Another variation on this theme involves scads, particularly round scads, which will often completely envelope adult Atlantic goliath groupers. Presumably this is more of a shelter-seeking move. Cleaning behavior may be a consequence of associating with larger fishes. At least two members of the family, leatherjacks and permits, engage in this behavior as juveniles. Both species have been observed cleaning external parasites from larger fishes, such as mullets and needlefishes.

Florida pompanos, permits, and palometas as a group, are found over sandy bottoms, sometimes associated with nearby structured habitats. Permit tend remain in one area, forming local aggregations, but they also migrate in schools. Florida pompanos migrate up and down the coasts of Florida, in response to water temperature and turbidity/visibility. They feed on snails, clams, worms, mole crabs (also known as sand fleas), calico crabs, and other swimming crabs. The young reside in the shallowest water, on coastal or inshore beaches. Here they feed on crustaceans, shrimps, and tiny clams, taking advantage of wave-generated water motion that periodically dislodges their prey from the sand.

Juvenile permits (*Trachinotus falcatus*) cleaning a white mullet (*Mugil curema*), Jupiter, Florida

Cobias
Family Rachycentridae

- **Cobia** (*Rachycentron canadum*)

Background

Cobias are the sole species in the family Rachycentridae. While cobia is the preferred common name, along the Gulf Coast, particularly in the Florida panhandle, they are often called "lings" or "lemonfishes." Off Texas they are known as "sergeantfishes," and off the Mid-Atlantic Bight they are referred to as "crab-eaters," an apt moniker. When swimming, the large pectoral fins are erect. This—coupled with the torpedo-shaped, dark-brown body; forked tail; high second dorsal fin; somewhat flattened head; small eyes; and broad, rounded snout—gives cobias a shark-like in appearance, an illusion not lost on divers observing one emerge from the bounds of visibility. The dorsum is dark; the belly is white; and there is a prominent to faded, dark midlateral stripe, edged in white. Individuals often display irregular patches of white along their flanks, head, and median fins. The first dorsal fin consists of 8 short, independent spines, a feature that separates cobias from similarly shaped and colored remoras in the genus *Echeneis* (Echeneidae), which have that fin modified into a sucking disc. Cobias rarely reach a length of 6 ft and a weight of 150 lb; they are normally seen in the under–50 lb size class.

Distribution and Habitat

The species is distributed worldwide; in the western Atlantic, cobias range from Massachusetts to Argentina. They occur in coastal and offshore waters throughout Florida and less commonly enter the lower, more saline reaches of estuaries.

Natural History

Reproduction occurs from April to September, with peaks in April–June in the Gulf of Mexico and mid-June–mid-August off Chesapeake Bay. Spawning probably occurs at night. Females spawn every 5–12 days. The eggs are pelagic and larvae grow quickly, reaching juvenile sizes of 4–4.75 in within 2 months and 2 ft in the first year. The lifespan of cobias extends to 9–11 years.

Crustaceans are a major part of their diet, along with fishes and, to a lesser extent, squids. We have found swimming crabs (Portunidae), searobins (*Prionotus* spp.), and flounders (Bothidae and Paralichthyidae), all bottom dwellers, in their stomach contents.

Cobia (*Rachycentron canadum*), Hobe Sound, Florida

Atlantic and Gulf of Mexico populations generally segregate, but tagging studies reveal that interchange between stocks does occur. Cobias are migratory and generally move in loose groups of less than 25. Florida east-coast cobias move southward in the winter into the Florida Keys, where they mix with the migratory, offshore Gulf of Mexico stock of the species. Individuals within the latter stock migrate northward and westward in the early spring, into offshore Gulf waters, and then return to the Keys in the fall. A second Gulf stock spends the summer inshore and overwinters a bit farther offshore. A similar two-stock situation—one more-or-less residential and the other highly migratory—may exist along the U.S. East Coast (e.g., cobias tagged in Florida have been recaptured off New Jersey).

Cobias frequent areas with structured bottom, such as artificial and natural reefs, but occasionally adults are encountered lying on sand near hardbottom. They exhibit the somewhat unusual behavior of closely associating with large marine animals, especially sharks and rays. In Florida waters we have observed cobias in the company of bull, tiger, and whale sharks; southern and roughtail stingrays; giant mantas; devil rays; and leatherback turtles (*Dermochelys coriacea*). They also have been accompanying Atlantic goliath groupers and cownose rays. The nature of this association does not appear to be obligate, but such near-contact behavior by cobias may represent an evolutionary precursor to the direct physical attachment that has evolved in the closely related remoras. One twist is that cobias often host remoras. We have seen cobias most frequently with large stingrays; the functional advantage of the association may be related to the bottom-hugging rays displacing meal-worthy demersal fishes and invertebrates.

Dolphinfishes
Family Coryphaenidae

- **Pompano dolphinfish** (*Coryphaena equiselis*)
- **Dolphinfish** (*Coryphaena hippurus*)

Background

In many respects, dolphinfishes are the ideal fish for offshore anglers (readily hooked, fight well, are beautifully colored, and are wonderful to eat). Dolphinfishes used to be referred to just as "dolphins" and have been marketed alternatively under the names "mahi-mahi" (the Hawaiian name) and "dorado" (the Hispanic name) to avoid confusion with the charismatic marine mammals.

Pompano dolphinfish (*Coryphaena equiselis*)

Dolphinfishes (*Coryphaena hippurus*), Jupiter, Florida

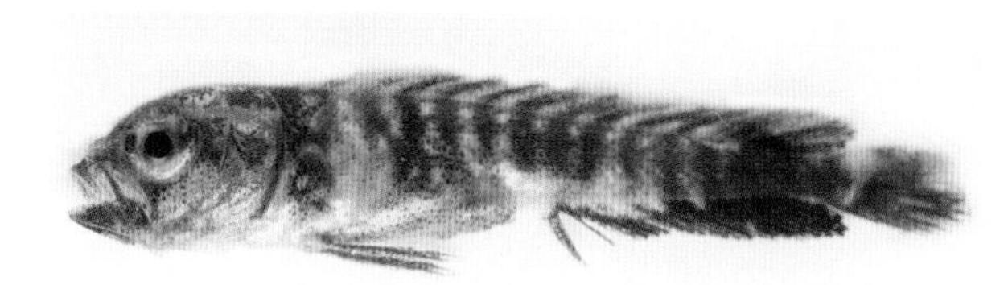

Juvenile dolphinfish (*Coryphaena hippurus*)

The dolphinfish species grows to a much greater size (6 ft) than the pompano dolphinfish (<2 ft), but at similar sizes the two species look very much alike. Mature male dolphinfishes (usually referred to as bulls) have steep, blunt heads. The smaller females (cows) have more-slender, rounded heads, as do both sexes of pompano dolphinfishes.

Distribution and Habitat

Dolphinfishes are found in oceanic and outer-shelf waters all around Florida and worldwide. They usually are found at and offshore of the shelf break, where water temperatures exceed 68°F. Dolphinfishes preferentially associate with sargassum patches and pelagic flotsam (trees and logs, shipping pallets and boards, hawser lines, and plastic buckets are dolphinfish magnets!).

Natural History

Few marine fishes grow as quickly and produce as many young as the dolphinfish. Both males and females attain sexual maturity during their first year of life. They spawn in oceanic waters multiple times during the year, but the peak time is spring and summer. Females produce 58,000–1.5 million eggs per spawning event. Eggs and larvae are pelagic, and juveniles associate with drifting sargassum. Members of this species grow very fast, up to 0.185 in per day and up to 5 ft in their first year of life. Individuals only live to a maximum of 4 years old, but most die before reaching age 2. A large adult can weigh 60–70 lb. Both dolphinfish species are gregarious, typically found in small schools. Dolphinfishes feed on a gamut of pelagic fishes, crustaceans, and squids, but they accidently ingest a variety of items (e.g., light bulbs, cigarettes, lighters, and other man-made debris) as they feed amongst flotsam. The fishes most commonly found in their diets are sargassum associates: triggerfishes, jacks, flyingfishes, pipefishes, puffers—and dolphinfishes. Dolphinfishes are preyed on by their own kind, as well as yellowfin tunas, blue marlins, white marlins, swordfishes, and wahoos.

Dolphinfishes prefer oceanic waters and are generally associated with the Loop Current, Florida Current, and Gulf Stream. In the western Atlantic, the dolphinfish species is believed to have two subpopulations that have discreet migratory circuits. Based on catch records, Florida is within the northern migratory route, where they are found near Puerto Rico during the winter (December–February) and move into Bahamian waters during March and April. They reach Florida during May and June, the Carolinas by June and July, move to Bermuda for July and August, and end up back at Puerto Rico by late fall and winter. There may also be a separate migratory pattern in the Gulf of Mexico.

Remoras
Family Echeneidae

- **Sharksucker** (*Echeneis naucrates*)
- **Whitefin sharksucker** (*Echeneis neucratoides*)
- **Slender suckerfish** (*Phtheirichthys lineatus*)
- **White suckerfish** (*Remora albescens*)
- **Whalesucker** (*Remora australis*)
- **Spearfish remora** (*Remora brachyptera*)
- **Marlinsucker** (*Remora osteochir*)
- **Remora** (*Remora remora*)

Background

Remoras are an odd lot that make their living as professional hitch-hikers. Using an oval-shaped sucker disc on top of their heads (a highly modified dorsal fin), remoras latch onto big fishes, small fishes, turtles, whales, and even an occasional inanimate object. Remoras form two natural groups. One group (consisting of the genera *Echeneis* and *Phtheirichthys*) is characterized by having long, slender bodies; pointed jaws, with the lower jaw protruding well past the upper; and narrow sucking discs. Shark-suckers, the largest (3 ft) and most often encountered of the remoras, and their look-alike, the somewhat smaller (2.5 ft) whitefin shark-suckers, have a white-edged, black lateral stripe that runs through the eye to the tail; this is best seen in younger individuals, as the stripe becomes increasingly obscure in adults. The dark dorsal, anal, and caudal fins have white-edged margins. The young of these two species greatly resemble young cobias of similar size. In subadults, the extent of the white on the tips of the tail is more pronounced in the whitefins, but unfortunately the white margins disappear in adults. Adult whitefins have a slightly chunkier body than shark-suckers and their base coloration is paler; that said, the best way to separate the two is to count the lamellae (transverse plates) of the sucking disc: usually 23 in the sharksucker and 21 in the whitefin. Slender suckerfishes (2.5 ft) are extremely thin; they have a much shorter head and a smaller sucking disc that does not extend past the pectoral-fin bases.

Members of the second group (genus *Remora*) have shorter, thicker bodies; rounded tails; rounded jaws; broad sucking discs; and lack white midbody stripes. White suckerfishes (12 in) are pale-gray to a bluish-tinged white, with a small disc that reaches just past the pectoral-fin origin. Whalesuck-ers (20 in) are brown, with narrow white edges on the dorsal and anal fins. They have a large head and a massive disc that extends well past

Sharksuckers (*Echeneis naucrates*), Florida Middle Grounds

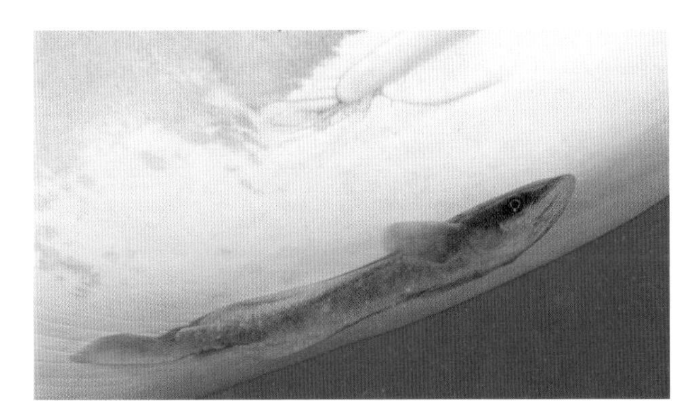

Remora (*Remora remora*), attached to a large tiger shark (*Galeocerdo cuvier*)

the tips of the pectoral fins and is about half the length of the whole fish. The dorsal and anal fins of spearfish remoras (15 in) are lighter in color than the brown to grey body; their short disc only reaches to the middle of the pectoral fins. The body and fins of marlinsuckers (15 in) are uniformly dark-gray, and the disc reaches well past the pectoral fins. Remoras are even darker (black), have a more robust body than spearfish remoras, and the disc extends posteriorly to near the end of the pectoral fins.

Distribution and Habitat

All members of this family are found in coastal and oceanic habitats throughout Florida's marine waters and, more specifically, wherever their primary hosts live. Adult sharksuckers, whitefin sharksuckers, and slender suckerfishes attach to a variety of hosts, including reef fishes, but they may be encountered as free livers. We have found sharksuckers on parrotfishes, spadefishes, trunkfishes, barracudas, Atlantic goliath groupers, African pompanos, cobias, sharks, and turtles. Whitefin sharksuckers are relatively uncommon, compared with sharksuckers, and probably attach to a similar range of hosts. Slender suckerfishes are even less common and are the one species frequently reported to attach to inanimate objects.

Species in the genus *Remora* are more selective in their choice of hosts, and all associate with oceanic pelagic species (mackerel sharks are a notable exception in the host category—remoras are unknown from white, porbeagle, and mako sharks). White suckerfishes specialize in rays, particularly mantas, and (to a lesser extent) sharks. Whalesuckers only latch onto marine mammals: whales, dolphins, and porpoises. Spearfish remoras mostly attach themselves inside the gill chambers of swordfishes and sharks, but they also ride on Atlantic sailfishes, blue marlins, white marlins, and spearfishes. Marlinsuckers specialize on Atlantic sailfishes and blue and white marlins. They also prefer the inside of the gill chamber and are difficult to detect unless you boat a billfish; generally following whatever struggle ensues in the boat, you will find a small gray fish stuck to the deck. Remoras almost exclusively attach to sharks but have been reported from turtles and billfishes.

Natural History

Little is known about reproduction in remoras. Presumably eggs are broadcast into the water column, and larvae develop in the pelagic realm. Other aspects—finding mates, courtship, and spawning behavior—are unknown.

Remoras consume plankton, invertebrates, and external parasites from their hosts. Remoras were long thought to be the beneficiaries of scraps emanating from feeding sessions by the host—usually sharks. But research indicates that they are more opportunistic, and feed when possible on a variety of invertebrates. Some species do eat external parasites, but rarely are these a major component of their diet. The sucker disc is a modified dorsal fin, moved drastically forward over the top of the head. The attachment mechanism is like venetian blinds: each of the lamellae have little hooks that facilitate suction.

The evolution of the sucker disc and hitchhiking behavior have been the subject of some research, as well as speculation. In one proposed evolutionary scenario, the remora lineage includes dolphinfishes and cobias. The physical resemblance between cobias and sharksuckers is striking and—when coupled with the cobias' propensity for associating with and following sharks, rays, and turtles and the dolphinfishes' love of hanging out under flotsam and sargassum—it looks like these two species have a logical genealogical linkage with the remoras' lifestyle. But the jury is still out.

Snappers
Family Lutjanidae

- **Black snapper** (*Apsilus dentatus*)
- **Queen snapper** (*Etelis oculatus*)
- **Mutton snapper** (*Lutjanus analis*)
- **Schoolmaster** (*Lutjanus apodus*)
- **Blackfin snapper** (*Lutjanus buccanella*)
- **Red snapper** (*Lutjanus campechanus*)
- **Cubera snapper** (*Lutjanus cyanopterus*)
- **Gray snapper** (*Lutjanus griseus*)
- **Dog snapper** (*Lutjanus jocu*)
- **Mahogany snapper** (*Lutjanus mahogoni*)
- **Lane snapper** (*Lutjanus synagris*)
- **Silk snapper** (*Lutjanus vivanus*)
- **Yellowtail snapper** (*Ocyurus chrysurus*)
- **Wenchman** (*Pristipomoides aquilonaris*)
- **Slender wenchman** (*Pristipomoides freemani*)
- **Vermillion snapper** (*Rhomboplites aurorubens*)

Background

Snappers are archetypal perch-like fishes, with well-developed teeth, spiny fins, and stout bodies. The various species—economically important components of the snapper-grouper commercial and recreational fisheries—can be distinguished by color, shape, and preferred microhabitat. The smallish (to 18 in) black snappers are uniformly dark-purplish to greenish-brown. They have long pectoral fins (the tips reach the origin of the anal fin) and a symmetrically ovoid body shape that differs from that of other snappers. On the other end of the spectrum are queen snappers, with a relatively slender body; pointed snout; low, straight-edged anal fin; and deeply lunate tail, with the upper lobe being almost filamentous and a bit longer than the lower one. Queen snappers reach 3 ft long and are pinkish-red (darker in color above the body midline), with a large yellow eye. The uniquely colored yellowtail snappers (2.5 ft; 10 lb) also look more streamlined (owing to the lunate tail). They have a broad, yellow midbody stripe that starts narrowly at the snout and gets progressively wider until it merges onto the yellow tail. The upper body has elongated yellow spots, and the dorsal fin is yellow. The anal fin has a rounded margin, a character that is shared by vermilion snappers (a.k.a. "beeliners," "mingos," or "California reds"). Vermilions (2 ft; 8 lb) are sleeker than most snappers, with a tail that is not quite lunate; the eyes, dorsal and caudal fins, and upper body are red, with the the latter transected by narrow, oblique, yellow-gold lines. Wenchmen and slender wenchmen are similar in size (9–12 in), shape, and color: pale-pink, narrow bodies, with a deeply forked tail and relatively large yellowish eyes. Slender wenchmen are (ahem!) a little more slender than wenchmen.

Queen snapper (*Etilus oculatus*). *Photo by George H. Burgess.*

Yellowtail snapper (*Ocyurus chrysurus*), Looe Key, Florida

Vermilion snappers (*Rhomboplites aurorubens*), Hobe Sound, Florida

The appearance of all of the above species differs from the most species-rich snapper genus, *Lutjanus*, which includes fishes that are most morphologically recognizable as snappers. Some ichthyologists have suggested that the dissimilar yellowtail snappers (genus *Ocyurus*) properly belong in the genus *Lutjanus* as well, in large part because they not infrequently hybridize with *Lutjanus* species. The most frequent cross appears to be with lane snappers, but we have seen hybrids with gray snappers as well. (Early naturalists recognized the yellowtail × lane hybrids as a distinct species and appropriately named it *Lutjanus ambiguus*.)

Lutjanus species can be separated by color pattern, but the differences between some species and life stages are subtle. Some species have a small dark spot on or near the lateral line, above the front of the anal fin. Lane and mutton snappers generally look alike, with pink to reddish bodies and tails, but the muttons' dark spot is much smaller than that of lane snappers; the anal fin is angular, with the longest rays in the middle of the fin, and the head profile is more rounded. Both often flash bars on the sides. Lane snappers have yellow horizontal stripes on the dorsal fin, head, and flanks; mutton snappers have bright-blue lines behind and under the eyes. Mutton snappers get much larger, reaching almost 3 ft and weighing about 30 lb, while lane snappers rarely exceed 2 ft and about 7 lb. These two are sometimes confused at small sizes (<4 in), but juvenile lanes have bright-yellow pelvic fins, while those of muttons are reddish-orange.

Three other species exhibit single lateral spots at times. Mahogany snappers (15 in; 4 lb) have the largest black spot (diameter greater than that of the eye), but the spot sometimes is faint or missing in adults. The dorsum is reddish-pink; the flanks are silvery, with faint, thin yellow stripes; and the eye is relatively larger than in other *Lutjanus* species. Juvenile mahoganies can be confused with young lane snappers, as both have bright-yellow pelvic fins, but the lateral spot on mahoganies is larger, darker, and more oblong. Juvenile red and silk snappers also show a spot below the soft dorsal fin, but this disappears with growth.

Red snappers are the kings of the group in reputation and value, hence any reddish snapper is sold and placed on a plate as such. The true, or genuine, red snapper (3 ft; 60 lb) is uniformly reddish-pink to brick-red, with red eyes and fins; the caudal fin bears a thin black margin. The anal fin is angular, a trait it shares with look-alike silk snappers. Silk snappers (commonly known as "yellow-eye snappers") are smaller in size (2.5 ft; 18 lb), and differ in having a yellow eye; a thin, dark-red caudal-fin margin; and faint, fine yellow lines on the flanks. Silk snappers might be confused with the similar-sized (2 ft; 12 lb) blackfin snappers (sometimes called "hambone snappers"), which also have yellow eyes and are uniformly red to red-orange. Blackfin snappers, however, have a distinctive black, comma-shaped spot on the base of the pectoral fin. The anal fin is rounded, and the caudal fin lacks a marginal band. Juveniles are pale-gray, with a bright-yellow tail and caudal peduncle.

Schoolmasters and grey, cubera, and dog snappers also have rounded anal fins. Adult schoolmasters (2 ft; 10 lb) are distinctive in having coppery-brown bodies, with 7–8 pale bars, or in being uniformly

Left: Lane snapper (*Lutjanus synagris*), Lake Worth Lagoon, Florida. *Right*: Juvenile, Lake Worth Lagoon, Florida.

Left: Mutton snapper (*Lutjanus analis*), Lake Worth Lagoon, Florida. *Right*: Juvenile, Lake Worth Lagoon, Florida.

Left: Mahogany snapper (*Lutjanus mahogoni*), Singer Island, Florida. *Right*: Juvenile, Jupiter Florida.

dull-yellow; the fins are golden-yellow, and the canine teeth on the upper jaw are especially long. Juveniles are colored much like adults and can be confused with small dog snappers, as both have blue lines under the eye. Gray snappers (about 2.5 ft; 18 lb)—known as "mangrove snappers" or "mangos" to most Floridians—have grayish-brown fins and base body coloration, with reddish-brown to brick-red centers on the midbody scales, forming indistinct stripes. They may display (especially at

Red snapper (*Lutjanus campechanus*)

night) 5–6 weak dark bars. A diagonal, reddish-brown bar bisects the eye and moves onto the snout; it is most obvious in juveniles and subadults and fades with growth. Juveniles look very much like miniature adults but have orange-red median and pelvic fins, and a thin blue line under the eye. Very small individuals are dark-brown (almost black), with clear median fins and tail. Possibly these newly settled individuals mimic bits of decaying plant material found in their favored habitat to avoid being eaten by visually oriented predators.

Cubera snappers somewhat resemble gray snappers, but the two can be separated at all but the smallest life stages (<1 in) by subtle patterns in shape and coloration. Young cubera snappers are coppery to dark-red, and there is a series of 9 dark bars along their flanks. These bars are not always visible on large adults, but the head profile slopes much less than that of gray snappers, and the body is relatively more slender. Size is a key separator, as cuberas are the largest of all snappers, growing to more than 5 ft long and weighing more than 125 lb. Dog snappers (3 ft; 30 lb) are dark–orangish-red dorsally, grading to lighter orange on the underside, with yellow-orange fins. Subadults and adults have a distinctive, pale triangular mark extending from under the eye to the upper jaw. There is a linear series of electric-blue dashes under the eye, extending onto the operculum.

Distribution and Habitat

The geographical distribution of snappers can be generalized into continental or insular, and shallow (<150 ft) or deep (>150 ft). All snappers are reef fishes, but many undergo developmental habitat shifts with age and growth. Red snappers are a distinctly continental species found around north-central Florida (but absent along the southeast coast, from Jupiter through the Keys), in water depths of 60–300 ft, the shallowest records being for young fishes in the Gulf. Insular mutton snappers have a somewhat anti–red snapper distribution; they are most abundant

Left: Blackfin snapper (*Lutjanus buccanella*), Jupiter, Florida. *Right*: Juvenile.

Left: Gray snapper (*Lutjanus griseus*), Crystal River, Florida. *Right*: Juvenile, Jupiter, Florida.

where red snappers are rare, in water depths from 1 to 200 ft. Gray and lane snappers are ubiquitous species found in shallow (<150 ft) habitats around the state. Both occur in seagrass beds, and gray snappers routinely ascend coastal rivers into freshwater reaches; they are a common sight in springs around the Homosassa and Crystal Rivers. Schoolmasters and dog and mahogany snappers are insular species found in abundance only in the Florida Keys and southeastern Florida, generally in depths of less than 150 ft. Cubera snappers are uncommonly found throughout Florida but reach peak abundance in the Florida Keys and southeastern Florida, generally occurring in depths of less than 200 ft; young are found near mangroves.

Vermilion and yellowtail snappers also have somewhat complementary distributions. Vermilions are continental fishes, found in water depths from 80 to 500 ft; like red snappers, the young are found in the shallow end of the depth range. Yellowtail snappers, on the other hand, are generally restricted to shallow waters (10–150 ft deep) in southeast Florida and the Florida Keys and are not common in the Gulf, only occasionally occurring near the shelf edge. Young yellowtails reside in seagrass meadows, patch reefs, and nearshore hardbottom in coastal waters. We have seen young-of-the-year yellowtail snappers in shallow water off the southwest coast of Florida, but they do not overwinter there. Blackfin and silk snappers and both species of wenchmen are deepwater snappers that occur around the state, in water depths between 100 and 1000 ft. Black and queen snappers also live in deep water (>400 ft) but are rare north of the Florida Keys.

Natural History

Depending on the species, snappers spawn in the water column, either over protracted periods or within narrow windows. For some species, when the time is right (spring and summer months for most snappers), individuals migrate from far

Cubera snapper (*Lutjanus cyanopterus*), Jupiter, Florida

Dog snapper (*Lutjanus jocu*)

Schoolmaster (*Lutjanus apodus*), Looe Key, Florida

and wide to converge at historically used aggregation sites on a segment of the reef. These sites are typically near the shelf edge and may be used by multiple species, including groupers and jacks. Dog, mutton, lane, and cubera snappers have been reported to aggregate in the Caribbean and the Bahamas, at reef sites used by multiple generations of fishes. Verifying snapper spawning aggregations off Florida has been elusive, with the exception of the Florida Keys and Dry Tortugas. One such area is Riley's Hump, a hardbottom feature west of the Dry Tortugas where mutton snappers aggregate to spawn in June, July, and August. Cubera snappers may also use this site, as do black groupers and scamps. The fundamental disparities between insular and continental environments possibly have resulted in the development of differing spawning patterns between continental populations of species known to form spawning aggregations in insular areas. In addition, decades of heavy fishing pressure along the narrow continental shelf of southeast Florida and the Florida Keys may have contributed to a decline in historical aggregations. In species that have been studied, spawning corresponds very well with the moon phase, enabling accurate predictions of its timing.

The planktonic duration of eggs and larvae ranges from 20 to 30 days for species that have been studied. A reasonable portion of the spawning products from aggregations in the Florida Keys are carried by the Gulf Stream into southeast and eastcentral nursery habitats. Several snapper species undergo a developmental migration similar to that described for grunts: planktonic young settle into habitats shallower than are typically used by the adults, and as these young grow, they move from shallow to deep water. The choice of nursery area varies with a species' preference for combinations of habitat structure, cross-shelf position (inshore, coastal, and shelf), and water depth. Gray, mutton, cubera, and dog snappers generally settle only in shallow (<15 ft) lagoon, bay, or estuarine waters, while others, such as schoolmasters and lane, mahogany, and yellowtail snappers, are more flexible and settle either there or in nearby coastal waters of similar depths. In contrast, vermilion, blackfin, red, silk, queen, and black snappers favor settling in deeper (>15 ft) offshore waters.

The newly settled young of most snapper species reside in structured habitats, such as seagrass meadows, mangrove prop roots, nearshore hardbottom, and oyster reefs. As individuals grow, they often move to deeper waters, occupying different habitats as they develop. A general pattern is that the young do not grow up side by side with the adults, the latter usually reside some distance away, on hardbottom, in deeper waters. Species that tend follow this life-history pattern are schoolmasters and gray, lane, mutton, dog, blackfin, and cubera snappers. Cross-shelf patterns represent more of a mosaic of habitats along a depth gradient in the Florida Keys. Shelf-edge reefs, fringing reefs, patch reefs, seagrass meadows, limestone-bottom habitats, and mangrove shorelines form a continuum. On the coasts of peninsular Florida, juvenile habitat tends to be in lagoons, estuarine environments, or nearshore reefs along the beach. Inlets and passes represent portals to the vast nursery areas of the Indian River Lagoon, Charlotte Harbor, Florida Bay, Tampa Bay, and St. Andrews Bay. The ubiquitous red snappers provide an exception to this pattern; their young settle on the inner continental shelf, on softbottom with shell-hash or other low-relief substrates. Vermilion, blackfin, silk, and queen snappers also disdain inshore nursery habitats, but their young are occasionally found in shallower water than the adults, usually where deep water is nearby (e.g., off southeastern Florida and the Keys).

Snappers have well-developed teeth that serve them well as top-tier predators. Schoolmasters and gray, dog, and cubera snappers prefer fishes, shrimps, and crabs. Mutton snappers forage over sandy bottoms and, in addition to fishes, feeds on worms, clams, snails, crabs, and other creatures that live in the sand. Yellowtail and vermilion snappers are epibenthic (above but still near to the bottom) feeders, harvesting the bounty carried by currents. Both species eat larger plankton, including small fishes. Silk, blackfin, red, queen, and black snappers consume fishes and invertebrates available in deeper waters. Although we have never observed a juvenile in Florida waters, predatory young black snappers bears a striking resemblance to planktivorous blue chromises, a case of mimicry that may help the the former approach unsuspecting prey.

Tripletails
Family Lobotidae

- **Atlantic tripletail** (*Lobotes surinamensis*)

Background

Atlantic tripletails are the sole representative of the family Lobotidae in the western Atlantic. Adult Atlantic tripletails are shaped a bit like a freshwater sunfish, with a deep body; spiny dorsal and anal fins; and broad soft dorsal and anal fins that contribute to the common name of this species. The head steeply slopes to a projecting mouth; the eyes are relatively small. Adults, which grow to 3.5 ft and weigh up to 40 lb, are dark-brown to olive-green, with splotches and flecks of darker pigment. A dark line projects from the posterior eye margin downward to the edge of the opercle. Juveniles (between 3 and 8 in) are bright-yellow to yellow-orange, with or without black flecks of pigment. In all individuals, regardless of background color, there are 2 dark lines radiating posteriorly from the eye, at right angles to one another.

Distribution and Habitat

Atlantic tripletails are distributed worldwide, in tropical and subtropical waters. In Florida they are known from pelagic (associated with floating sargassum and flotsam), coastal, and inshore environments.

Natural History

Atlantic tripletail spawning has not been observed, but females are ripe from June to August. Spawning probably occurs in deep offshore waters, as larvae and early juveniles are found primarily in offshore waters. Small individuals (1–3 in) reside within patches of drifting sargassum. Juveniles make their way to inshore waters, where they are found around mangrove shorelines. It is here that yellow-phase individuals hide in plain sight by imitating a floating leaf. Yellowing mangrove leaves, so prevalent in their preferred habitat, are the model for this mimicry. We have seen Atlantic tripletails up to twice the size of a floating mangrove leaf lying on their sides at the water surface. This ruse facilitates access to unsuspecting prey and probably also fools birds or larger predatory fishes seeking a meal.

Atlantic tripletails grow relatively fast, apparently reaching 18 lb in less than 3 years. Adults habitually associate with buoys, pilings, and flotsam. We once retrieved a discarded wooden fish box floating in a weed line, only to find a full-sized Atlantic tripletail nestled inside. It is not unusual to find adult Atlantic tripletails associated with drifting objects well offshore (ca. 25 mi) of the east coast. From Flamingo (at the southern end of the Everglades) northward along the Gulf coast, fishers seek Atlantic tripletails around the floats for stone crab and lobster traps. On the east coast, Atlantic tripletails appear to migrate from South Carolina and Georgia to the area between Cape Canaveral and Ft. Pierce. During the cold winter of 2010, there was an unusual influx of adult Atlantic tripletails into the Loxahatchee River, presumably individuals seeking relief from the colder temperatures of the northern and central Indian River Lagoon.

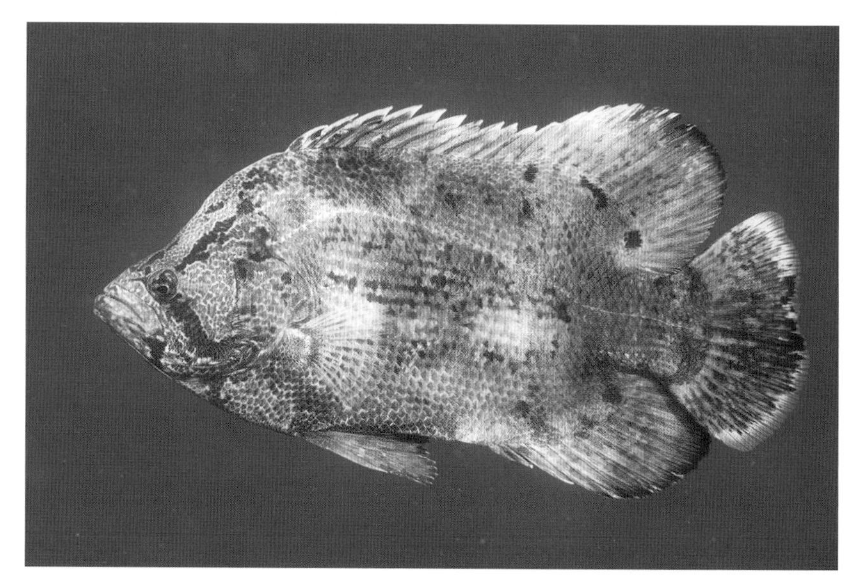

Atlantic tripletail (*Lobotes surinamensis*), Palm Beach, Florida

Juvenile Atlantic tripletail (*Lobotes surinamensis*), Loxahatchee River, Florida

Mojarras
Family Gerreidae

- **Irish pompano** (*Diapterus auratus*)
- **Rhombic mojarra** (*Diapterus rhombeus*)
- **Spotfin mojarra** (*Eucinostomus argenteus*)
- **Silver jenny** (*Eucinostomus gula*)
- **Tidewater mojarra** (*Eucinostomus harengulus*)
- **Bigeye mojarra** (*Eucinostomus havana*)
- **Slender mojarra** (*Eucinostomus jonesii*)
- **Mottled mojarra** (*Eucinostomus lefroyi*)
- **Flagfin mojarra** (*Eucinostomus melanopterus*)
- **Striped mojarra** (*Eugerres plumieri*)
- **Yellowfin mojarra** (*Gerres cinereus*)

Background

Mojarras are abundant shallow-water, bottom-dwelling fishes that occupy a wide variety of habitats. With a few exceptions, identification is difficult, as most simply appear as small (maximum length 6–9 in), slim, silvery fishes with deeply forked tails. Four species stand out from the rest in being larger (14–16 in); possessing long, robust dorsal- and anal-fin spines; and having deeper bodies. Irish pompanos and rhombic mojarras are silver laterally, with yellow pelvic fins, but they can be separated by counting the large anal-fin spines: Irish pompanos have 3 and rhombic mojarras have 2. Striped mojarras are shaped similarly to these two species, but the upper sides and fins of adults are yellowish, with a series of thin brown stripes on their flanks. Young striped mojarras do not have stripes and are easily confused with young Irish pompanos. Yellowfin mojarras are not quite as deep bodied, have 8 dusky bars on their sides, and only the pelvic fins are yellow.

The seven species of *Eucinostomus* are quite similar in appearance, with very subtle differences among the species. The two species that are easiest to identify have distinctive black blotches on their dorsal fins. The black fin margin on flagfin mojarras is wide, and there is a thinner white band just below it, while the dorsal fin of bigeye mojarras has a black triangular wedge and lacks white bordering. Bigeye mojarras also have dusky, oblique bars on the flanks that are missing on the all-silver flagfins. These two species—plus tidewater, slender, and mottled mojarras—are the most-slender members of the family, separating themselves from both the four deeper-body species and two species having intermediate body depths, spotfin mojarras and silver jennies. Silver jennies generally have a slightly

Striped mojarra (*Eugerres plumieri*), Loxahatchee River, Florida

Irish pompano (*Diapterus auratus*), Loxahatchee River, Florida

Yellowfin mojarra (*Gerres cinereus*), Lake Worth Lagoon, Florida

deeper body than spotfin mojarras, and adults do not have the oblique dusky bars found on the latter. Both have a dusky to black margin on the dorsal fin, a character shared by tidewater, slender, and mottled mojarras. The second and third oblique dusky bars of tidewater mojarras have an opposing short bar, resulting in two Y-shaped marks laterally, under the dorsal fins. Slender and mottled mojarras are almost identical in having faint, oblique body bars, but slender mojarras have a V-shaped mark on the snout, and mottled mojarras have a square spot of black pigment in the eye, above the iris. To be honest, juvenile mojarras all look alike, and even trained professionals pull out their hair trying to identify them.

Distribution and Habitat

The distribution of mojarras in Florida follows one of two basic patterns: either found off both coasts or confined to the east coast. The species found along both coasts include spotfin mojarras, silver jennies, tidewater mojarras, and striped mojarras. Those occurring only along the east coast are Irish pompanos and rhombic, bigeye, slender, mottled, flagfin, and yellowfin mojarras. Rhombic mojarras are a rarely observed, tropical peripheral species reported only from the Indian River Lagoon. All mojarra species are relegated to shallow water, and some (e.g., striped mojarras, Irish pompanos, and tidewater mojarras) regularly ascend coastal rivers into freshwater. Spotfin mojarras and silver jennies are regular estuarine residents, while slender and mottled mojarras are encountered more often in higher salinities. Mojarras occupy a variety of habitats, numerically dominating seagrass beds and mangrove areas, as well as on sandy beaches, in estuaries, and on low-relief, mixed-substrate bottoms containing gorgonians and sponges. They are absent on reefs.

Left: Flagfin mojarra (*Eucinostomus melanopterus*), Lake Worth Lagoon, Florida. *Right*: Bigeye mojarra (*Eucinostomus havana*), Lake Worth Lagoon, Florida.

Left: Tidewater mojarra (*Eucinostomus harengulus*), Loxahatchee River, Florida. *Right*: Mottled mojarra (*Eucinostomus lefroyi*), Jupiter, Florida.

Left: Silver jenny (*Eucinostomus gula*), Lake Worth Lagoon, Florida. *Right*: Spotfin mojarra (*Eucinostomus argenteus*), Crystal River, Florida.

Natural History

Mojarras are water-column spawners, but details about this activity are not well known. Irish pompanos and striped mojarras migrate downstream from brackish or freshwater reaches of coastal rivers to spawn near or in the ocean. Most species have prolonged spawning seasons or spawn multiple times during a year; hence early juveniles are found year round in Florida.

Mojarras are daylight predators that feed on sediment-dwelling organisms, their greatly protrusible mouths offering a superb adaptation for picking invertebrates out of the bottom sediment. Popular food items include copepods, ostracods, amphipods, isopods, mysids, crustaceans, bivalves, nematodes, and (especially) polychaete worms. The well-developed swim bladder apparently has no sound-producing function, but its connection with the auditory capsule presumably helps mojarras locate buried food items by serving as an amplifier. Mojarras are often found in loose aggregations and are among the most abundant inshore fishes on both coasts of Florida. They are sought out by a veritable who's who of predators, including fishes, wading and diving birds, porpoises, and aquatic snakes. Their high abundance suggests that they play an important role in nearshore food webs.

Grunts
Family Haemulidae

- **Black margate** (*Anisotremus surinamensis*)
- **Porkfish** (*Anisotremus virginicus*)
- **Barred grunt** (*Conodon nobilis*)
- **Bonnetmouth** (*Emmelichthyops atlanticus*)
- **Margate** (*Haemulon album*)
- **Tomtate** (*Haemulon aurolineatum*)
- **Caesar grunt** (*Haemulon carbonarium*)
- **Smallmouth grunt** (*Haemulon chrysargyreum*)
- **French grunt** (*Haemulon flavolineatum*)
- **Spanish grunt** (*Haemulon macrostomum*)
- **Cottonwick** (*Haemulon melanurum*)
- **Sailors choice** (*Haemulon parra*)
- **White grunt** (*Haemulon plumierii*)
- **Bluestriped grunt** (*Haemulon sciurus*)
- **Striped grunt** (*Haemulon striatum*)
- **Boga** (*Haemulon vittata*)
- **Pigfish** (*Orthopristis chrysoptera*)
- **Burro grunt** (*Pomadasys crocro*)

Background
Grunts are similar to snappers but lack the latter's prominent teeth. Grunts get their name from their ability to produce grunting sounds by grinding their pharyngeal teeth (located in the throat). The different species can be readily separated by body shape and color pattern. Porkfishes and black margates are easily distinguished from all others by their deep bodies, sharply sloping heads, and distinctive color patterns. Adult porkfishes have yellow and blue stripes on their sides and 2 distinguishing, vertical black bars on the head: the first running diagonally from the top of the head through the eye to the jaw, and the second vertically from the dorsal-fin origin

Top left: Black margate (*Anisotremus surinamensis*), Jupiter, Florida. *Top right*: Juvenile black margate, Palm Beach, Florida. *Bottom left*: Porkfish (*Anisotremus virginicus*), Marquesas Keys, Florida. *Bottom right*: Juvenile porkfish, Lake Worth Lagoon, Florida.

Burro grunt (*Pomadasys crocro*), Loxahatchee River, Florida

straight down. Juvenile porkfishes have golden- to bright-yellow heads and silvery-white bodies, with 2 black stripes: 1 above and parallel with the lateral line, and 1 on top of the lateral line (midlateral stripe). Adult black margates have a single wide bar that runs diagonally behind the gill and above the pelvic fin. Each scale on the flanks has a black spot, and the fins and tail are black. Juveniles have 2 thick, black longitudinal stripes and a squarish caudal spot.

Burro grunts are uniformly golden-brown, with dusky pigment in the fins and tail and no stripes or obvious markings. Juveniles and adults have a similar coloration. Burro grunts do have a conspicuous morphological feature that distinguishes them from the others—an enlarged anal spine.

The *Haemulon* species are colorful in a subdued way; each one characterized by a particular striped pattern that is the primary way to separate the species visually. Early *Haemulon* juveniles (2 to 5 in) also exhibit species-specific patterns of lateral stripes (the lengths of the midlateral stripe and the upper eye stripe emerging from the upper posterior margin of the eye are key to separating early juveniles) and caudal spots that aid in their identification. At first glance, some of the species look very similar. For example, white grunts and bluestriped grunts both

Top left: White grunt (*Haemulon plumierii*), Lake Worth Lagoon, Florida. *Bottom left*: Juvenile white grunt, Jupiter, Florida. *Top right*: Bluestriped grunt (*Haemulon sciurus*), Lake Worth Lagoon, Florida. *Bottom right*: Juvenile bluestripe grunt, Jupiter, Florida.

Left: Smallmouth grunt (*Haemulon chrysargyreum*), Lake Worth Lagoon, Florida. *Right*: Juvenile.

have blue stripes, but white grunts only have blue stripes on the head; bluestriped grunts have blue stripes running the length of the body, and the tail is dark-brown. Juvenile white grunts are golden-tan to yellowish; the upper eye stripe is long (extending past the gill cover) and curves downward to the midlateral stripe. Bluestripe juveniles are brown to yellowish; the upper eyestripe is long but does not curve downward, instead paralleling the midlateral stripe.

French, smallmouth, and striped grunts all have yellow stripes against a white background; the stripes on French grunts are diagonal, below the midline, while smallmouth and striped grunts have straight longitudinal stripes. The latter species two can be separated by body shape: striped grunts are more slender and have a more deeply forked tail than smallmouth grunts. Juvenile smallmouth grunts are pale, with a long upper eye stripe that parallels the midlateral stripe, which is connected with the caudal spot. French and striped grunts both have short upper eye stripes (not extending past the gill cover), but French grunt juveniles,

Top: Striped grunt (*Haemulon striatum*), Singer Island. *Bottom*: Juvenile, Lake Worth Lagoon.

like adults, have diagonal scale rows beneath the midlateral stripe, and striped grunts do not. Juvenile striped grunts have only a faint caudal spot.

Other species are more easily recognized. Cottonwicks have black and yellow-gold stripes, with 1 thick black stripe running along the lateral line from the tail forward through the eye, and another black stripe running along the back and into the lower lobe of the tail. Juvenile cottonwicks have long upper eyes stripes that curve downward to the midlateral stripe,

Left: French grunt (*Haemulon flavolineatum*), Ft. Lauderdale, Florida. *Right*: Juvenile, Lake Worth Lagoon, Florida.

Left: Cottonwick (*Haemulon melanurum*), Lake Worth Lagoon, Florida. *Right*: Juvenile, Juno, Florida.

Left: Caesar grunt (*Haemulon carbonarium*), Pecks Lake, Florida. *Right*: Juvenile, Singer Island, Florida.

which is continuous with the caudal spot. Caesar grunts have a coppery-colored background, with yellow stripes, and black pigment on the soft dorsal, soft anal, and tail fins. Juveniles have a long upper eye stripe that parallels the mid-lateral stripe, rather than curving downward; the caudal spot is large and rectangular. Margates—the largest of the *Haemulon* group found in Florida—reach 2.4 ft and weigh 8 lb. They generally lack conspicuous stripes and are silvery-gray, with a black back, fins, and

Left: Margate (*Haemulon album*), Lake Worth Lagoon, Florida. *Right*: Juvenile, Lake Worth Lagoon, Florida.

Left: Spanish grunt (*Haemulon macrostomum*), Singer Island, Florida. *Right*: Juvenile, Singer Island, Florida.

tail. Anglers sometimes call them "silver snappers." Juvenile margates have a long, parallel upper eye stripe, and the midlateral stripe connects with a heavily pigmented caudal area.

Spanish grunts are distinctive, with bold black stripes; a black underside against a silvery background, with pale-yellow along the back through the soft dorsal fin; and a bright-yellow smudge at the posterior base of the soft dorsal fin. Spanish grunt juveniles have a long, downward-curving upper eye stripe and a large, wedge-shaped caudal spot. Tomtates have a coppery-brown midlateral stripe and thin, yellow, lateral stripes over a white background, with a large dark spot on the caudal peduncle. Tomtate juveniles are whitish-tan, with a short upper eye stripe and a small, distinct caudal spot. The scale rows beneath the midlateral stripe are parallel. Sailors choice have small black dots on each scale that form diagonal stripes along the flank, above the lateral line. Juveniles have a whitish-silver body, with a long, black, downward-curving upper eye stripe and a distinct caudal spot.

Bonnetmouths and bogas can be readily recognized from the other *Haemulon* species by their slim bodies and elongated snouts. Bogas are blue dorsally, with a white belly and thin, dark stripes. Bonnetmouths are thin and silver, with a light-brown back and faint dark stripes. These two species were only recently placed in the grunt family, based on an analysis of DNA sequences.

Pigfishes are shaped similarly to the *Haemulon* species, but their bodies are silvery to blue-gray and covered with broken golden stripes that extend from head to tail. The stripes tend to be more continuous and straight on the caudal peduncle. Juveniles have a series of stripes against a silvery background, as do *Haemulon*, but the mid-dorsal stripe is much wider than the stripes in juvenile

Left: Tomtate (*Haemulon aurolineatum*), Palm Beach, Florida. *Right*: Juvenile, Jupiter, Florida.

Haemulon. Juvenile pigfishes differ from *Haemulon* juveniles in having 2 thick, brownish lateral stripes: 1 along the dorsal midline, and 1 on the lateral line. These stripes are much thicker than the black stripes found on the early juveniles in the *Haemulon* species. Barred grunts are more slender, with a smaller head than the *Haemulon* species, and they have a series of dark, broad bars.

Top: Sailors choice (*Haemulon parra*), Lake Worth Lagoon, Florida. *Bottom*: Juvenile, Jupiter, Florida.

Distribution and Habitat

Although some grunts occur throughout Florida's coastal and shelf waters (to a maximum of 150 ft deep), most species are concentrated in southeast Florida and the Florida Keys, in water depths less than 100 ft. Grunts are tropical species that use habitats ranging from seagrasses and mangroves to reefs and wrecks during their life history. Black margates occur throughout the Keys and along the east coast to northeast Florida, but not off Florida's west coast. Porkfishes reside mostly in waters in the southeastern portion of the state and will occasionally appear on the west Florida shelf. Pigfishes, white grunts, and tomtates are distributed widely around the state, from the shelf to inshore waters. Sailors choice, caesar grunts, cottonwicks, French grunts, Spanish grunts, bluestriped grunts, smallmouth grunts, striped grunts, margates, bonnetmouths, and bogas are restricted to southeastern Florida, eastcentral Florida, and the Florida Keys. Burro grunts are a tropical peripheral species that normally live in the brackish waters of Central America and the Greater Antilles and are generally limited to coastal rivers in Florida, such as those draining into the Indian River Lagoon. Barred grunts are another tropical peripheral species rarely seen in Florida waters.

Natural History

Grunts spawn in the water column, in coastal and shelf habitats. Some may aggregate at specific locations to spawn, but this not been well documented in Florida, and reproduction is poorly known in most species. Their larvae are planktonic for only about 2 weeks. The larvae of most species settle in shallow (<15 ft) nearshore waters throughout the year, but their numbers peak during summer and fall. Species of *Haemulon* and *Anisotremus* exemplify the phenomenon of ontogenetic (developmental) migration: settling in shallow water and, as they grow, moving to deeper waters, utilizing a range of habitats along the way. Patterns vary in species-specific fashion, with some species being flexible and others somewhat rigid in how they use different habitat/depth combinations. Some species may be exclusively found in particular habitats, and they may also show a preference for inshore (inside barrier islands) versus coastal habitats (outside barrier islands). For example, black margates appear to settle only on nearshore hardbottom habitats (<15 ft) in coastal waters, rarely (if ever) inside barrier islands. Smallmouth, Spanish, and caesar grunts follow a similar pattern but will use habitats inside barrier islands, where salinity is high (>25 psu). Young bluestriped grunts prefer to settle in inshore seagrass meadows (<15 ft deep). Others—such as sailors choice, white grunts, and French grunts—are more flexible,

Top: Boga (*Haemulon vittata*). *Bottom*: Bonnetmouth (*Emmelichthyops atlanticus*), Juno, Florida.

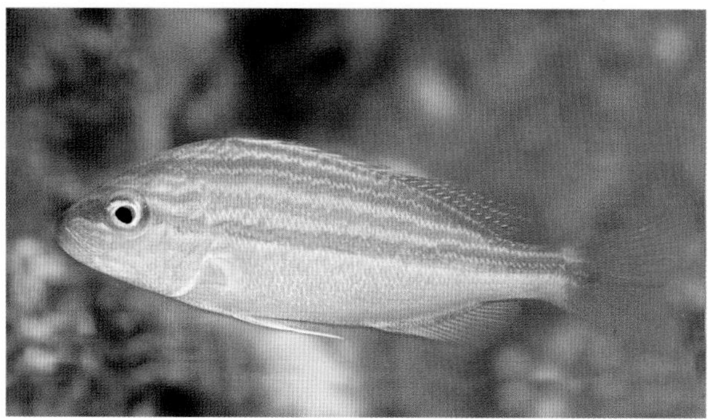

Top: Pigfish (*Orthopristis chrysoptera*), Lake Worth Lagoon, Florida. *Bottom*: Juvenile, Lake Worth Lagoon, Florida.

Juvenile porkfish (*Anisotremus virginicus*) removing external parasites from a sheepshead (*Archosargus probatocephalus*), Lake Worth Lagoon, Florida

French grunts (*Haemulon flavolineatum*) displaying, Ft. Lauderdale, Florida

settling on structural habitat in either inshore or coastal waters, including on seagrasses, sponges and other invertebrates, hardbottom, or artificial structures. Porkfishes and tomtates are very flexible, settling on almost any structured habitat, across a range of water depths, in inshore, coastal, or shelf waters. At the opposite extreme are bogas, bonnetmouths, cottonwicks, and striped grunts, which preferentially settle on habitat in deeper (>15 ft, and usually much deeper) shelf waters.

Adult grunts tend to school and can represent most of the fish biomass in a particular habitat. When swimming on a shallow nearshore reef in southeast Florida, you will notice that grunt schools are often composed of several different species (e.g., French, white, bluestriped, and caesar grunts). These groups hover lazily above their home structure by day and then disband at dusk, spreading out over adjacent sand flats to feed on sand-dwelling invertebrates that emerge at night. Because of this feeding habit, grunts are important distributors of organic material on reefs, sand flats, and seagrass meadows (e.g., feeding off of a reef and then discharging the nutrient-laden wastes onto the reef).

Other species mostly consume plankton during daylight hours. Tomtates are one of the most abundant, reef-associated species in Florida waters, often forming huge schools around natural and artificial structures. Tomtates feed primarily in the water column, but we have watched individuals break off from the larger school to forage briefly over the sand and then rejoin the school. Striped grunts mix with bonnetmouths, tomtates, and (occasionally) round scads into schools that expand above the reef, feeding on plankton during daylight hours. Behavior and schooling patterns in these groups vary, depending on the speed and direction of local currents. Mixed-species schooling is undoubtedly an adaptation that lessens the risk of being eaten. Predation is an ever-present risk—we have watched Atlantic trumpetfishes, great barracudas, gags, and black and tiger groupers quickly but nonchalantly pluck individual grunts from a large group. As much as grunts contribute to the cycling of nutrients and materials among reefs and seagrass meadows, they also serve as an important resource for many fish-eating predators. Thus grunts, by their numbers and actions, are

Sheepsheads (*Archosargus probatocephalus*), Lake Worth Lagoon, Florida

Top: Silver porgy (*Diplodus argenteus*), Jupiter, Florida. *Bottom*: Spottail pinfish (*Diplodus holbrookii*), Venice, Florida.

ecologically important to shallow-water food webs.

Pigfishes are found in estuarine, coastal, and shelf waters, usually over softbottom but, more often than not, adjacent to structured habitat. Pigfishes feed on invertebrates during the day, grabbing mouthfuls of sand and filtering any prey organisms out through the gills. In the Indian River Lagoon. pigfishes are used as bait for spotted seatrouts and snooks. Presumably their grunting sounds attract the desired quarry. Some porkfish juveniles pick external parasites from larger fishes, including sheepsheads, great barracudas, doctorfishes, snooks, and yellowfin mojarras. Our observations suggest that not all individuals engage in cleaning behavior. Cleaners may be solitary or gather in groups of 3–5 and will maintain a cleaning station that is revisited by clients.

The ecology and habits of burro grunts are not well known in Florida, but they are a common demersal and riverine species in Mexico. We have collected this species in rivers, lagoons, and canals along the southeast coast of Florida.

Grunts generally seem indifferent, but on closer inspection you might see two individuals of the same species facing one another with their mouths agape, a phenomenon known as kissing. The biological significance of this behavior—a favorite of underwater photographers—is unknown, but it probably involves male-male aggression; therefore kissing may not be an accurate descriptor.

Porgies
Family Sparidae

- **Sheepshead** (*Archosargus probatocephalus*)
- **Sea bream** (*Archosargus rhomboidalis*)
- **Grass porgy** (*Calamus arctifrons*)
- **Jolthead porgy** (*Calamus bajonado*)
- **Saucereye porgy** (*Calamus calamus*)
- **Whitebone porgy** (*Calamus leucosteus*)
- **Knobbed porgy** (*Calamus nodosus*)
- **Sheepshead porgy** (*Calamus penna*)
- **Littlehead porgy** (*Calamus proridens*)
- **Silver porgy** (*Diplodus argenteus*)
- **Spottail pinfish** (*Diplodus holbrooki*)
- **Pinfish** (*Lagodon rhomboides*)
- **Red porgy** (*Pagrus pagrus*)
- **Longspine porgy** (*Stenotomus caprinus*)

Background

Porgies are perch-like fishes with fairly deep bodies, sloping heads, and pronounced front teeth. Sheepsheads are the most recognizable of the clan, with alternating black and white bars along the flanks; a gray, sloping head; and bad front teeth. They are the largest of the porgies, reaching a total length of almost 3 ft and weighing up to 18 lb.

The two *Diplodus* species have a rounded profile and a large, conspicuous black spot at the base of the tail, which set them apart from all other porgies. Both species are called "spots" by anglers in southeastern Florida. Even when in hand, the two species can be difficult to tell apart. Subtle differences in the size and shape of the caudal spot can help separate the two, and geography also helps. Spottail pinfishes have a larger caudal saddle, covering the entire caudal peduncle, while silver porgies have a spot rather than a full saddle. Spottail pinfishes have a dusky

Pinfish (*Lagodon rhomboides*), Indian River Lagoon, Florida

Sea bream (*Archosargus rhomboidalis*), Key West, Florida

Top: Grass porgy (*Calamus arctifrons*), Lake Worth Lagoon, Florida. *Bottom*: Juvenile, Lake Worth Lagoon, Florida.

Littlehead porgy (*Calamus proridens*), Ft. Lauderdale, Florida

margin on the soft dorsal and anal fins that is not evident on silver porgies. Both species have 9 faint bars, as well as dark pigment on the opercular margin.

Pinfishes, one of the most common nearshore fishes in the state, are silvery, with alternating yellow and blue stripes, faint grayish bars, and a large black spot above the pectoral fin. Sea breams are very similar in appearance to pinfishes. On close inspection, a black spot behind the pectoral fin and thin, yellow-gold stripes on the body (no blue stripes, as with pinfishes) identify this species. Sea breams grow to just over 12 in.

The seven *Calamus* species look superficially very similar in color and shape: silvery, with the deepest part of the body forward of a vertical midline. Grass porgies are small (<12 in), have 6 jagged bars on flanks that are flecked with black, and have thin dark lines on the tail, all of which distinguish them from other members of the genus. Juveniles have a similar but more pronounced pattern of lines and bars. Littlehead porgies (≤18 in) have a steeply sloping head, with a series of wavy blue lines under the eye and a yellow wash over the head. Knobbed porgies (24 in) have a steeply sloping head that is pale-yellow (like littlehead porgies), but instead of wavy lines, it is covered with yellow spots. Jolthead porgies are the largest member of this genus, growing to over 24 in. They have a sloping head; a silver body, with a light-bronze cast; and rusty-orange corners on the mouth. Saucereye porgies reach about 16 in long; have an iridescent-

Knobbed porgy (*Calamus nodosus*), Florida Middle Grounds

Jolthead porgy (*Calamus bajonado*)

blue crescent mark under the eye; and yellow corners on the mouth. Whitebone porgies have a variegated or blotched pattern on the flanks. Sheepshead porgies have bars similar to sheepsheads but can turn them on and off. They normally are plain silver, with a slightly curved head profile and a distinctive black spot on the axil of the pectoral fin.

Red porgies are shaped similarly to the *Calamus* species, but the body is not as deep, and they are pinkish on the head and dorsal portion of the body, with iridescent-blue flecks on the face; this species can reach a total length of 28 in. Longspine porgies are small (<12 in), silver, and unmarked, but with distinctive, elongated dorsal spines (third through fifth spines).

Distribution and Habitat

Porgies are widely distributed around Florida, where they reside in inshore and offshore structured habitats. Sheepsheads are found from freshwater springs to offshore reefs (≤100 ft). Young sheepsheads occur only in estuaries, usually in low to medium salinity (0–18 psu). Pinfishes are one of the most abundant species in shallow inshore habitats around the state. Young pinfishes numerically dominate fish assemblages in seagrass habitats off both coasts. Sea breams are found on reefs, under mangrove canopies, and within seagrass meadows in southeast Florida and the Florida Keys. Spottail pinfishes and silver porgies have complementary distribution patterns. Silver porgies only occur from southeast Florida through the Florida Keys and are absent in the Gulf of Mexico. They are most common in southeast Florida, where they mix with spottail pinfishes. Spottail pinfishes are found along the west coast and from southeast Florida northward. Within this range, both species associate with nearshore hardbottom, jetties, piers, and other structures. With the possible exception of grass porgies, which are mostly found in seagrass meadows, the various *Calamus* species associate with reefs and hardbottom around the state, in water depths from 1 to up to 330 ft. Red porgies occur around the state and associate with deep reefs (75–600 ft). Longspine

Left: Saucereye porgy (*Calamus calamus*). *Right*: Whitebone porgy (*Calamus leucosteus*), Hobe Sound, Florida.

Sheepshead porgy (*Calamus penna*), Lake Worth Lagoon, Florida

porgies are only found in northeastern and northwestern Florida, usually over softbottom, in depths ranging from 60 to 400 ft.

Natural History

Porgies are water-column spawners, and some species aggregate at certain times of the year to mate. Sheepsheads aggregate to spawn during winter months (November–March) off both Florida coasts. Aggregations form in coastal and shelf waters, usually over natural or artificial hardbottom.

Young sheepsheads usually settle in estuaries, where salinity is low; individuals less than 5 in long are rarely found in salinities higher than 12 psu. Pinfishes also aggregate to spawn, but apparently they gather over level sandy bottoms in coastal waters. Young pinfishes settle in inshore areas during spring, when they are abundant. Spawning by spottail pinfishes or silver porgies has not been observed but probably occurs in coastal waters. From February to April, young from these two species settle in coastal and high-salinity inshore waters along the east coast. Little is known about spawning in the *Calamus* species, but adults may aggregate on outer-shelf reefs.

Sheepsheads feed on attached and motile invertebrates, particularly crabs, isopods, and shrimps. Pinfishes and sea breams change their diets as they age and grow. Small individuals feed on zooplankton, intermediate-sized fishes primarily eat small crustaceans, and larger individuals consume plant matter. Silver porgies and spottail pinfishes undergo a feeding progression similar to that described for pinfishes, with the notable addition of ectoparasite feeding by some intermediate-sized individuals. We have observed juvenile silver porgies cleaning parasites from barracudas, gray angelfishes, and snooks on nearshore hardbottom habitats in southeastern Florida. The *Calamus* species

Sheepshead (*Archosargus probatocephalus*) aggregation, Pecks Lake, Florida

generally feed over sandy areas adjacent to reefs or hardbottom. You can watch them move in characteristic start-stop fashion as they root through sand or vegetated bottoms, searching for polychaete worms, crustaceans, and mollusks. Longspine porgies feed on bottom-dwelling invertebrates over open sandbottom or mudbottom.

Threadfins
Family Polynemidae

- **Atlantic threadfin** (*Polydactylus octonemus*)
- **Littlescale threadfin** (*Polydactylus oligodon*)
- **Barbu** (*Polydactylus virginicus*)

Background

Threadfins are curious fishes, so named because of their highly filamentous pectoral-fin rays. They are small (usually less than 1 ft long), silvery fishes, with pointed and somewhat translucent snouts. The latter characteristic garners the Florida local names "glass noses" or "glass-nose whitings." The three Florida species look very much alike, but the pectoral fins of barbus are pale, with a dusky center, while Atlantic threadfins and littlescale threadfins have dark, almost-black pectoral fins. The two latter species are tough to quickly identify (definitive identification requires scale counts), but the 2 dorsal and caudal fins of littlescale threadfins are darker than those of Atlantic threadfins.

Distribution and Habitat

Atlantic threadfins occur statewide, over mud- and sandbottom, from the surf zone to the outer shelf, but they are most commonly encountered in depths of 15–90 ft. Their

Atlantic threadfin (*Polydactylus octonemus*), Jupiter, Florida

pelagic larvae are found over deep water (to almost 9000 ft). The less-widespread barbus are found on the east coast of Florida, south to Biscayne Bay. Littlescale threadfins are confined to the southeast coast, from Palm Beach to Biscayne Bay.

Natural History

Only Atlantic threadfins have been studied in any detail. In the Gulf of Mexico, they mature at 7–9 months of age and spawn from mid-December to mid-March. Larvae settle in water depths less than 50 ft at 2–4 months of age and gradually move to deeper waters during the summer. Reproduction occurs along the outer continental shelf or upper continental slope. They apparently don't live much more than a year. All three Florida threadfin species feed on worms, shrimps, crabs, and mollusks living in the substrate. Their elongate pectoral rays have sensory properties that are used by these fishes to locate prey in the turbid and turbulent environments in which they live.

Drums and Croakers
Family Sciaenidae

- **Silver perch** (*Bairdiella chrysoura*)
- **Blue croaker** (*Corvula batabana*)
- **Striped croaker** (*Corvula sanctaeluciae*)
- **Sand seatrout** (*Cynoscion arenarius*)
- **Spotted seatrout** (*Cynoscion nebulosus*)
- **Silver seatrout** (*Cynoscion nothus*)
- **Weakfish** (*Cynoscion regalis*)
- **Jackknife-fish** (*Equetus lanceolatus*)
- **Spotted drum** (*Equetus punctatus*)
- **Banded drum** (*Larimus fasciatus*)
- **Spot** (*Leiostomus xanthurus*)
- **Southern kingfish** (*Menticirrhus americanus*)
- **Gulf kingfish** (*Menticirrhus littoralis*)
- **Northern kingfish** (*Menticirrhus saxatilis*)
- **Atlantic croaker** (*Micropogonias undulatus*)
- **Reef croaker** (*Odontoscion dentex*)
- **High-hat** (*Pareques acuminatus*)
- **Blackbar drum** (*Pareques iwamotoi*)
- **Cubbyu** (*Pareques umbrosus*)
- **Black drum** (*Pogonias cromis*)
- **Red drum** (*Sciaenops ocellatus*)
- **Star drum** (*Stellifer lanceolatus*)
- **Sand drum** (*Umbrina coroides*)

Background

Drums and croakers are a species-rich and numerically dominant assemblage of bottom-dwelling fishes that inhabit shallow, near-shore waters. They use specialized muscles connected their taut swim bladders to produce the drumming and croaking sounds that give the family its common names. The most-streamlined members of the group, the seatrouts and weakfishes, are elongated, with pointed snouts and protruding lower jaws, and they have a pair of large canine teeth near the tip of the upper jaw. The 2–3 ft long weakfish and spotted seatrout adults have caudal fins with straight or near-straight edges (lanceolate as juveniles) and dark body spotting, while the smaller (1–1.5 ft) sand and silver seatrouts have lanceolate tails at all sizes and are silvery in coloration. Spotted seatrouts are the easiest to identify,

Spotted seatrout (*Cynoscion nebulosus*), Tampa Bay, Florida

having distinct, pupil-sized black spots on the rear two-thirds of the silvery, iridescent upper body and tail and less–well-defined spots on both dorsal fins. Large individuals (over 10 lb) are referred to as "gator trouts." The entire grey–blue-green body of weakfishes (also called "grey seatrouts") has a series of indistinct dark spots and splotches that morph into short, oblique, upward-trending rows; only the caudal fin has any (indistinct) spotting. The inside of mouth is yellow-gold. Silver seatrouts, the smallest of the seatrouts (16 in), are pretty much uniformly silver, except for faint yellow fins; their distinctive character is a mouth often rimmed with dark pigment. They have a shorter anal fin (8–11 rays) than the slightly larger and similarly dull (pale-gray) sand seatrouts (10–11 rays). The latter also have a small dark blotch on the pectoral-fin bases and the most-angular caudal fin of all the seatrouts. To confuse things, weakfishes and sand seatrouts hybridize along Florida's east coast, producing offspring that are not easily distinguishable, and sand and silver seatrouts may do the same in the Gulf.

Gulf kingfish (*Menticirrhus littoralis*), Jupiter, Florida

Sand drum (*Umbrina coroides*), Palm Beach, Florida

The three species of kingfishes (referred to by Floridians as "whitings" or "grey mullets," and are not to be confused with king mackerels, also called "kingfishes") and sand drums have a single, stout mental barbel on the tip of the lower jaw. Kingfishes are very slender, with long snouts; inferior mouths; tall, shark-like, spinous dorsal fins; and lower caudal lobes that are longer than the top lobes, resulting in an S-curve margin to the tail. Gulf kingfishes, the most readily identified member of this group, are totally grey-silver, except for a dipped-in-black-ink upper caudal tip and a dusky tip to the spiny dorsal fin. Southern and northern kingfishes can be confusing, as both have dark–golden-brown, oblique bars on the body and head. Southern kingfishes have 7 or 8 parallel diagonal bars on the flanks, and one of opposite orientation on the nape, resulting in the first 2 bars forming a V behind the operculum. Northern kingfishes have similar but less–fully formed diagonal dark bars on the flanks, but there are 2 bars under the nape, with the first 3 bars forming a V. Northerns also have a brown, horizontal lateral stripe posteriorly that extends out onto the tail, but only the lower caudal lobe is brown (the entire tail of southern kingfishes is uniformly brown). All kingfishes attain sizes of 18–24 in. The smaller sand drums (14 in) show the more typical shape of the family (short

Left: Jackknife-fish (*Equetus lanceolatus*), Lake Worth Lagoon, Florida. *Right*: Juvenile, Sarasota, Florida.

Left: Spotted drum (*Equetus punctatus*), Jupiter, Florida. *Right*: Juvenile, Jupiter, Florida.

and dumpy), have a straight caudal-fin margin, and have 8–9 dark bars on the body.

The most strikingly marked drums are five small (10–13 in) reef dwellers that have similar patterns of dark stripes against a light background, an identification challenge compounded by pattern changes that occur with age and growth. Adults are readily distinguished, but identifying the young is much more difficult. Jackknife-fishes and spotted drums are among the most handsome fishes found on hardbottom habitats (adults have steep, almost-vertical foreheads; very elongated first dorsal fins; and 3 conspicuous, dark body bands (bordered by thin white bands in jackknife-fishes). The anterior-most band runs from the lower jaw through the eyes; the second runs vertically from the tip of the pelvic fin to the top of the head; and the third extends from the tip of the pointed tail (from the base of the tail in spotted drums) forward along the flank and then bends upward to the elongated tip of the first dorsal fin. The base coloration of adult jackknife-fishes is creamy white, and that of adult spotted drums is white. Adult spotted drums differ in having 2 additional pairs of narrow, dark-brown, near-horizontal bands above and below the large central band; dark-brown soft dorsal and anal fins, with white spots (those fins are light-colored in jackknives), and by having dark pectoral fins and a dark smudge on the tip of the snout.

Adult high-hats and cubbyus are easily confused. The body and fins of cubbyus are grey–chocolate brown–bronze, with alternating thin and thinner dark stripes. In contrast, high-hats have alternating wide, dark-brown and silver-white stripes and a spiny dorsal fin that is vertically bicolored, dark anteriorly and white posteriorly. The base color of blackbar drums is silver-white, and they have a distinctive wide, oblique, dark-brown bar extending from the tip of the pelvic fin upward to the anterior portion of the dorsal fin. They also have a single wide, perfectly horizontal, dark stripe running from that bar to the midtip of the tail; a dark, diagonal eye-mask running from the nape to the mouth; and a broad, light-brown band running from the base of the anal fin upward onto the rear two-thirds of the soft dorsal fin. Dorsal-fin elongation in juveniles of all of these species is pronounced, making them popular aquarium fishes. Juvenile jackknife-fishes and spotted drums bear dark bar patterns similar to those of adults, but jackknives are golden-yellow, with a thin vertical line on the snout, versus white, with a smudge, in spotted drums. Cubbyu and high-hat juveniles have alternating wide and thin black stripes over a white background. The snout of cubbyus has a V-shaped mark that morphs into an 0 shape with growth; at equal sizes, high-hats have a small spot between the eyes that develops into an inter-ocular bar.

Red and black drums are the largest members of the family, both reaching a length of about 5.5 ft, and are important fishery species. Red drums (known widely as "redfishes") have a long, tapering body; thick shoulders; a terminal mouth; and lack chin barbels. But coloration is the key to their identification: the upper two-thirds of the body in red drums is pale- to coppery–brown-grey; the belly is white; and there are 1 or more variably shaped, ocellated black spots near the upper base of the tail.

Black drums are much stockier

Left: High-hat (*Pareques acuminatus*), Jupiter, Florida. *Right*: Juvenile, Jupiter, Florida.

Left: Cubbyu (*Pareques umbrosus*), Jupiter, Florida. *Right*: Juvenile, Jupiter, Florida.

Red drum (*Sciaenops ocellatus*), Tampa Bay, Florida

than red drums; have an inferior mouth, with 10–15 chin barbels; and the body is bronze-black, except for the white belly. There are 4 or 5 wide, dark bars on the body that are most pronounced in the young. Atlantic croakers, which grow to 20 in, also have multiple chin barbels. They have an inferior mouth, a dusky spot on the base of the pectoral fin, and are silver-golden in color (but can appear steely–purple-pink when viewed from the side), with a series of thin, oblique bars on the flanks. The caudal fin is 3-lobed in adults and pointed in juveniles, distinguishing them from spots, which share a silver-golden base coloration and diagonal bars on the sides, and have a straight or slightly forked tail margin at all sizes. Also, as their name correctly notes, spots (called "mazukies" on Florida's east coast) have an eye-sized black spot located above the upper corner of the operculum and lack the barbels and dusky pectoral spot of Atlantic croakers.

The remaining group of half a dozen species are small (1 ft or less) and mostly nondescript in coloration. Banded drums have 7–9 faint, brownish bars on the sides of the body and might be confused with sand drums, but they lack the latter's chin barbel, have an lanceolate tail, and clearly have an underbite. A distinctive large black spot on the base of the pectoral fin identifies reef croakers, which have a terminal mouth, a straight-edged tail, and are bronze–light-brown in color. Star drums and silver perches are small, silvery fishes with terminal mouths. Silver perches have a moderately deep body, a straight-margined to moderately lanceolate tail, and yellow fins, leading Gulf anglers to call them "yellowtails" (not to be confused with the snapper species of that name). Star drums have a pointed, lanceolate caudal fin that resembles an old-fashioned canoe paddle; a slightly concave forehead; and are silvery-gray, with dusky fins. The body shapes of blue and striped croakers are similar, but they differ slightly in coloration and decidedly in distribution. Blue croaker have a gray–brassy-brown fin and body coloration, with thin stripes along the flanks, formed by rows of dark spots. The tail is more rounded than that of striped croakers, which have light-colored fins and golden-brown stripes, formed by rows of dark spots on the grey-silver body.

Spot (*Leiostomus xanthurus*), Lake Worth Lagoon, Florida

Atlantic croaker (*Micropogonias undulatus*), Loxahatchee River, Florida

Striped croaker (*Corvula sanctaeluciae*), Jupiter, Florida

Distribution and Habitat

Most of the drums and croakers are continental forms that are widely distributed around Florida, but some have restricted, tropical peripheral distributions or are tied to the presence of reefs. Spots, Atlantic croakers, red and black drums, spotted seatrouts, and silver perches are primarily estuarine species, found statewide. Spots and Atlantic croakers prefer open sand- or mudbottom in estuaries. Red drums favor shallow water. They roam between seagrass meadows, oyster reefs, and mangrove shorelines, and can also be found in shallow shelf waters. Star drums inhabit estuarine conditions at the mouths of rivers and in shallow nearshore waters, to depths of about 60–70 ft, primarily off northeast Florida and the panhandle, where they are found with banded drums (which range deeper, to about 300 ft). Gulf, northern, and southern kingfishes occur in coastal and inshore waters around the state, but the best place to find these species is in the surf zone. On calm days, you can watch kingfishes from the beach as they move back and forth with the waves in the wash zone, searching for prey. Spotted seatrouts occur around the state and are generally thought to spend their entire life-cycle within estuaries. Their presence and abundance generally depends on the availability and extent of their preferred habitat: seagrass meadows. In Florida, we have seen adult spotted seatrouts in seaward, nearshore waters off Florida Bay, Venice, Crystal River, and Satellite Beach—all areas adjacent to estuarine seagrass hotspots. Weakfishes are basically restricted to Florida's east-coast waters, and sand seatrouts are confined to Gulf waters, with both species inhabiting estuaries and nearshore coastal waters. Silver seatrouts are less of an estuarine fish, occurring along beaches in nearshore waters throughout peninsular Florida.

Sand drums occur in the surf zone of sandy beaches and nearshore hardbottom along the entire east coast of Florida and are absent from the west coast. Striped and blue croakers are tropical peripheral species, with very restricted distributions. Striped croakers are found only on the east coast, from Palm Beach Inlet north to about Satellite Beach, and are usually

Reef croaker (*Odontoscion dentex*), Pecks Lake, Florida

associated with structured habitat, in water depths from nearshore to about 40 ft. Blue croakers are largely confined to softbottom seagrass beds in and on the northwestern margin of Florida Bay. Reef croakers occur on shallow reefs (<100 ft), from Stuart around to the Big Bend ecoregion on the west coast. The five reef-dwelling species of *Pareques* and *Equetus* occur around the state, with blackbar drums being restricted to deep reefs, in water depths exceeding 150 ft.

Natural History

This family is divided ecologically into mud- or sandbottom dwellers of estuarine and shallow shelf waters, and bottom-associated reef dwellers. All are water-column spawners. Several of the general bottom dwellers—spots, Atlantic croakers, and banded, red, and black drums—migrate from their estuarine haunts to shelf waters to spawn. Most of these offshore migrators spawn during winter months, and temperature is probably an important factor in the timing of their spawning. A massive influx of juvenile Atlantic croakers and spots into coastal lagoons and estuaries occurs in the early spring. Surprisingly, the red drum population in Mosquito Lagoon appears to spawn within the lagoon, while on the west coast and in northeastern Florida, red drums migrate en masse to shelf waters to spawn. We have caught spots in bottom trawls in water as deep as 600 ft, offshore of southeast Florida.

Sonic communication in drums and croakers is important, so their drumming and croaking sounds play an important part in their ecology. Thus far, all drums that have been investigated have characteristic sounds. Most (probably all) species produce sounds in a collective chorus during spawning seasons. Fishery managers are developing techniques to locate spawning sites using underwater microphones that identify chorusing or reproducing groups in inshore water bodies, such as Indian River Lagoon or Tampa Bay.

Drums and croakers are archetypical bottom feeders that use their subterminal mouths and

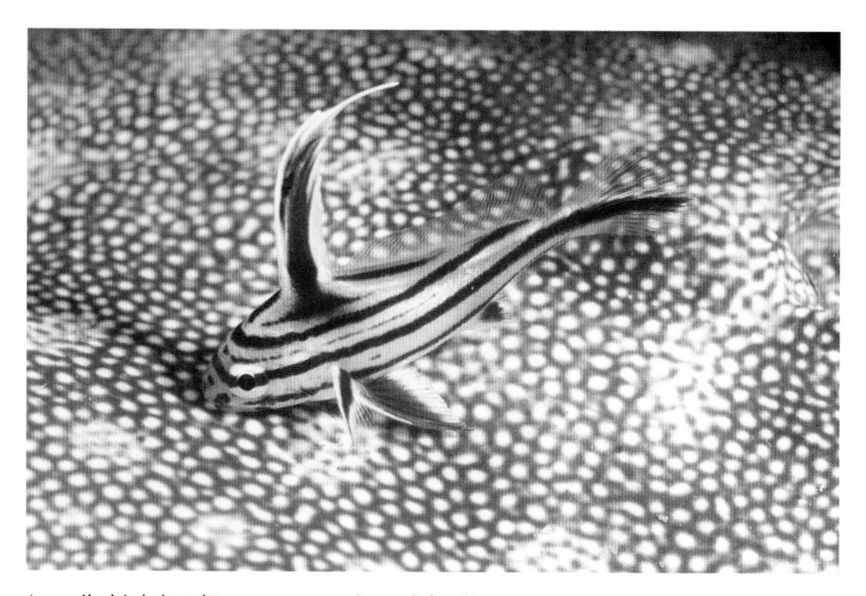

Juvenile high-hat (*Pareques acuminatus*) feeding on mucous or organic matter from the back of a yellow stingray (*Urobatis jamaicensis*), Lake Worth Lagoon, Florida

sensory barbels to find worms, clams, shrimps, and other invertebrates hiding in soft sediments. The size and type of prey consumed depends on the size and species of the predators. Black drums have large, triangular, knob-covered crushing plates in their throats, capable of breaking open oyster and clam shells. Seatrouts, equipped with large terminal mouths and fangs, feed on more-mobile fishes, crabs, and shrimps. The reef-dwelling drums consume benthic invertebrates. We have observed young high-hats cleaning external parasites from the sides of adult common snooks. Another curious behavior exhibited by high-hats is what appears to be the ingestion of either mucous or organic material sticking to the mucous layer on resting yellow stingrays (chapter 19). We have also observed juvenile high-hats pick at organic material clinging to the small hairs on spiny lobster antennae. This sort of mucous-feeding behavior has not been reported in the scientific literature for the species.

Most drums are preyed on by larger fishes. Spots and sand drums are favorite live baits for snooks on the southeast coast. We have taken spots from the stomachs of greater amberjacks caught from 300 ft water depths off eastcentral Florida. Spots and croakers happily eat each other and (occasionally) their own. Weakfishes and seatrouts also munch on these two species.

Goatfishes
Family Mullidae

- **Yellow goatfish** (*Mulloidichthys martinicus*)
- **Red goatfish** (*Mullus auratus*)
- **Spotted goatfish** (*Pseudupeneus maculatus*)
- **Dwarf goatfish** (*Upeneus parvus*)

Background

Goatfishes are readily recognized by their 2 long chin barbels. The four goatfish species have similar body shapes but differ in color patterns and habitats. Yellow goatfishes are pale overall, with a deeply forked tail, yellow fins (all but the pectorals), and a yellow lateral stripe, but when attended by a cleaner fish or shrimp, they change their color to a dark-red. Red goatfishes are reddish-orange, with 2 or more yellow lateral stripes, and 2 red and orange bands trisecting the spinous dorsal fin; juveniles and subadults are pale, with a dark-red lateral line. Spotted goatfishes vary from light-brown to almost-white, typically with 3 squarish blotches along the flanks. These fishes become reddish, and the blotches merge into irregular bars, at night and while resting during the day. There are faint blue spots running lengthwise along the flanks, and neon-blue lines on the snout. Dwarf goatfishes are pinkish to brown, with a faint yellow lateral stripe; diagonal bands run across both of the dorsal fins and each caudal-fin lobe.

Distribution and Habitat

The yellow and spotted goatfishes are tropical species, normally associated with insular habitats. They are most common in south Florida,

Yellow goatfish (*Mulloidichthys martinicus*), Looe Key, Florida

Juvenile red goatfish (*Mullus auratus*), Lake Worth Lagoon, Florida

Spotted goatfishes (*Pseudupeneus maculatus*), Palm Beach, Florida

where there are appropriate water temperatures and suitable habitat. Dwarf and red goatfishes have a more continental distribution. Yellow goatfishes occur around shallow-water reefs and seagrass beds. Red goatfishes prefer silty, sedimentary bottoms, in somewhat deeper waters (30–400 ft). Spotted goatfishes are found around reefs and in seagrass meadows, from nearshore to 120 ft. Dwarf goatfishes are widespread in Florida, in deeper waters (ranging from 120 to 300 ft).

Natural History

Goatfishes are water-column spawners; otherwise little is known about reproduction in this family. Spotted and yellow goatfishes aggregate to spawn in Caribbean locations, but such events have not been reported for Florida waters. Goatfishes feed on crustaceans, worms, clams, and other critters living in or on silt- or sandbottom. The long chin barbels have taste-sensitive cells that facilitate foraging in sediments. Watching a foraging goatfish is very entertaining. The barbels are like two twitching digits, probing the upper layers of sediment for morsels. The inferior mouth allows easy plucking of any prey item flushed out by the vigorous barbel movements. Schools of yellow goatfishes serve as nuclear predators that often attract followers—wrasses, mojarras, jacks, yellowtail snappers, coneys, and other species—looking for prey items flushed out of the sand by the foraging goatfishes. On the other hand, individual yellow goatfishes join schools of smallmouth grunts, which they resemble, in an association known as protective mimicry, where both species gain an increased level of protection by forming a larger school of look-alikes. Goatfishes are eaten by reef and seagrass predators, including cornetfishes, trumpetfishes, red hinds, Nassau groupers, blue runners, and dog and mutton snappers. Pelagic young fall prey to tunas.

Sweepers
Family Pempheridae

- **Glassy sweeper** (*Pempheris schomburgkii*)

Background
Glassy sweepers, sometimes locally referred to as "hatchetfishes" (erroneously—marine hatchetfishes [Sternoptychidae] are bioluminescent deepwater fishes), are the only member of this family in Florida waters. They are small, laterally compressed, deep-bodied fishes, with relatively large eyes and a large mouth. Coloration is generally reddish to coppery in adults; juveniles under 0.75 in are transparent. Adults reach a total length of about 5 in.

Distribution and Habitat
Glassy sweepers are generally restricted to southeastern Florida and the Florida Keys, where they are found in reefs and hardbottom habitats, at depths from 3 to about 120 ft.

Natural History
Reproduction has not been observed in glassy sweepers. In the early summer, planktonic larvae settle en masse on nearshore hardbottom habitats in southeast Florida, and they are often associated with newly settled grunts. These transforming young settle along the sand-rock edges of natural and artificial structures, in water depths as shallow as 3 ft.

During daylight hours, adult glassy sweepers generally hide beneath ledges and overhangs or in the recesses of caves. Just after sunset, individuals emerge synchronously (all at the same time) from their hiding places and follow regular routes to midwater feeding areas up to a half mile from their home shelter, where, as individuals or small groups, they feed on larval crabs and shrimps passing by in the water column. Feeding continues through the night, until about an hour before sunrise, when they return to their daytime refuges.

Glassy sweeper (*Pempheris schomburgkii*), Palm Beach, Florida

Sea Chubs
Family Kyphosidae

- **Darkfin chub** (*Kyphosus bigibbus*)
- **Highfin chub** (*Kyphosus cinerascens*)
- **Bermuda chub** (*Kyphosus sectatrix*)
- **Yellow chub** (*Kyphosus vaigiensis*)

Background
Florida's four sea chub species are nearly indistinguishable, with their definitive identification based on counting dorsal- and anal-fin rays and gill rakers. Obviously these are not very convenient characters while on the deck of a boat or underwater, so tough-to-denote subtleties of body coloration, fin shape, and fin coloration will have to do. Highfin chubs can be distin-

Highfin chub (*Kyphosus cinerascens*), Palm Beach, Florida

Juvenile chub (*Kyphosus* sp.), Jupiter, Florida

Yellow chub (*Kyphosus vaigiensis*), Jupiter, Florida

guished by a small but noticeable bump on the nape that is lacking in the other three species, as well as In having longer soft dorsal- and anal-fin rays than Bermuda and yellow chubs. The latter species possesses a series of thin, gold to bronze horizontal stripes on the body and 2 wide, gold to bronze stripes on the head. In darkfin chubs, the soft dorsal and anal fins have dark margins and angular profiles. Bermuda chubs look very similar to darkfin chubs, but the soft dorsal and anal fins are uniformly pale gray, lacking dark margins. Frankly, differences between the latter two species are minor, and field identification is extremely difficult. Juveniles of all species are dark-brown to greenish-brown, with white spots (about eye-sized) over the entire body.

Distribution and Habitat

Recent DNA work revealed that the four Atlantic species are found worldwide, in tropical waters. In Florida, all species associate with structured habitat, in coastal and shelf waters, to depths of 300 ft, but adults are most commonly encountered in less than 100 ft of water, often in seagrass beds, on rocky bottoms, and on coral reefs. Juveniles associate with floating sargassum, over deeper water.

Natural History

Sea chub reproduction occurs in both mass and pair-spawning events in the water column, usually over reef habitat. The eggs, larvae, and juveniles of all species are epipelagic, with the young living under drifting flotsam and sargassum. Juveniles raft for an unknown period of time, until they reach a length of about 5 in. While pelagic, they eat a lot of copepods, as well as algae. Adult sea chubs are almost totally herbivorous, feeding on attached algae and floating sargassum, leading to foul-smelling internal organs (viscera). We have observed groups of sea chubs attacking drifting sargassum, thrashing the surface waters with a fury that frequently fools anglers into thinking more-highly desired piscivores are involved. Although sea chubs are herbivores, they will take a baited hook and are good fighters for their size. By day, sea chubs school (often with fellow members of the genus) in the water column, above reef promontories or high-relief features, where they graze on algae. A tagging study outside of Florida demonstrated that each Bermuda chub has a discrete round or ovoid home range that often overlaps with the home ranges of others in this species, but it frequents one or two favorite spots each day. At night the schools break apart, and individuals move to their favored nighttime resting spots under ledges, in crevasses, or in caves.

Butterflyfishes
Family Chaetodontidae

- **Foureye butterflyfish** (*Chaetodon capistratus*)
- **Spotfin butterflyfish** (*Chaetodon ocellatus*)
- **Reef butterflyfish** (*Chaetodon sedentarius*)
- **Banded butterflyfish** (*Chaetodon striatus*)
- **Longsnout butterflyfish** (*Prognathodes aculeatus*)
- **Bank butterflyfish** (*Prognathodes aya*)
- **Guyana butterflyfish** (*Prognathodes guyanensis*)

Background

Butterflyfishes are the archetypical tropical-reef fishes. They are perpetually puckering up; bear prominent dorsal spines; and have deep, laterally compressed bodies, with bold diagnostic color patterns. There are two natural groups, the shallow-water *Chaetodon* species and the deeper-dwelling *Prognathodes* group, the former being round to oblong in shape and the latter more triangular in appearance, with a more pronounced snout. All *Chaetodon* species reach 6–8 in long, and all have a prominent, sickle-shaped, dark stripe that starts in front of the dorsal fin, crosses the eye, and then bisects the operculum. The evolutionary value of this is tied to having false eye spots at

Left: Foureye butterflyfish (*Chaetodon capistratus*), Looe Key, Florida. *Right*: Juvenile.

Top: Spotfin butterflyfish (*Chaetodon ocellatus*), Jupiter, Florida. *Bottom*: Juvenile, Lake Worth Lagoon, Florida.

the other end of the body, seeding just enough doubt in the mind of a predator for a butterflyfish to scoot away.

The median fins of foureye butterflyfishes are straw-colored; the sides are white; and a series of dark, thin lines parallel the scale row above and below the lateral line, in opposing directions. The posterior flank has a large ocellated spot (black outlined with white) just below the soft dorsal fin. Juveniles have vertical eye bars, like the adults, but bear 2 ocellated spots on the soft dorsal fin and 2 broad, dusky bands on the body. Spotfin butterflyfishes have uniformly white sides and bright-yellow median and pelvic fins. A small black spot is present on the rear margin of the soft dorsal fin, and at night a dusky spot appears on the body, near the rear base of the dorsal fin. Juveniles are colored like the adults, but they have a second vertical black bar emanating from a diffuse spot at the rear base of the dorsal fin, the net result looking like a pair of parentheses: ( ). Reef butterflyfishes have a larger, diagonal black band running along the posterior margin of the soft anal fin, over the caudal peduncle, and over most of the soft dorsal fin, resulting in an oblong blotch on that fin. The sides of the body are yellow and white, with 6–7 diagonal, parallel, faint brown lines; the caudal fin and parts of the dorsal and anal fins are golden-yellow; and the paired fins are clear. Young look like the adults, except that they have a small dark spot in lieu of the blotch on the rear dorsal fin. Banded butterflyfishes are distinctive, with color shades of brown over a white base coloration. Two dark, diagonal bars cross the midbody, the rearmost connecting on the rear dorsal and anal bases with a third that crosses the caudal peduncle, producing a posterior O-shaped mark. Juveniles resemble the adults, adding a white-ringed black spot on the soft dorsal fin.

The three *Prognathodes* species are separated by habitat depth as well as by appearance. Longsnout butterflyfishes lack any vertical bars, and the rear dorsolateral third of the body is rusty brown, grading to orange laterally, and then white on the belly. There are 5 yellow-orange head bars, with 3 originating at the base of the dorsal fin—1 going between the eyes onto the snout and a pair leading to each eye—and a second pair, with each going from an eye onto the snout. Bank butterflyfishes have a narrow black bar extending from the base of the dorsal fin, through the eye, to near the corner of the mouth, with the subocular portion being more dusky than dark. A second wide, black bar begins at the middle of the dorsal-fin base and angles downward and backward onto the base of the mid-anal fin. All fins except the pectorals are predominantly yellow, and the sides between the dark bars are white.

Reef butterflyfish (*Chaetodon sedentarius*), Jupiter, Florida

Top: Banded butterflyfish (*Chaetodon striatus*), Jupiter, Florida. *Bottom*: Juvenile.

Guyana butterflyfishes are similar in pattern to bank butterflyfishes, but the bars extend to the margins of the dorsal and anal fins, and they have a third bar running parallel with and behind the second, ending at the base of the caudal peduncle.

Distribution and Habitat

Butterflyfishes are found on hard-bottom and reef habitats, and all species except Guyana butterfly-fishes have been found statewide. Like other reef fishes, adults are most abundant in the Keys and on offshore reefs to the north. *Chaetodon* adults are usually found in shallow-water habitats, including reefs and areas with man-made debris; young frequent seagrass beds and any available inshore hardbottom, including bridge and dock pilings, jetties, mooring lines, and sunken boats. Juveniles of the *Prognathodes* species are usually found alongside the adults, in deeper reef areas. Depth is a big ecological separator in this family. Although they occasionally get to depths greater than 160 ft, banded butterflyfishes prefer shallow water (<20 ft). Foureye and spotfin butterflyfishes range from the shoreline to depths of 200 ft but are most common at 25–75 ft and 50–120 ft, respectively. Reef butterflyfishes replace spotfins at greater depths and are most abundant at 80–200 ft, but range to almost 350 ft. Longsnout butterfly-fishes, although uncommonly seen in shallow waters, are most often encountered at depths of 50–180 ft and have been observed as deep as 475 ft. They, in turn, are replaced by bank butterflyfishes, ranging from 115 to 400 ft deep and dominant at depths greater than 260 ft, and, ultimately, by Guyana butterflyfishes, most common at 500–650 ft (range = 130–750 ft). The latter have been taken by mixed-gas divers at the limits of diving depths in the Keys, and Guyanas probably are found on reefs at greater depths along Florida's east coast.

Natural History

Butterflyfishes usually keep to a particular home range on the reef but do not aggressively guard it. They are well known for traveling in twos, and it is often assumed that these are mated pairs. Spawning has been observed for some species and happens at dusk. Although mating usually occurs in pairs, groups of individuals can be involved. Courtship includes various displays between individuals, and eggs and sperm are released into the water column. The larval stage of butterflyfishes looks nothing like a butterflyfish: the bones of the skull are enlarged and cover the head like armor. These pelagic larvae are so odd looking that early naturalists often described them as different species, formally named *Tholichthys*. Fossils reveal that this larval stage was in existence in the Oligocene epoch, more than 30 million years ago, demonstrating its longstanding evolutionary value. This stage lasts about 30 days, followed by the settlement of miniature adults (for most species) into off-reef habitats.

Butterflyfishes consume a variety of invertebrates, including attached and mobile forms.

Longsnout butterflyfish (*Prognathodes aculeatus*)

Bank butterflyfish (*Prognathodes aya*)

Their long, forceps-like jaws are well adapted for picking up small, motile invertebrates and nipping off the soft parts of attached invertebrates (and, on occasion, large drifting jellyfish). Coral polyps and worm tentacles are favorite tidbits. It is not uncommon to see reef butterflyfishes feeding on plankton in the water column. Butterflyfishes are frequent customers at shrimp-cleaning stations.

Angelfishes
Family Pomacanthidae

- **Cherubfish** (*Centropyge argi*)
- **Blue angelfish** (*Holacanthus bermudensis*)
- **Queen angelfish** (*Holacanthus ciliaris*)
- **Rock beauty** (*Holacanthus tricolor*)
- **Gray angelfish** (*Pomacanthus arcuatus*)
- **French angelfish** (*Pomacanthus paru*)

Background

The brightly colored angelfishes are signature reef species in Florida's tropical and subtropical waters. They are deep bodied, laterally compressed, and each species bears the family trademark: a long, conspicuous spine on the gill cover. That preopercular spine and the much shorter dorsal spines separate angelfishes from the equally attractive butterflyfishes, with which they are closely allied.

Cherubfishes are the smallest member of the family, reaching little more than 2 in long at maturity. They are oval shaped, with a golden-yellow head; the body and fins are royal-blue, with the median fins rimmed with thin, electric-blue margins. Rock beauties also are easily distinguished by their color pattern: bright-yellow, with a black body patch that increases in size with growth. Early juveniles are mostly yellow, with a blue-ringed black spot below the soft dorsal fin. That spot grows into an all-black polygon occupying the rear upper body and base of the dorsal fin in subadults. At adult size (about 12 in), rock beauties are more black than yellow, the patch now covering most of the rear two-thirds of body and the dorsal and anal fins. The head, tail, and paired fins are bright-yellow, as are most margins of the dorsal and ventral fins and the tail; the jaws are deep-blue or black.

Queen angelfishes, also reaching 1 ft long, are arguably one of the most spectacularly colored reef fishes in the region. Words can't adequately describe all of their subtleties of color and shading—they look like fantasy fish, painted by someone trying to use all the colors in a new paint set. Each blue-green flank scale is broadly edged in yellow; the tail and paired fins are bright-yellow; and the rear tips of the dorsal and anal fins are orange-yellow. There is an ocellated spot, with a cobalt-blue center surrounded by electric-blue rings, on the forehead behind and above the eyes (the nape) and a large blue spot on each pectoral-fin base. The somewhat larger (to 16–18 in) blue angelfishes look very similar to queen angelfishes, but the blues and yellows are more muted and pastel-like. Importantly, they lack the ocellated spot on the blue nape and the bold blue spots on the pectoral bases. The pelvic fins and the rear margins of the dorsal, anal, and caudal fins are yellow; the pectoral fins are blue basally and yellow-margined. Blue and queen angelfishes commonly hybridize, and the resultant mixes were once described as a unique species: Townsend's angelfish (*Holacanthus townsendi*). The intermediate characters arising from this hybridization vary considerably but usually include a nonexistent to partially formed ocellated spot on the nape.

Cherubfish (*Centropyge argi*), Juno, Florida

Left: Rock beauty (*Holacanthus tricolor*), Jupiter, Florida. *Right*: Juvenile, Looe Key, Florida.

Left: Queen angelfish (*Holacanthus ciliaris*), Singer Island, Florida. *Right*: Juvenile, Jupiter, Florida.

Juvenile queen and blue angelfishes are difficult to identify, the chief difference being that the thin, blue body bars are straight in blue angelfishes and comma-shaped in young queen angelfishes.

The less colorful French and gray angelfishes have grey faces, with white jaws. Each body scale of an adult gray angelfish is gray-brown, with a light edge. The inner sides of the pectoral fins are yellow; the rear edges of the dorsal, anal, and caudal fins have thin, light margins. The tails of gray angelfish adults (to 15 in) are straight edged, while those of adult French angels (to 16 in) are more rounded and lack the light margin. The base color of French angelfishes is black, with each flank scale edged in yellow. The eye is ringed with yellow, as are the edges of the operculum and preopercular spine; there is a yellow triangular spot on the pectoral-fin bases. Juveniles of gray and French angelfishes are easily confused, both being uniformly black, with 3 thin, yellow body bars, and having royal-blue on the pelvic fins. But French angelfishes have a yellow ring around the tail, including its posterior margin, which is lacking in gray angelfishes. As angelfish juveniles grow, their color patterns go through transitional phases between the juvenile and adult patterns described above.

Distribution and Habitat

Cherubfishes occur sparingly statewide, mostly on low-relief ledges off southeastern Florida and the Keys, in depths of 80–350 ft, and, in the Gulf, only on offshore

(150–300 ft) reefs, often in association with the rock mounds of sand tilefishes. Although blue and queen angelfishes occur together in southeastern Florida and the Florida Keys, in depths less than 400 ft, the two species exhibit continental and insular distribution patterns, respectively. As such, blue angelfishes are common and widespread on hardbottom habitats throughout Florida, while queen angelfishes are primarily found in abundance only in southeastern Florida and the Keys. In the Gulf, both species occur on offshore reefs, in depths of 60–300 ft. In southern Florida, rock beauties are found on deep reefs as adults and in shallow to intermediate water depths on rock and coral bottoms as juveniles, but, being an insular form like queen angelfishes, they are rare in the deep Gulf waters of Florida. Gray and French angelfishes also represent a species-pair having zoogeographical affinities and distributions mirroring those of blue and queen angelfishes; the continental gray angelfishes are common throughout Florida waters, while the insular French angelfishes are most plentiful in southern Florida. Like blue and queen angelfishes, grey and French angels prefer hardbottom habitats bearing coral and sponges, and in the Gulf they are predominantly found on deepwater (60–200 ft) reefs. Juveniles of both species settle in a wide range of water depths and structured habitats.

Left: Blue angelfish (*Holacanthus bermudensis*), Lake Worth Lagoon, Florida. *Right*: Juvenile, Palm Beach, Florida.

Left: Gray angelfish (*Pomacanthus arcuatus*). *Right*: Juvenile, Jupiter, Florida.

Left: French angelfish (*Pomacanthus paru*), Hillsboro, Florida. *Right*: Juvenile, Lake Worth Lagoon, Florida.

Natural History

Angelfishes are pair-spawners that reproduce throughout the lunar cycle. Courtship and spawning occur in the water column, usually at dusk. Temporary, sex-specific coloration occurs during courting and spawning in both sexes of grey angelfishes and in male cherubfishes. Males usually are larger than females. The social aspects of angelfish reproduction are diverse, ranging from promiscuous breeding groups in grey angelfishes, to male-controlled harems (queen angelfishes, rock beauties, and cherubfishes), to monogamy in French angelfishes. Adult gray and French angelfishes and cherubfishes are regularly encountered in opposite-sex pairs; blue and queen angelfishes pair up occasionally; and rock beauties tend to fly solo. The pelagic larvae of angelfishes are planktonic for 17–40 days (depending on the species) before setting on appropriate bottom.

All *Holacanthus* and *Pomacanthus* species feed heavily on sponges and, to a lesser degree, on other attached invertebrates. French and grey angelfishes even consume poisonous zoanthid corals. Cherubfishes differ from the other species in feeding on algae, cyanobacteria (blue-green algae), and organic material. Juveniles of *Holacanthus* and *Pomacanthus* are territorial and clean parasites from client species. Young French and gray angelfishes dance to attract prospective clients. Young rock beauties have been reported to feed on the mucous covering of larger fishes. Adult angelfishes often become clients themselves, serviced by cleaning gobies.

Hawkfishes
Family Cirrhitidae

- **Redspotted hawkfish** (*Amblycirrhitus pinos*)

Background

Redspotted hawkfishes, the only member of this family in our waters, are small (<3 in), shy fishes, with 4 dark-green bars interacting with 4 thinner bars, and their whole body covered with small red spots. They are the only fish species in Florida with short cirri on the dorsal-fin spines.

Distribution and Habitat

Redspotted hawkfishes prefer structured habitats and are often found around living stony coral and fire coral colonies, in water depths from nearshore to about 100 ft. They are found from southeastern Florida through the Florida Keys and northward to the panhandle.

Natural History

Little is known about the life history of redspotted hawkfishes, but

Redspotted hawkfish (*Amblycirrhitus pinos*), Hillsboro, Florida

they are thought to be a territorial species. Individuals rest on their thick pectoral fins, usually under a ledge or within the branches of a coral colony. They feed mostly on small planktonic organisms, especially copepods and crustacean larvae, but will also take small benthic critters, such as worms and isopods.

Damselfishes
Family Pomacentridae

- **Sergeant major** (*Abudefduf saxatilis*)
- **Night sergeant** (*Abudefduf taurus*)
- **Blue chromis** (*Chromis cyanea*)
- **Yellowtail reeffish** (*Chromis enchrysura*)
- **Sunshinefish** (*Chromis insolata*)
- **Brown chromis** (*Chromis multilineata*)
- **Purple reeffish** (*Chromis scotti*)
- **Yellowtail damselfish** (*Microspathodon chrysurus*)
- **Dusky damselfish** (*Stegastes adustus*)
- **Longfin damselfish** (*Stegastes diencaeus*)
- **Beaugregory** (*Stegastes leucostictus*)
- **Bicolor damselfish** (*Stegastes partitus*)
- **Threespot damselfish** (*Stegastes planifrons*)
- **Cocoa damselfish** (*Stegastes variabilis*)

Background

Damselfishes are small (<8 inches), colorful, but pugnacious reef fishes. Color pattern is usually sufficient to distinguish the species in this family, but you have to be mindful of variations within species, depending on the mating condition or life stage. The genus *Abudefduf* is represented by sergeant majors and night sergeants. Sergeant majors are a relatively deep-bodied form, with distinctive black and yellow alternating bars and a gray face. Reproductively active males develop a bluish cast over their normal color pattern. Night sergeants are a bit chubbier

Sergeant major (*Abudefduf saxatilis*), Lake Worth Lagoon, Florida

Night sergeant (*Abudefduf taurus*)

than sergeant majors and have dark bars over a light-tan to cream-colored body.

Yellowtail damselfishes are another large, thick-bodied species, growing to 8 in long. Their body color ranges from dark-blue (a female's courtship color pattern) to tan or gray (a male's courtship color), with a bright-yellow tail. Juveniles are dark-blue, with a transparent tail fin, and are covered with electric-blue spots. The striking coloration of juveniles provides the basis for the sometimes-used common name "jewelfishes."

Members of the genus *Chromis* are small (<4 in) and hover in the water column. Blue chromises are deep-blue, with black margins on the upper and lower caudal lobes; the back can also be black. Brown chromises have a similar profile to blue chromises but are gray- to brown-colored, with a black spot at the base of the pectoral fin and a pale-yellow spot at the base of the posterior dorsal fin. Yellowtail reeffishes are dark-blue above and white below, with the tail and posterior median fins colored bright-yellow. Smaller individuals have an electric-blue line running through the eyes. Sunshinefishes are strikingly bicolored as juveniles, with golden-yellow on the upper third and bright-purple on the lower two-thirds of the body. Adults are dark-green to olive above and white on the belly, with an electric-blue mark on the upper eye. Purple reeffishes are uniformly cobalt-blue, with small electric-blue spots covering the body. Courting male purple reeffishes have pale blotches on the flanks, with the head and fins retaining their normal blue coloration.

The genus *Stegastes* is composed of site-attached, bottom-dwelling species. Dusky damselfishes are uniformly dark-brown. Juveniles have a red-orange swath over the face and back, against a steely-blue to lavender body; an ocellated spot located at the base of the dorsal

Left: Yellowtail damselfish (*Microspathodon chrysurus*). *Right*: Juvenile, Marquesas, Florida.

Top: Brown chromis (*Chromis multilineata*), Palm Beach, Florida. *Bottom*: Blue chromis (*Chromis cyanea*).

Yellowtail reeffish (*Chromis enchrysura*), Jupiter, Florida

fin; a small black spot on the caudal peduncle; and the face peppered with electric-blue spots. Adult longfin damselfishes superficially resemble dusky damselfishes—uniformly dark-brown, with the same shape. The distinguishing feature, which is subtle in the field, is that in longfin damselfishes the soft dorsal and anal fins extend beyond the base of the caudal fin (hence the vernacular name "longfin"). Juveniles are distinctive: a yellow body, with blue (or sometimes purple) lines extending from the head, over the back, and into an ocellated spot on the dorsal fin.

Beaugregory adults are dark-brown, with olive to golden coloration on the flanks; the tail is often golden. Small blue dots sparsely cover the face, and the median fins often have thin blue margins. Juveniles are royal-blue above and yellow below, with an ocellated spot on the dorsal fin. Cocoa damselfishes look very similar to beaugregories: adults range from dark- to golden-brown, with tiny blue dots on the face and body. There is black spot on the caudal peduncle that separates this species from beaugregories, but in dark-colored adults the spot is difficult to see. Juveniles are bicolored—blue above, yellow below—with the black spot on the caudal peduncle that distinguishes them from juvenile beaugregories. Bicolor damselfishes are black anteriorly and white posteriorly. Although usually half black and half white, the proportion of the body covered by black or white varies. Some individuals have golden-yellow on the belly. Courting adults are pale all

Left: Sunshinefish (*Chromis insolata*), Jupiter, Florida. *Right*: Juvenile, Jupiter, Florida.

Left: Purple reeffish (*Chromis scotti*), Jupiter, Florida. *Right*: Breeding male, Jupiter, Florida.

over, with bits of black around the eyes and near the tail. Threespot damselfishes are caramel-brown, with a black spot on the caudal peduncle and another black spot on the base of the pectoral fin. Courting adults become pale-gray, with a dark band running through the eyes and down to the base of the anal fin. Juveniles are bright-yellow, with a black spot on the caudal peduncle, at the base of the dorsal fin, and at the base of the pectoral fin. In addition, there are thin, dusky bars on the flanks and thin, bright-blue margins on the median fins.

Distribution and Habitat

Damselfishes are found on reefs and structured habitats in coastal, shelf, and high-salinity inshore waters. Most species occur in southeastern Florida and the Florida Keys. Sergeant majors are found around the state, in shelf, coastal, and inshore waters, but they are most common in southeastern Florida and the Keys. Night sergeants inhabit very shallow, surge-prone hardbottom, from Cape Canaveral southward around the state. Yellowtail damselfishes are found from Stuart southward on the east coast, and throughout the Florida Keys, but are rare to absent on the west coast. This species associates with elkhorn, staghorn, and fire coral colonies, in water depths from 3 to 30 ft.

Blue and brown chromises are found on shallow reefs, to about 120 ft depths, from Cape Canaveral on the east coast around to the Florida Keys and Dry Tortugas;

blue chromises have been recorded from the Florida Middle Grounds. Brown chromises occur from Cape Canaveral south through the Florida Keys and to the Dry Tortugas but are rare on the west Florida shelf. Yellowtail reeffishes, sunshinefishes, and purple reeffishes are found around the state, in water depths ranging from 70 to 300 ft.

Several of the *Stegastes* species—including dusky damselfishes, longfin damselfishes, beaugregories, and threespot damselfishes—are most common off southeastern Florida and the Florida Keys. Beaugregories and dusky damselfishes do occur sparsely off the west coast of Florida. Cocoa damselfishes are the only *Stegastes* damselfish found in the northern waters of both coasts. Bicolor damselfishes occur around the west coast, through the Florida Keys, and north to about Cape Canaveral. Most of the *Stegastes* species prefer shallow water (<30 ft), but bicolor damselfishes are the exception, being most common from 50 to 200 ft depths.

Natural History

All of the damselfishes lay adhesive eggs on the substrate, usually on a patch of hardbottom cleaned by the male. The seasonal, lunar, and daily timing, as well as the degree to which nesting sites are defended, varies among species. The egg clusters (nests) are guarded, at times fiercely, by males. The sometimes-complicated social aspects of spawning, nesting, and territoriality vary among species.

In southeastern Florida and the Florida Keys, sergeant majors

Left: Dusky damselfish (*Stegastes adustus*), Palm Beach, Florida. *Right*: Juvenile, Palm Beach, Florida.

Left: Longfin damselfish (*Stegastes diencaeus*). *Right*: Juvenile.

Left: Beaugregory (*Stegastes leucostictus*). *Right*: Juvenile.

Left: Male cocoa damselfish (*Stegastes variabilis*), Lake Worth Lagoon, Florida. *Right*: Juvenile, Jupiter, Florida.

Left: Threespot damselfish (*Stegastes planifrons*). *Right*: Juvenile, Lake Worth Lagoon, Florida.

spawn year round; in other areas, such as the westcentral shelf or the panhandle, spawning is restricted to warmer spring and summer months. Females deposit as many as 20,000 eggs on any hard substrate, including natural rock, limestone, old tires, plastic debris, or fiberglass boat hulls. The eggs are reddish-purple when first laid, and they gradually pale as they develop. During nonspawning periods, sergeant majors are not territorial, forming groups or schools above structured habitat; at spawning times, the male actively defends his nest against all comers.

Attending males don a dark-blue cast over their normal coloration. Eggs hatch, and the larvae are planktonic for some time, after which the juveniles associate with flotsam and sargassum. Young are commonly observed under fixed buoys, moorings, mangrove roots, or other structures that break the water surface. Sergeant majors feed on plankton that is carried by currents over the structured habitats where they live. The natural history of night sergeants is not well known, but reproduction and feeding are thought to be similar to those described here for sergeant majors.

All of the *Stegastes* species hold territories, where they feed, advertise for mates, and fight with rival males. The spawning period begins when the male expands his territory, clears a patch of bottom to serve as a nest, and changes his color pattern. Males advertise to females by bobbing their heads, and several species produce sounds to communicate with mates or rivals.

Bicolor damselfish (*Stegastes partitus*), Lake Worth Lagoon, Florida

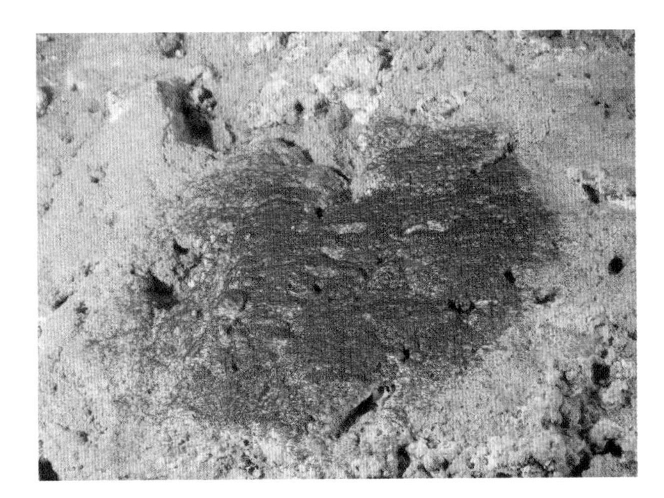

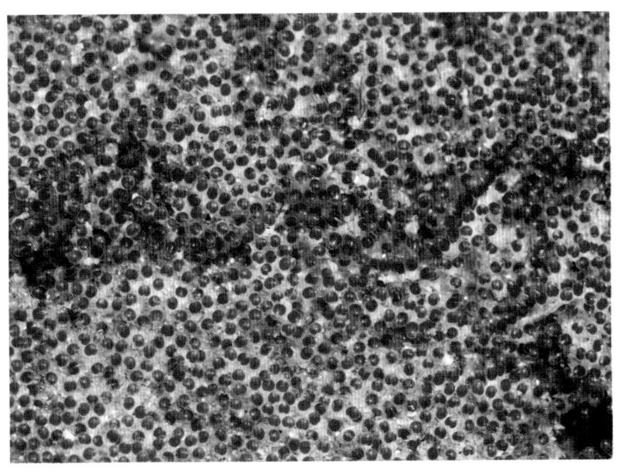

Top: Sergeant major (*Abudefduf saxatilis*) nest, Jupiter, Florida. *Bottom*: Individual eggs, Palm Beach, Florida.

Eggs are fertilized on the nest site, where they adhere to the bottom for 4–5 days before hatching.

Reproduction in the *Chromis* species is less complicated—usually pairs spawn, with the male fertilizing eggs as they are being released by the female. The males signal females by way of color changes. Purple reeffishes become dark above the midline, with pale blotches below. Their eggs are deposited on small areas of hard substrate that have been cleared by the male.

The *Stegastes* damselfishes are omnivorous, leaning mostly toward the algae section of the spread. Threespot damselfishes are called "farmers," because of their propensity for killing portions of live staghorn coral colonies to allow algae to grow in their place (an underwater version of slash-and-burn). Although these fishes do feed on algae, the tiny crustaceans and mollusks attracted to the algal lawn are also on the menu. Threespot damselfishes are by far the most-staunch territory defenders of the bunch. When diving in areas around adult threespot damselfishes, expect to be nipped on any expose surface: face, flanks, and toes. The *Chromis* species hover above reefs and wrecks, feeding on plankton suspended in the passing water column.

Wrasses and Parrotfishes
Family Labridae

- **Spotfin hogfish** (*Bodianus pulchellus*)
- **Spanish hogfish** (*Bodianus rufus*)
- **Creole wrasse** (*Clepticus parrae*)
- **Bluelip parrotfish** (*Cryptotomus roseus*)
- **Red hogfish** (*Decodon puellaris*)
- **Dwarf wrasse** (*Doratonotus megalepis*)
- **Greenband wrasse** (*Halichoeres bathyphilus*)
- **Slippery dick** (*Halichoeres bivittatus*)
- **Painted wrasse** (*Halichoeres caudalis*)
- **Yellowcheek wrasse** (*Halichoeres cyanocephalus*)
- **Yellowhead wrasse** (*Halichoeres garnoti*)
- **Clown wrasse** (*Halichoeres maculipinna*)
- **Rainbow wrasse** (*Halichoeres pictus*)
- **Blackear wrasse** (*Halichoeres poeyi*)
- **Puddingwife** (*Halichoeres radiatus*)
- **Hogfish** (*Lachnolaimus maximus*)
- **Emerald parrotfish** (*Nicholsina usta*)
- **Midnight parrotfish** (*Scarus coelestinus*)
- **Blue parrotfish** (*Scarus coeruleus*)
- **Rainbow parrotfish** (*Scarus guacamaia*)
- **Striped parrotfish** (*Scarus iseri*)
- **Princess parrotfish** (*Scarus taeniopterus*)
- **Queen parrotfish** (*Scarus vetula*)
- **Greenblotch parrotfish** (*Sparisoma atomarium*)
- **Redband parrotfish** (*Sparisoma aurofrenatum*)
- **Redtail parrotfish** (*Sparisoma chrysopterum*)
- **Bucktooth parrotfish** (*Sparisoma radians*)
- **Yellowtail parrotfish** (*Sparisoma rubripinne*)
- **Stoplight parrotfish** (*Sparisoma viride*)
- **Bluehead** (*Thalassoma bifasciatum*)
- **Rosy razorfish** (*Xyrichtys martinicensis*)
- **Pearly razorfish** (*Xyrichtys novacula*)
- **Green razorfish** (*Xyrichtys splendens*)

Background

Ichthyologists recently combined the closely related wrasses and parrotfishes into a single megafamily, the Labridae. Both groups are colorful, large-scaled fishes, with a single long (and, with few exceptions, low) dorsal fin and pectoral fins with oblique bases. Wrasses are slender, laterally compressed fishes, with protrusible mouths armed with separated (and notably canine) teeth. Most species are small (all but three species are <1 ft), while about half the parrotfishes are more than 1 ft long. In contrast, parrotfishes are a stout-bodied, moderately compressed group of species, with nonprotusible mouths bearing teeth fused into solid upper and lower plates, resembling a parrot's beak. In most wrasses and parrotfishes, color patterns vary with age and gender, resulting in a confusing array of types. Males tend to be more brightly colored than females, and juveniles are colored differently

Left: Hogfish (*Lachnolaimus maximus*). *Right*: Juvenile, Indian River Lagoon, Florida.

than adults; these changes in coloration make wrasses and parrotfishes some of the more-difficult species to identify among Florida fishes. All species propel their stiff bodies by employing pectoral-fin sculling, essentially rowing their way through life. And then there is sex—most are born as females and, depending on the species, change sex with age, growth, and social structure.

Terminal phase male hogfish (*Lachnolaimus maximus*), Marquesas, Florida

Hogfishes, Spanish hogfishes, and spotfin hogfishes (also known in Florida as "Cuban hogfishes") are a related group that can easily be distinguished by body shape, coloration, and size. Hogfishes, often erroneously called "hog snappers," are the largest of the wrasses, at 3 ft long and weighing about 20 lb. They have very deep bodies; long, straight foreheads (napes) and snouts, and large jaws. The trailing tips of the caudal fin and the free anterior dorsal spines are elongated. The head profile of terminal males changes from the relatively steep angle seen in immature individuals to a more curving profile that, along with their prominent tusks, gives the snout a hog-like appearance. Terminal males are pale-red laterally, with dark-purplish-red on the top of the head down to the midline of the eye (which is bright-red) and the base and leading edges of the caudal fin. Small juveniles have wavy, thin, pale bars against a reddish-brown background; larger individuals range from pale-gray to dark-red, with indistinct bars that vary with the environmental background. At all sizes, there is a small dark spot located at the rear base of the dorsal fin.

Spanish and spotfin hogfishes have similar body shapes, pointed snouts, and extended tips of the dorsal, anal, and caudal fins, but they differ in coloration. The anterior upper-two-thirds of Spanish hogfishes (18 in) is bluish-purple, and the area below and posterior to the eyes and pectoral fins is bright-yellow. Early juveniles are bluish-purple on the anterior half

Left: Terminal phase spotfin hogfish (*Bodianus pulchellus*), Jupiter, Florida. *Right*: Juvenile, Hobe Sound, Florida.

Left: Terminal phase Spanish hogfish (*Bodianus rufus*), Palm Beach, Florida. *Right*: Juvenile, Palm Beach, Florida.

of the body and bright-yellow on the posterior half. Some older individuals are almost-completely bluish-purple, with only a smattering of yellow remaining. Tusk-like teeth project from the lower jaws of these larger males. Spotfin hogfishes are smaller in size (8 in) and are easily distinguished by their fire-engine-red body; a midlateral, whitish stripe that trails off posteriorly; and a bright-yellow rear part of the dorsal fin, caudal peduncle, and tail (the leading edge of the lower caudal lobe is edged in red). The upper tip of the pectoral fin has a black spot, and the dorsal and anal fins are edged in blue or black. Early juveniles are uniformly bright-yellow, with a black spot on the front of the dorsal fin.

Dwarf wrasses (3 in) differ from other wrasses in having the dorsal fin clearly demarcated into two parts, with the anterior portion taller than the posterior one. They are relatively deep bodied and laterally compressed; have a fairly pointed snout; and range in color from pale-green (with pink stripes) to dark–emerald-green (with flecks of white and orange). Creole wrasses, with their deeply forked tails and streamlined bodies, look nothing like the other members of the group (they actually resemble damselfishes in the genus *Chromis*). Juveniles and initial-phase individuals are uniformly deep purplish-blue, with a series of light spots or miniblotches on the upper body, under the dorsal fin. Terminal males are bicolored: the front half is similarly purple, and the rear half predominantly golden-yellow. The snout is dark-purple to black.

Razorfishes are extremely laterally compressed, with steep, vertical foreheads and small eyes, placed high up. Pearly and green razorfishes have rounded caudal fins. Pearly razorfishes (8.5 in) have multiple vertical, electric-blue lines on the head. Terminal males are

Dwarf wrasse (*Doratonotus megalepis*), Indian River Lagoon, Florida

Top: Terminal phase male creole wrasse (*Clepticus parrae*), Palm Beach, Florida. *Bottom*: Initial phase, Palm Beach, Florida.

pearly white, with a pinkish cast laterally, and have pinkish dorsal and anal fins, with thin, variegated blue lines. There is a diffuse, oblique, reddish-orange blotch under and behind the tip of the pectoral fin. Intermediate-phase individuals are pale-bluish to green; the head has several thin blue bars; and each body scale has a thin blue line. Young pearly razorfishes are pale, with 4 diffuse body bars. Green razorfishes (5.5 in) are greenish-brown, with about 5 vertical orange bars on the head and elongated pelvic fins. Terminal males are lime-green to orangish, with 1 or more small, conspicuous, ocellated black spots at midbody, near the tip of the pectoral fin. The dorsal, anal, and caudal fins have thin, wavy, light-blue lines and pink to brick-red margins. Females are pearly whitish to green, with 5–6 grayish to orangish-brown bars and a series of small white spots along the base of the dorsal fin. The dorsal and anal fins are pale-maroon, with thin, light-blue lines. The first 2 dorsal spines are elongated in juvenile green razorfishes; in combination with their olive-green to dark-brown bodies, they might be confused with dwarf wrasses. Rosy razorfishes (6 in) have straight margins on their caudal fins, and small dark spots in the pectoral-fin armpits. Terminal males are reddish-pink; their pale-yellow heads have faint blue bars and a dark blotch on the gill cover. Individual body scales have blue centers. In females, the area behind and below the pectoral fin is pale, with thin, vertical red lines and black spots; this nearly transparent abdominal window may indicate the stage of ovarian development to interested males. There is a dark vertical blotch on the operculum.

Small, cigar-shaped members of the genera *Halichoeres* and *Thalassoma* undergo some degree of developmental color change, resulting in multiple (minimally three) phases: juveniles, immature adults (female or male), and terminal males (also known as supermales). Adding to the complexity are transitional phases between these three life stages. Puddingwives, the largest (18 in) in this group, have a deeper body than other members of these two genera. Both sexes, as well as subadults, have blue spots on the fins and body; blue lines on the head; yellow margins on the caudal fin; and a small dark spot on the upper base of the pectoral fin. Terminal males are blue-green to golden-yellow, usually with a broad, light, midbody bar off the tip of the pectoral fin; when in this color phase, they might be confused with parrotfishes. Initial-phase puddingwives have bright-orange bodies, with green central pigmentation on each flank scale; 5 wide white bars cross the upper back. Juveniles have a rusty-orange body, interrupted by blocky white bars and stripes; there is a large, blue-ringed black spot on the upper midbody and lower portion of the dorsal fin.

Yellowcheek wrasses (12 in) have yellow heads, upper flanks, and caudal fins; a broad, blue-black to purple lateral stripe extends from the operculum posteriorly onto the tail. Terminal males have a blue crest on the nape; the basal portion of the posterior part of the dorsal fin is blue-black; and there is a tilde-shaped (~), black-edged yellow blotch immediately behind the eye. Juveniles are bright-purple below the yellow head and shoulders.

Yellowhead wrasses (7.5 in) have thin dark lines and spots radiating from the eyes, and thin blue margins on the dorsal and anal fins. Terminal males have golden-yellow heads, separated from a grayish-green body by a diagonal, jet-black bar that turns upward and forms a dorsal stripe running posteriorly toward the caudal fin. The upper part of the head and the shoulders are grayish-brown in

Left: Terminal phase pearly razorfish (*Xyrichtys novacula*), Lake Worth Lagoon, Florida. *Right*: Initial phase, Lake Worth Lagoon, Florida.

Left: Terminal phase green razorfish (*Xyrichtys splendens*), Lake Worth Lagoon, Florida. *Right*: Initial phase, Lake Worth Lagoon, Florida.

Left: Initial phase puddingwife (*Halichoeres radiatus*), Lake Worth Lagoon, Florida. *Right*: Juvenile, Jupiter, Florida.

Left: Yellowcheek wrasse (*Halichoeres cyanocephalus*), Juno, Florida. *Right*: Juvenile, Jupiter, Florida.

the initial phase; the remainder of the head and the body usually are pale–yellow-orange. Early juveniles are orange, with a thin, bright-blue, midlateral lateral stripe.

Although not a member of the *Halichoeres* or *Thalassoma* genera, red hogfishes (6 in) share their look (slender bodies, with a gently sloping forehead), but the tips of the caudal fin are slightly elongated (vs. truncate). The back is pale-red; the belly is whitish; yellow spots mark the upper head, flanks, and tail; and yellow lines radiate from the eyes. Greenband (9 in) and painted (8 in) wrasses also have a pale-red to salmon base body coloration. Greenbands have a bright-yellow midlateral stripe that is continuous in the initial phase and broken in terminal males. A green band extends from the tip of snout to the eye in both phases, and the line continues to the upper edge of the operculum in males. Males have an elongated dark spot over the pectoral fin, and a tiny black spot is found on the midbase of the tail in initial-phase individuals.

Painted wrasses (8 in) have bright-blue lines on the head and usually show a tiny black spot on the base of the mid-dorsal fin. Terminal males also have a dusky opercular spot behind the eye and are pale-green dorsally, grading to pink laterally, with a series of blue spots running the length of the body. Two dusky lateral stripes—1 running through the eye and 1 below, running through the pectoral fin—occur in the initial phase. Juveniles are whitish to pale-pink, with a dark lateral stripe; a series of light-blue spots is just above the stripe.

Several wrasses have black or dusky dorsal-fin blotches or spots in one or all color phases. Blueheads (7 in) change color con-

Top: Terminal phase male yellowhead wrasse (*Halichoeres garnoti*), Jupiter, Florida. *Bottom*: Juvenile, Jupiter, Florida.

Red hogfish (*Decodon puellaris*)

siderably as they grow, but they always have an elongated black blotch at the front of the dorsal fin. Newly settled young are bright-yellow or white, with or without a pronounced black lateral stripe. In the initial phase, 7–8 mostly bicolored (lighter above, darker below) saddles overlay the yellow coloration, and a purplish stripe extends from the snout through the eye. Mature males have deep-blue heads, separated from a blue-green body by 2 broad black bars that almost form a V and border a thick white wedge.

The black spot on the dorsal fin of clown wrasses (5.5 in) is rectangular in shape and located about a quarter of the way back from the fin origin. Reddish-orange lines run through and around the eyes, and a small blue spot is found on the base of the pectoral fin. The anterior part of the upper body of terminal-phase males is greenish to greenish-blue, fading to greenish-yellow posteriorly; a black spot occurs in the center of the body, just posterior to the pectoral fin. Initial-phase individuals have interrupted, reddish-orange to pink lines laterally. Juveniles are brown (sometimes roan) dorsally and white below, with a broad, black, lateral stripe.

Many (but not all) juvenile and initial-phase slippery dicks (9 in) have a spot on the mid-dorsal fin; the variability of this character is emblematic of the species, which has more color variation than most wrasses. A bicolored (dark- and bright-colored) spot is found on the opercular margin in the initial and terminal phase, and all phases may have a dark midlateral stripe, composed of series of diffuse spots. Juveniles are either white, with a dark lateral stripe, or dark–reddish-brown dorsally and pale below, with a dark lateral stripe. Larger juveniles similarly vary between white and dark-brown, but they may develop a second, ventrally

Left: Initial phase greenband wrasse (*Halichoeres bathyphilus*). *Right*: Juvenile, Jupiter, Florida.

Left: Terminal phase painted wrasse (*Halichoeres caudalis*), Lake Worth Lagoon, Florida. *Right*: Juvenile painted wrasse, Venice, Florida.

placed lateral stripe as they grow into the initial phase. The light and dark color patterns may indicate a geographical dispersal, with the darker individuals more prevalent in northern, continental habitats of the peninsula, and the lighter individuals more common in southeast Florida and the Florida Keys. Initial-phase females are greenish, with 2 lateral stripes. Terminal males are variably pale-green and have orangish-pink lines on the face, body, and tail.

Juvenile and initial-phase blackear wrasses (8 in) occasionally have a dark spot on the mid-dorsal fin, and all phases have a small dark spot at the rear base of that fin. Early juveniles are brown-orange on the head, have a white lateral stripe, and have a series of black blotches running along the flank. Later juvenile and initial-phase individuals are uniformly lime-green, with a tricolored (black–yellow–reddish orange) spot behind the eye (the "black ear"). Males are drab-green, with some pale-orange blotches and spots; the head is pale-orange, with a dark-orange to purplish spot, rimmed in electric-blue, just behind the eye.

Rainbow wrasses (4.5 in) lack dorsal spots or blotches but, in the terminal phase, do have an elliptical blotch at the base of the caudal fin. The dorsal fin is orange; an orange streak, rimmed in neon-blue, bisects the tail. Horizontal, neon-blue lines are found above and below the eyes. Initial-phase individuals and juveniles have a midlateral stripe that is darkest from the snout to about the tip of the pectoral fins and then fades posteriorly, and a dark-brown stripe under the base of the dorsal fin.

As with the wrasses, color pattern, body shape, and size provide clues to recognition for the parrotfish species. Bluelip parrotfishes are the smallest (≤5 in) species in the family, and their slender bodies

Left: Terminal phase bluehead (*Thalassoma bifasciatum*), Palm Beach, Florida. *Right*: Initial phase, Lake Worth Lagoon, Florida.

Left: Terminal phase male clown wrasse (*Halichoeres maculipinna*), Lake Worth Lagoon, Florida. *Right*: Initial phase clown wrasse, Hobe Sound, Florida.

give them more of a wrasse-like appearance. They are the only parrotfish species without fully fused teeth; only the bases are fused, and the distal teeth appear to be separate. Early juveniles are pale, with reddish-pink to olive-green backs, with or without a wide, white, lateral band. Initial-phase individuals range from light-brown to red, with white undersides. The terminal phase is darker green, with a wide, pinkish, lateral stripe; 2 pink lines run diagonally from the base of the eye to the mouth; and the upper portion of the eye has a bright-blue dash. Terminal males also have blue on the lower jaw and inside the mouth (hence the common name). Emerald parrotfishes (8 in) are a relatively drab species. Juvenile and initial-phase individuals are pale-green to brown, often with 2 white, lateral stripes that may fade to a more variegated pattern. The median fins and tail may be reddish, with thin, wavy lines. A reddish line runs diagonally from the eye to the origin of the jaw, and 2 small, light-blue marks occur behind the eye socket. The terminal phase is dark- to emerald-green, often with 6 darker, irregular bands and a red-orange iris.

Parrotfishes of the genus *Scarus* are united in having an upper jaw that overhangs the lower jaw, and teeth that are are completely united. Adult blue, midnight, and rainbow parrotfishes have elongated extensions of the caudal fins. Initial- and terminal-phase blue parrotfishes (4 ft) are uniformly dark-blue. In the terminal phase, they develop a prominent, squared-off head profile. Juveniles are pastel-blue, with yellow skullcaps over the back from just forward of the eyes, tapering posteriorly to the front part of the dorsal fin.

Midnight parrotfishes (3 ft) also

Left: Initial phase slippery dick (*Halichoeres bivittatus*), Florida Middle Grounds. *Right*: Initial phase, Lake Worth Lagoon, Florida.

are easy to recognize, being indigo-blue overall (although lateral scales often have lighter-blue centers), with royal-blue patches on the head at all ages and sizes. Rainbow parrotfishes (>4 ft) are the largest of the parrotfishes. Initial-phase rainbows and juveniles have orange heads and bodies; the scales on the flanks have blue-green centers; the outer margin of the pectoral fins is orange; and the dorsal and anal fins have electric-blue margins. Terminal males are dull orange over the front half of the body and dark–emerald-green posteriorly; the tail is orange, with a narrow, green, outer central edge. The face often develops a deep purplish-red hue, and the blue-colored, fused teeth frequently sport tufts of filamentous algae.

In their initial phase, queen parrotfishes (19.75 in) have a whitish head and a dark-gray to brown body, with a broad, white, lateral stripe. Terminal males are blue-green, with dark-pink edges on the scales; darker blue-green and orange bands are found above and below the mouth. The base and outer portions of the lobes of the lunate caudal fin are orange to pink, with the leading edges and central section being blue-green; the pectoral fins are blue-green, with an orange-pink, upper central streak. The dorsal and anal fins are orange-pink basally, with blue-green margins.

Terminal-phase princess parrotfishes (11.75 in) differ in having the orange-pink stripe sandwiched by 2 blue-green stripes; a blue, truncate caudal fin, with thin, submarginal caudal stripes; and a broad, rectangular yellow blotch under and behind the pectoral fin. They are blue-green laterally, with an orange lower head and 2 broad, blue, lateral stripes above and below the eye. Terminal-phase male striped parrotfishes (11 in) are the ones most likely to be confused with terminal-phase male princess parrotfishes. The former are distinguished by having broader, blue leading-edge margins on the

Terminal phase male slippery dick (*Halichores bivittatus*), Lake Worth Lagoon, Florida

Left: Blackear wrasse (*Halichoeres poeyi*), Lake Worth Lagoon, Florida. *Right*: Juvenile, Jupiter, Florida.

Left: Terminal phase bluelip parrotfish (*Cryptotomus roseus*), Lake Worth Lagoon. *Right*: Initial phase, Lake Worth Lagoon, Florida.

truncate caudal fin, and a central, blue section, with thin, wavy, orange-pink stripes; they also often have a broad, gray-yellow-pink midlateral stripe. Terminal-phase male princess parrotfishes are blue-green laterally, with the body scales edged in orange; have 2 dark lines running through the eye; and bear a rectangular, pink-orange–yellow-white blotch under and behind the pectoral fin. Initial-phase princess and striped parrotfishes are tricky to distinguish. Both species are white, with 3 dark-brown lateral stripes. The yellow snouts of young princess parrotfishes are lacking in striped parrotfishes.

Members of the other major genus of parrotfishes, *Sparisoma*, have a lower jaw that overhangs the upper jaw, and fused teeth that still retain individual integrity. The most-distinctive species are stoplight parrotfishes (2 ft), which, in the initial phase, have a deep-red to pink belly; a dark-brown back, with regularly spaced white spots; and a light-gray to brown head. All fins (except the lunate caudal fin) are deep red; the white, lunate, caudal fin has a broad red marginal band and narrow red leading edges. Terminal males are blue-green laterally, with pink-edged scales. There is a yellow blotch on the caudal peduncle; a yellow, submarginal crescent on the tail; and a small bright-yellow spot on the upper operculum.

Initial-phase yellowtail parrotfishes (18 in) have grayish-brown bodies, often with 4–5 irregular dark bars; yellow caudal peduncles; and almost-truncate tails. Terminal-phase males are blue-green overall, with yellow tails; the base of the upper pectoral fin has a black spot. Juveniles are a drab, mottled gray-brown. Regardless of phase, the dorsal, anal, and pelvic fins and the central part of the tail are red in redtail parrotfishes (18 in); a dark spot is always found on the

Left: Adult emerald parrotfish (*Nicholsina usta*), Lake Worth Lagoon, Florida. *Right*: Initial phase, Indian River Lagoon, Florida.

Left: Adult blue parrotfish (*Scarus coeruleus*), Looe Key, Florida Keys. *Right*: Juvenile, Lake Worth Lagoon, Florida.

Adult midnight parrotfish (*Scarus coelestinus*), Lake Worth Lagoon, Florida

base of the upper pectoral fin. Initial-phase redtail parrotfishes are mottled, from pink to pale-red. Terminal-phase males are blue-green, with an elongated blue patch on the flank, just behind the pectoral fin. In both redtail and redband parrotfishes, the caudal fin is almost truncate in initial-phase individuals, and lunate in terminal-phase ones. Juvenile redband parrotfishes (11 in) are a mottled brown, with a conspicuous white spot at the posterior base of the soft dorsal fin. That white spot is retained in the initial phase, in which the body is a mottled blue-green above and a mottled dark-red below; all fins except the clear pectorals also are a

Rainbow parrotfish (*Scarus guacamaia*), Lake Worth Lagoon, Florida

Top: Terminal phase queen parrotfish (*Scarus vetula*). *Bottom*: initial phase, Looe Key, Florida Keys.

mottled red. Terminal-phase males have green heads, with a bright-red line running from the corner of the mouth to the upper corner of the operculum; a small yellow blotch, with 1 or more small black spots, is found above the base of the yellow pectoral fin.

Bucktooth parrotfishes (7 in) are a mottled olive-brown in the initial phase; the base of the pectoral fin and the margin of the operculum are metallic blue-green. Terminal-phase males are green-brown, with an elongated black blotch on the base of the pectoral fin and a black margin on the rounded tail. Greenblotch parrotfishes (4 in), in their initial phase, are red on the back, with a series of discontinuous red stripes on the flanks. Terminal-phase males are green, with a reddish hue on the midflanks, caudal peduncle, and tail; the belly is bluish-grey. A rectangular green blotch is found above the pectoral fin.

Distribution and Habitat

Wrasses and parrotfishes are cornerstone species on tropical reefs, and Florida has the full complement of Caribbean species inhabiting its reefs, wrecks, and seagrass, mangrove, and hardbottom habitats, from the shelf edge to high-salinity inshore regions. Wrasses segregate loosely by water depth and substrate preference. Slippery dicks, puddingwives, and clown, rainbow, yellowhead, creole, yellowcheek, and blackear wrasses prefer structured habitats. Slippery dicks do not commonly occur in the Keys; rather, they are a continental form that is most common north of Biscayne Bay and the Ten Thousand Islands regions, including off the Florida panhandle, in depths of 1–25 ft. Puddingwives and clown, rainbow, yellowhead, creole, and yellowcheek wrasses are found on the southeast coast, from Jupiter to the Keys, in, respectively, 2–55 ft, 1–85 ft, 15–77 ft,

Top: Terminal phase striped parrotfish (*Scarus iseri*), Lake Worth Lagoon, Florida. *Middle*: Initial phase, Lake Worth Lagoon, Florida. *Bottom*: Juvenile, Palm Beach, Florida.

3–85 ft, 20–100 ft, and 60–150 ft depths, demonstrating the vertical gradation of depth chosen within a single ecogroup. Blackear wrasses are more widespread; they have been recorded from waters 8–40 ft deep, in the area between Cape Canaveral to Choctawhatchee Bay in the panhandle. Their wider distribution may reflect a more-plastic lifestyle, since they are equally at home in shallow seagrass meadows, particularly those near reefs, such as in the Florida Keys. Blueheads also have a wider distribution; they are known from Jupiter (in the southeast) to St. Andrews Bay (in the panhandle). Greenband wrasses are similarly distributed, but from a deeper perspective; they are documented from Daytona Beach to Cape San Blas, but only on deep, mesophotic (areas of limited sunlight) reefs, in depths of 120–433 ft. Painted wrasses also can occur in deeper depths (15–968 ft) in Florida; they are known from the Keys and the Gulf of Mexico north to Destin in the panhandle.

Dwarf wrasses reside in seagrass beds from Vero (on the east coast) southward through the Florida Keys. They are absent or rare on the western peninsular coast, but they are slightly more common along the panhandle. Spanish hogfishes are rare north of Jupiter Inlet on the east coast, but they are more common in the Keys. Similarly, spotfin hogfishes can be found from off of Daytona southward, but they are not as plentiful there as in the Keys. While both species are known from the Gulf, their abundances must be low, given that we have no museum specimens from Florida's Gulf waters. Both Spanish and spotfin hogfishes are found on reefs and wrecks in deeper waters (>30 ft). Spotfin hogfishes are known to occur on deeper reefs (to depths just under 400 ft), but our Florida depth records for the species are from 1 to 145 ft. Spanish hogfishes are said to inhabit somewhat shallower depths, but our records from Florida mirror the depth range of the spotfins (5–150 ft). Like greenband wrasses, red hogfishes are only found on deep reefs, in depths of 182–607 ft. Adult hogfishes tend to be found in mid- to outer-shelf waters (30–250 ft deep) these days, but the young reside in shallower seagrass meadows and on sponge bottom; our depth records are from

3 to 85 ft. The outer part of the west Florida shelf (including the Florida Middle Grounds) harbors a concentration of very large adult hogfish.

Razorfishes occur in groups (colonies) over sandbottom, usually near reefs or other structures, in shelf and saline inshore waters. Two species are predominantly insular in distribution, being most common off southeast Florida and the Florida Keys: green razorfishes (Stuart to the Keys; 8–95 ft deep) and rosy razorfishes (Biscayne Bay to the Keys; 20–25 ft deep). Pearly razorfishes, with continental affinities, are more widespread in Florida's shelf waters (3–643 ft deep), from Ft. Pierce to Choctawhatchee Bay.

Unlike wrasses, which have deeper-water representatives, parrotfishes are strictly shallow-water dwellers. While some wrasses tolerate cooler, deeper waters and nonstructured habitats, parrotfishes need be in the immediate vicinity of algae, seagrasses, or (in some instances) sponges, all of which only grow in well-lit, shallow waters. Juveniles of most species can be found in seagrass meadows or other shallow habitats. Young rainbow parrotfishes prefer mangrove areas throughout the Florida Keys, Florida Bay, and Biscayne Bay, but they are also associated with rock jetties and rip-rap. Redtail and yellowtail parrotfishes live on reefs, in clear water, but they are also among the most-common species on nearshore reefs and inshore seagrass meadows, where water clarity is more variable. Adult bucktooth, yellowtail, bluelip, and emerald parrotfishes commonly reside in seagrass meadows. Species that prefer the clear waters of reefs and adjacent hardbottom habitat include blue, midnight, princess, and rainbow parrotfishes. Some species—such as greenblotch, redband, redtail, stoplight, and yellowtail parrotfishes—are equal-opportunity types that can be found in both habitats, their

Left: Terminal phase princess parrotfish (*Scarus taeniopterus*), Hillsboro, Florida. *Right*: Initial phase.

Left: Terminal phase stoplight parrotfish (*Sparisoma viride*), Lake Worth Lagoon, Florida. *Right*: Initial phase, Lake Worth Lagoon, Florida.

Left: Terminal phase yellowtail parrotfish (*Sparisoma rubripinne*), Lake Worth Lagoon, Florida. *Right*: Initial phase, Lake Worth Lagoon, Florida.

Left: Terminal phase redtail parrotfish (*Sparisoma chrysopterum*), Lake Worth Lagoon, Florida. *Right*: Initial phase, Lake Worth Lagoon, Florida.

choices often dictated by their life-history stage.

Habitat preferences and availability have resulted in two major distributional patterns among Florida's parrotfishes. One group, largely confined to southeastern Florida (from about off of Vero to Biscayne Bay) and the Keys, includes blue, greenblotch, midnight, princess, and queen parrotfishes. These are dyed-in-the-wool tropical, insular, reef-dwelling species. The second distributional pattern covers southeastern Florida (from about Sebastian Inlet to Biscayne Bay) and the Keys, but also extends into the northeastern Gulf of Mexico, off the Florida panhandle. This group consists of bluelip, bucktooth, emerald, rainbow, redband, redtail, stoplight, striped, and yellowtail parrotfishes, including some species with continental affinities (e.g., emerald parrotfishes) and preferences for continental-like habitats, namely seagrass beds.

Natural History

Wrasses have complex mating systems, involving sex reversals, harems, supermales, sneaky males, pair spawning, and group spawning. On the surface, this sounds like potential material for daytime television scripts, but the mating behavior in wrasses illustrates the variety of ways individuals from one family contribute their offspring to future generations. Once all the gyrations and mating displays are over with, wrasses and parrotfishes spawn in the water column, usually in an upward rush toward the surface, either in groups or as pairs. Reproductive styles differ slightly among the wrasse genera. Shallow-dwelling *Halichoeres*—such as slippery dicks, puddingwives, and clown, blackear, and yellowhead wrasses—may be born as males or females. Many of the females change sex with age, size, and social structure, and some

Top: Terminal phase redband parrotfish (*Sparisoma aurofrenatum*), Lake Worth Lagoon, Florida. *Bottom*: Initial phase, Jupiter, Florida.

males will remain subdominant for their entire lives.

For the *Halichoeres* species that have been studied, dominant males (also known as terminal males or supermales) patrol plots on the seafloor that represent their home ranges, but these are not defended territories. Within a plot, a terminal male stays busy courting and trying to mate with all females in sight, as well as incessantly harassing subdominant males. Courting involves fin flarings and open-mouth displays; similar displays, coupled with rapid rushes, are used to run off any trespassing males. The subdominant males often get the last laugh, however, as they mate with females in a clandestine fashion, whenever possible; they have been dubbed sneaky males for their backdoor style. Less is known about mating in the deeper-dwelling yellowcheek, painted, rainbow, and greenband wrasses, but it is probable that pair spawning is more prevalent, as these deeper-dwelling species are less abundant, and mates are harder to find. In general, spawning occurs in the late afternoon, but tidal flow also has an influence, with outgoing tides being the trigger for some species.

The mating system of blueheads expands on the *Halichoeres* theme with two notable additions: fiercely defended spawning territories, and organized group spawning—no sneaking required. Not all blueheads are born as females, and subdominant males participate in group spawns with multiple individuals. You can observe this phenomenon over patch reefs in the Florida Keys. A group of similarly colored (yellow), initial-phase fishes (usually multiple males per female) mill around above a reef; suddenly, all individuals rapidly dart upward a few feet and then return immediately to the reef. This is an impressive feat of synchronization, and if you blink you will miss it. Terminal males mate with females that enter their territories, which are established along the margin of a reef or structured habitat. Spawning sites used by dominant males tend to be employed repeatedly over time. Spawning occurs daily during summer months and is often related to the tidal phase. Little is known, however, about mating in dwarf wrasses in Florida. Observations made in the Caribbean indicate that males mate with multiple females, in territories within seagrass beds.

The spawning products of true wrasses transform into planktonic larvae, which spend a relatively brief period (average of 25 days) in the plankton. Several of the *Halichoeres* species settle from the plankton to the seafloor and spend their first postsettlement week or so buried in the sand. *Halichoeres* and *Thalassoma* wrasses feed opportunistically on small invertebrates, such as crustaceans, worms, and brittle starfish. They obtain easy meals by following stingrays, flying gurnards, snake eels, yellow goatfishes, and anything else that flushes out potential prey from the sediment. Some slippery dick, bluehead, and yellowcheek wrasse individuals pick external parasites (worms and crustaceans) from

Bucktooth parrotfish (*Sparisoma radians*), Lake Worth Lagoon, Florida

various fish clients, including groupers, barracudas, and moray eels.

Creole wrasses spawn after some courtship, which involves bicolored males brushing against available females. The willing female is pushed towards the surface by the courting male until the pair release gametes into the water column. Courting males can observed zooming around the reef near dusk in pursuit of mates.

Hogfishes are protogynous hermaphrodites: all individuals in this species are born as females and change sex with age and social status. The hogfish social system can be called haremic, as a single dominant male defends and mates with only a particular group of females. Once the male dies or leaves, one of the females, usually an older, larger individual (presumably the one with the highest social status), will transform into a male and assume that role within the harem. Spanish hogfishes also form harems, with a larger, distinctively colored male presiding. Spawning tends to occur from late in the afternoon to dusk. Hogfishes aggregate to spawn during winter months in the Caribbean Sea and the Bahamas; Florida hogfishes probably also aggregate. Some researchers have asserted that the size of hogfish transitional males has gotten progressively smaller, in response to fishing pressure. Hogfishes rarely take baited hooks (but will readily eat a live shrimp). This species is a favorite among spearfishers, not only because of its taste and texture, but also because of its trusting nature, making hogfishes an easy target. On the west Florida shelf, the largest adult hogfishes are found in deeper, offshore waters, probably a consequence of higher spearfishing pressure in shallow water. The same can be said about the east coast, where the largest individuals are found in water depths greater than 100 ft. Larval hogfishes settle in shallow waters, particularly in seagrass meadows. Adults forage for molluscs over sand-, shell-, and seagrass bottoms that usually are not far from artificial or natural reefs. Hogfishes have well-developed muscles and

Left: Terminal phase greenblotch parrotfish (*Sparisoma atomarium*), Hobe Sound, Florida. *Right*: Initial phase, Jupiter, Florida.

crushing teeth in their throats that form a second set of jaws—the pharyngeal jaws. Spanish hogfishes, spotfin hogfishes, and hogfishes can crush small snails and clams with ease. Juvenile Spanish and spotfin hogfishes are known to remove ectoparasites from client fishes. Little is known about the life history of red hogfish. On mesophotic reefs in the northeastern Gulf of Mexico, we have observed individuals swimming head-down at a 45 degree angle, presumably foraging.

Razorfishes generally form harems, where a larger, brightly colored male will squire a group of 5–10 females within a defined territory or area. Males advertise to passing females and chase off intruders of different species. All individuals apparently are born female; a transition to the male sex occurs when social pressures dictate (e.g., loss of the dominant male). Razorfishes feed opportunistically on sand-dwelling invertebrates, but rosy razorfishes also feed on plankton. As a consequence of inhabiting featureless sand plains, razorfishes are very adept at diving head-first into the sand when approached too closely by potential predators. Pearly razorfishes even mark their favorite sand-diving spot with a small pile of shells and rocks. The other two species will dive in at any time and place.

Parrotfishes spawn in the water column, in a rush to the surface. Similarly colored intermediate males and females spawn in groups, although brightly colored males only mate with single females. Terminal male parrotfishes patrol seafloor areas, warding off rival males and courting passing females. Some species in this group aggregate to spawn, others spawn only within their home range, and some may do both. Parrotfishes are primarily herbivores, with fused scraping teeth and long intestines that allow the gradual digestion of plant material—mostly algae. Feeding styles vary among the species and include browsers, scrapers, and excavators. Scrapers use their fused teeth to scrape algae off limestone substrates. Excavators, such as stoplight and queen parrotfishes, feed on algae embedded in limestone rock or living coral by breaking off chunks of it with their powerful jaws and then pulverizing the hard parts with their pharyngeal mills (grinding plates in their throats). The underwater sounds of rock-munching parrotfishes are very impressive. Algae are digested, along with any other organic material, and the sedimentary byproduct is passed into the water column, creating a minor turbidity plume of newly milled sand. Scientists have estimated that an adult stoplight parrotfish can produce 26 lb of sand per square yard per year.

Although herbivory is the norm, some parrotfish species (including striped, princess, redband, redtail, yellowtail, and rainbow parrotfishes) are not averse to sponges. In the Florida Keys, young rainbow parrotfishes feed on sponges that grow on mangrove prop roots. Emerald and bluelip parrotfishes are browsers, eating seagrass blades and nibbling algae off living or nonliving surfaces. Bucktooth parrotfishes nip seagrass blades and seem to prefer older blades that are covered with encrusting calcareous organisms. Presumably this aids in maintaining balanced pH levels in the alimentary tract (an antacid treatment of sorts).

Parrotfishes are tireless feeders, spending more than 90 percent of their waking hours feeding; individuals may completely turn over the material in their digestive tracts 10 times per day. All of this has important consequences for the ecology of the reefs. Through their feeding activities, parrotfishes create space for new attached organisms, manufacture sediment, promote coral growth, and prevent overgrowths of fleshy algae.

If you go night diving on reefs in the Florida Keys or southeast Florida, you will very likely observe snoozing parrotfishes. They lie around the bottom, looking a bit like decorative parade floats. Many are enveloped in mucous cocoons while sleeping. The adaptive value of the mucous cocoon is unknown, but it may deter predators, or even act as a mosquito net, keeping external parasites off the sleeping fishes.

Stargazers
Family Uranoscopidae

- **Northern stargazer** (*Astroscopus guttatus*)
- **Southern stargazer** (*Astroscopus y-graecum*)
- **Lancer stargazer** (*Kathetostoma albigutta*)
- **Freckled stargazer** (*Xenocephalus egregius*)

Background

Stargazers, like frogfishes and scorpionfishes, are grumpy looking, lie-in-wait predators, with large, upturned mouths that bear fleshy tabs. Northern and southern stargazers grow to about 22 and 18 in, respectively. Pretty much look-alikes, they have small white spots over a base coloration of brown to light-gray; the underside is white. The spots on northern stargazers are a bit smaller than those on southerns. The tails have a black dorsal and ventral margin, and a central black band; the latter extend onto the body in northerns and stays on the tail in southerns. Unlike these species, lancer and freckled stargazers lack a spinous dorsal fin and scales. Lancer stargazers are brown on the back, with small white spots, but they also differ in having a pair of stout spines, projecting upward from the shoulder girdles; 2 spots on the dorsal fin; and 3 spots on the tail margin. Freckled stargazers are brown to green, with a squared-off head profile; have bony ridges on the top of the head; and have a large, dark, rectangular blotch on the tail.

Distribution and Habitat

Northern stargazers occur from Cape Canaveral northward, and southern stargazers are found around most of the state. They mainly inhabit nearshore waters, but range to depths of about 250 ft. Most of the time both species live buried in the sand, with little more than their eyes showing. Lancer and freckled stargazers occur statewide, in depth ranges of 100–1400 ft and 80–1600 ft, respectively.

Southern stargazer (*Astroscopus y-graecum*), Singer Island, Florida

Natural History

Not much is known about reproduction in stargazers. Eggs and larvae are pelagic, and young of northern and southern stargazers settle to sandy bottoms, in shallow water. Stargazers possess an interesting adaptation—they have electric organs just behind their eyes, capable of delivering a respectable 50 volts. This may not be enough to kill prey or knock waders off their feet, but the electric field that is created possibly could attract small fishes into the vicinity of the waiting mouth or stun them just enough to distract them. The electric organs are derived from muscles formerly used in eye movement. To compensate for the loss of these muscles, southern stargazers use fluid-filled cavities under the eyes to push them upward, above the bottom, allowing the fish to see.

Stargazers are superbly adapted for life buried in the sand. Their nostrils and mouth are covered with fleshy tabs that prevent sand from entering, and the gill openings expel water behind the head. When buried, stargazers pass incoming water through the nostrils, over the gills, and out through small, vertically oriented openings. Their lie-and-wait predatory style requires some help in getting potential prey near the huge, lightning-fast maw. Southern stargazers appear to employ subtle tricks in this regard. The generated electric field may attract prey near the mouth. Water expelled from the small gill openings may move the surface sand enough to attract curious fishes, crabs, or shrimps. When one of these potential food items approaches, the eye—distant from the mouth—rotates back and forth, perhaps further attracting the unwitting prey over the strike zone. Juvenile stargazers must swallow their fish prey head-first, leading to some comedic great-circle-route swimming when they feed off the bottom in the nonstealth mode.

Individuals emerge at night, and we have seen them lying on top of the sand in the early morning. They bury themselves by shaking the entire body rapidly over the sand. We've watched bowling pin–shaped adults disappear in a few seconds. The best way to find a stargazer during the day is to search for areas where the sand has been disturbed or is a different color. Look closely for the eyes or a small patch of the spotted color pattern.

Triplefins
Family Tripterygiidae

- **Lofty triplefin** (*Enneanectes altivelis*)
- **Roughhead triplefin** (*Enneanectes boehlkei*)
- **Redeye triplefin** (*Enneanectes pectoralis*)

Background

Triplefins are tiny (1.5 in), large eyed, short-snouted relatives of sand stargazers, but they are readily distinguished from their brethren by the presence of 3 dorsal fins (the first 2 with spines). Redeye and lofty triplefins have scales on their bellies and pectoral-fin bases. Redeye triplefins are tan to reddish, with 5 dark bars—the last, on the caudal peduncle, is the broadest and darkest. The dorsal fin is barred; the tail has a reddish bar at the base, followed by (in sequence) white, brown, and white bars. The snout slopes downward, at a more acute angle than in the other triplefin species, and the anal fin is evenly pigmented. Lofty and roughhead triplefins also have 5 dark bars on a light-tan to red background, but the last bar is narrower and darker than the others and is not bordered by red caudally. Their anal, dorsal, and caudal fins are barred. A high first dorsal fin, notably taller than the second and third dorsal fins, separate out lofty triplefins and give the species its vernacular name. Roughhead triplefins lack scales on the belly and pectoral-fin bases and have a white bar, formed by 2 elongate white spots, on the base of the caudal fin.

Distribution and Habitat

Triplefins reside on reefs, from nearshore waters to depths of about 80 ft, ranging from southeast Florida through the Florida Keys. They are often found on shallow, hardbottom habitats, where wave surge can be high, and occasion-

Roughhead triplefin (*Enneanectes boehlkei*), Hillsboro, Florida

ally observed perched on sponges, such as *Aegelus*. The three species differ slightly in their distributions. Redeye triplefins occur off southeast and southwest Florida and in the Keys, while lofty triplefins are confined to the southeastern coast, from Jupiter through the Florida Keys. Roughhead triplefins are more widespread. They are found on the Gulf coast, from the Big Bend ecoregion south into the Florida Keys, and are rare off southeast Florida.

Natural History

Little is known about the biology of triplefins. Pairs have been observed standing on their long pelvic fins in what may be courtship behavior. Eggs are demersal and are deposited on hard substrate, in a hole or cave. Males differ from most other non-blenny fishes in having an accessory testicular organ, located between the 2 testes. Triplefins eat benthic invertebrates.

Sand Stargazers
Family Dactyloscopidae

- **Bigeye stargazer** (*Dactyloscopus crossotus*)
- **Speckled stargazer** (*Dactyloscopus moorei*)
- **Sand stargazer** (*Dactyloscopus tridigitatus*)
- **Arrow stargazer** (*Gillellus greyae*)
- **Masked stargazer** (*Gillellus healae*)
- **Warteye stargazer** (*Gillellus uranidea*)
- **Saddle stargazer** (*Platygillellus rubrocinctus*)

Background

Sand stargazers are superbly adapted for a life nestled into the sand. They are tiny (all species <3 in), with a long, tapered body; tan to pale-brown coloration; a flat head, with eyes set on top (in some species the eyes are on retractable stalks); an upturned mouth, and fleshy fringes on the lips and gill openings. This group is seldom seen, except by avid divers who concentrate on bottom-dwelling microfauna. Speckled and sand stargazers differ from the others in having stalked eyes and, often, in lacking dark to dusky lateral bars or saddles (when these appear, they occur only dorsally). Speckled stargazers are tan to light-brown, with darker marks on the scales, forming horizontal rows. Sand stargazers have limited brown mottling on the head and back, and the first 3–4 dorsal-fin spines are separate (no membrane between them). Saddle stargazers have 4 broad brown bars—1 on the head and 3 on the body—that cross from the dorsal to the anal fins. The first 3 dorsal spines form a brown mini-fin, and a weak notch bisects the remaining portion of the fin, characteristics this species shares with warteye stargazers. The latter have 4–5 dark-brown, V-shaped saddles, the one on the head extending down to mid-operculum, and the posterior trio reaching down to the lateral line; the head has dark blotches or spots.

Masked stargazers have 7 dark, inverted V-shaped bars on the flanks that extend down to the lateral line, and a dark bar between and under the eyes. Arrow stargazers have 6 narrow, V-shaped, black bars or saddles that terminate at the lateral line, and a 3-spined, reddish-brown dorsal finlet. The remaining dorsal fin is bisected by a pronounced notch. Bigeye stargazers can have 8–12 light, V-shaped bars or can be pallid; the first few dorsal spines are free.

Saddle stargazer (*Platygillellus rubrocinctus*), Singer Island, Florida

Distribution and Habitat

Sand stargazers live buried in the sand, some on wave-washed, sandy beaches and others on the periphery of reefs or hardbottom features. Bigeye stargazers live along the sandy beaches of eastcentral Florida, and arrow stargazers are restricted to the east coast. Warteye stargazers are found only in the

Arrow stargazer (*Gillellus greyae*), Palm Beach, Florida

Florida Keys, where they live in the sandy areas adjacent to reefs, in depths of 1–40 ft. Speckled, sand, and masked stargazers are found along shallow (<10 ft), sandy beaches around the state.

Natural History

Reproduction by sand stargazers is not well known. Individuals from several species have been found brooding a mass of eggs under each pectoral fin; in one species, it was the male holding the eggs. Sand stargazers, from their hiding places in the sand, ambush small invertebrates. Arrow stargazers lie partially buried, with their bodies in an S-curve prior to striking.

Combtooth Blennies
Family Blenniidae

- **Striped blenny** (*Chasmodes bosquianus*)
- **Stretchjaw blenny** (*Chasmodes longimaxilla*)
- **Florida blenny** (*Chasmodes saburrae*)
- **Pearl blenny** (*Entomacrodus nigricans*)
- **Barred blenny** (*Hypleurochilus bermudensis*)
- **Zebratail blenny** (*Hypleurochilus caudovittatus*)
- **Crested blenny** (*Hypleurochilus geminatus*)
- **Featherduster blenny** (*Hypleurochilus multifilis*)
- **Oyster blenny** (*Hypleurochilus pseudoaequipinnis*)
- **Orangespotted blenny** (*Hypleurochilus springeri*)
- **Feather blenny** (*Hypsoblennius hentz*)
- **Tessellated blenny** (*Hypsoblennius invemar*)
- **Freckled blenny** (*Hypsoblennius ionthas*)
- **Highfin blenny** (*Lupinoblennius nicholsi*)
- **Mangrove blenny** (*Lupinoblennius vinctus*)
- **Redlip blenny** (*Ophioblennius macclurei*)
- **Seaweed blenny** (*Parablennius marmoreus*)
- **Molly miller** (*Scartella cristata*)

Background

Combtooth blennies have long, continuous first and second dorsal fins, big eyes and lips, comb-like front teeth, and, in most species, fleshy cirri (frilly tufts) on their heads. Combtooth blennies, with their sinuous body movements, are easily distinguished from stiff-bodied gobies and dragonets. Color patterns often are the key to identification for combtooth blennies, but frequently there is extensive variability within a species, including sexual dimorphism, making things difficult.

Most combtooth blennies are slender, with blunt heads; the eyes are located at the top of a steep hill, descending to the tip of the snout. This look easily says "blenny," but it can be a real problem to identify the little guys to the species-level in the field, because often only their heads are seen peering out of a hole or a barnacle. Members of the genus *Chasmodes*, however, have more-traditional head profiles, with the eyes located about halfway down the hill. They

Redlip blenny (*Ophioblennius macclurei*)

have deeper, somewhat laterally compressed bodies, and the continuous dorsal fins are equal in height. Zoogeography really helps in identifying these three species, which look a lot alike and reach similar sizes (4 in). Florida blennies are distinguished by having a short upper jaw that usually ends below the center of the eye and never reaches the rear of the eye. The forehead is rounded, giving them a humped appearance, and the snout is short. Coloration is olive-green to brown, with an irregular mosaic of horizontal, lighter lines on the body. Stretchjaw and striped blennies are similarly colored, but they have longer jaws and snouts and differing head profiles. Stretchjaw blennies have the longest upper jaw, extending well past the rear of the eye, and the longest and most-pointed snout. The forehead is nearly straight. Striped blennies have upper jaws that reach the rear edge of the eye, and they have slightly humped foreheads. All three species have a small, bright-blue spot located between the first 2 dorsal spines.

Redlip blennies are the largest (5 in) of the combtooth blennies. They look like they hit a wall at high speed, as their head profile is nearly vertical. The body is either uniformly dark (reddish-brown, brown, or black) or pale-white. Regardless of base color, the lips, the edges of dorsal and anal fins, and pectoral fins are red. The more-elongated pearl blennies (4 in) have a similar (but not quite as pronounced) head profile but are unique in having a well-defined notch between the 2 dorsal fins. They are light-brown, with several large dark spots on the midline and numerous small white spots along the flanks. The similar-sized molly millers are easy to recognize, with a mohawk (a linear comb of red- and white-banded cirri) on top of their heads. The body ranges from light-brown to greenish, with 6 irregular dark bands and a light, wavy line on the midrear part of

Left: Seaweed blenny (*Parablennius marmoreus*), Lake Worth Lagoon, Florida. *Right*: Yellow phase, Venice, Florida.

Top: Male molly miller (*Scartella cristata*), Palm Beach, Florida. *Bottom*: Subadult, Lake Worth Lagoon, Florida.

the body that looks like the output of a heart monitor. There usually are small red spots on the head. Mangrove blennies are small (2 in) and have a series of dark squares at the base of the dorsal fin that meld into dark body bars; the paired cirri are finger-like, not branched. Highfin blennies (2.5 in) are brown, with 2 dark bars extending from the eye through the jaw. The front of the dorsal fin is much higher than the rear portion (markedly so in males). Seaweed blennies (3.5 in) are an extremely variable species. They ranges in color from brown or tan to orange to yellow, with spotting and mottling; telltale blue lines radiate under and in front of the eyes.

Unlike the rest of the group, tessellated blennies (2.5 in) are very colorful. The head, body, and dorsal and anal fins are densely covered by moderately large, dark-edged, bright–reddish-orange polygons and spots over a blue to steely-gray base color. There is prominent, dark ear-blotch behind the eye, and 3 blue stripes cross the throat. The pair of short cirri above the eyes are multibranched (up to 4 branches) in adults. Very long, multibranched cirri identify feather blennies (4 in). Their color is a mottled tan to brown, with 5–6 ill-defined, oblique body bars; the head is marked with a small dark blotch behind the eye; the dorsal fin has a blue spot at the front. The similar-looking freckled blennies (4 in) usually have a pair of moderately long, unbranched cirri (sometimes with only a single branching) above the eyes, and a pair of dark lines connecting the eye to mouth.

Species of *Hypleurochilus* are a confusing group that share a character requiring a microscope: a backward-directed canine tooth at the corner of the jaw. Color pattern, habitat choice, and distribution help in their identification. Barred blennies (4 in) are the easiest of this group to identify. They are tan to light-brown, with 5 or 6 irregular, dark-brown saddles that tend to merge at the midline of the body; the fins are banded; and the head and body are peppered with small bronze to rusty spots. Zebratail blennies (2.5 in) are uniformly dark-brown, except for translucent windows on the rear portions of the dorsal and anal fins; clear pelvic- and anal-fin margins; and their distinctive, zebra-striped tail, with 3–5 brown, vertical bands. There usually is a blue spot on the fleshy membrane between the first 2 dorsal spines. Orangespotted blennies (2 in) are tan to bluish-gray, with orange spots on the head, the front of the body, and the first dorsal fin. In the young, these body spots tend to get darker (approaching brownish) on the rear half of the trunk. Similar dark spots form bands on the second dorsal, caudal, and anal fins, and the spots coalesce to form bars on the body. The dorsal fin does not have a spot between the first 2 dorsal spines. The related crested and oyster blennies always have a dark spot there, and featherduster blennies often do, as well. Oyster blennies (2.5 in) are gray to brown, with 5 irregular, dark bars on the flanks; small orange spots on the upper body; and—diagnostically—dark chin bands. Crested blennies are gray to dark-brown, with orange brown spots on the head; oblique body bars often are present. Males are darker than females and have a fleshy crest on the nape. Featherduster blennies are variably colored (without recognizable distinctions), looking a lot like crested and oyster blennies but lacking their ear spots and (often) missing the dorsal-fin spot. Frankly, their distribution is the best clue to their identity.

Distribution and Habitat

Combtooth blennies, for the most part, dwell in holes, crevices, barnacles, and discarded bottles that are found in shallow water, close to shore; the reef-dwellers among this family range to greater depths where suitable habitat exists. *Chasmodes* species prefer nearshore seagrass, oyster, or mangrove habitats and have complementary distribution patterns, facilitating species identification. Striped blennies are confined to northeastern Florida, stretchjaw blennies are known only from the farthest-western part of the panhandle, and Florida blennies occur from Cape Canaveral south and west throughout the panhandle.

Happily, distributions also help us separate the similar-looking crested, oyster, and featherduster blennies. Crested blennies are known from the Northeast to Eastcentral ecoregions, in shallow, usually estuarine, hardbottom habitats, such as pilings, jetties, and oyster reefs. Oyster blennies frequent oyster bar, mangrove, and other hardbottom habitats, including pilings, on the southeastern coast and Florida Keys.

Barred blenny (*Hypleurochilus bermudensis*), Florida Middle Grounds

Orangespotted blenny (*Hypleurochilus springeri*), Lake Worth Lagoon, Florida

Featherduster blennies occur in shallow water, but only on the west coast, from the middle of the panhandle westward. Barred and orangespotted blennies frequent rocky hardbottom and reef habitats in south Florida. Barred blennies are found from nearshore to at least 150 ft depths, and they often are associated with tube sponges. Orangespotted blennies prefer the rock-rubble surge zone and may be found on barnacle-encrusted pilings or buoys. Zebratail blennies are a Florida endemic, occurring only in the Gulf, from St. Andrews Bay to Sarasota, on hardbottom habitats over the shelf.

Tessellated blennies were originally described from northern

South America, but there is a disjunct population thriving in empty barnacle shells that encrust the legs of oil platforms in the northern Gulf of Mexico, off Louisiana and Texas. In this location, they are only found in the surge-dominated, upper 10 ft or so of offshore (blue) waters. Some people have asserted that this species colonized the northern Gulf by riding in the ballast tanks of or in barnacles attached to ships heading from South America into the northern Gulf of Mexico, but a more likely explanation is that some tenacious larvae simply rafted north with the Loop Current, settling on the first shallow-water hardbottom (although artificial) habitat they encountered. Regardless of how they got here, tessellated blennies have been reported from the upper legs of what are known as "Air Force towers" off southwest Florida, where the combination of blue ocean water and shallow depths creates a habitat like no other. Feather blennies are found on softbottom habitat, such as in seagrass beds and near oysters, primarily in northeast Florida and the Gulf, from Cedar Key westward across the panhandle. Freckled blennies are similarly distributed.

Zebratail blenny (*Hypleurochilus caudovittatus*), Venice, Florida

Redlip blennies are found in coral-reef and rock-rubble habitats, along the southeastern coast and in the Florida Keys. Barred, orangespotted, and highfin blennies occur from Cape Canaveral and Tampa south through the Florida Keys. Seaweed blennies are known from around the state; they are the deepest dwelling (to at least 200 ft) and most widespread of all the blennies. Molly millers inhabit the shallowest, surge-washed shoreline from about Cape Canaveral southward through the Florida Keys. Other species are also confined to southern Florida and are rare in the Gulf. Pearl blennies are found from Boca Raton to the Keys, in shallow, rocky habitats and are common in tidepools. Highfin blennies (occurring from off of Stuart and Sarasota southward) and mangrove blennies (found in the Keys) prefer mangrove, oyster, and other hardbottom habitats.

Natural History

Combtooth blennies are favorites of fish watchers, because these fishes inhabit shallow water, have a curious behavior, and pose outright for cameras. Combtooth blennies deposit demersal eggs, usually in or near a crevice, hole, or empty barnacle shell. For those species that have been studied, the number of eggs varies but generally ranges from 500 to 1000 per nest. A dominant male normally guards the nest until hatching—usually 7–10 days after fertilization, depending on the water temperature. Most species are territorial, and details differ between species in how each holds and defends its territory. The mating process in molly millers involves alternative reproductive tactics. Large, colorful, aggressive males fight with rival males, court females, and guard nests. But hiding in nearby barnacles are other, less-conspicuous males that sneak in during mating, fertilize eggs, and then sneak out, leaving the big guy as the foster parent for the young. Paternity studies suggest that, on average, 12 percent of the progeny in a nest are unrelated to the guardian male. It is not clear how prevalent the alternative reproductive mode is among the other members of this family. Most species of combtooth blennies feed on algae, but some are omnivorous, throwing in a few invertebrates here and there along with their favorite green and red algae.

Labrisomid Blennies
Family Labrisomidae

- **Puffcheek blenny** (*Labrisomus bucciferus*)
- **Masquerader hairy blenny** (*Labrisomus conditus*)
- **Mock blenny** (*Labrisomus cricota*)
- **Palehead blenny** (*Labrisomus gobio*)
- **Mimic blenny** (*Labrisomus guppyi*)
- **Longfin blenny** (*Labrisomus haitiensis*)
- **Downy blenny** (*Labrisomus kalisherae*)
- **Spotcheek blenny** (*Labrisomus nigricinctus*)
- **Hairy blenny** (*Labrisomus nuchipinnis*)
- **Goldline blenny** (*Malacoctenus aurolineatus*)
- **Rosy blenny** (*Malacoctenus macropus*)
- **Saddled blenny** (*Malacoctenus triangulatus*)
- **Threadfin blenny** (*Nemaclinus atelestos*)
- **Coral blenny** (*Paraclinus cingulatus*)
- **Banded blenny** (*Paraclinus fasciatus*)
- **Horned blenny** (*Paraclinus grandicomis*)
- **Bald blenny** (*Paraclinus infrons*)
- **Marbled blenny** (*Paraclinus marmoratus*)
- **Blackfin blenny** (*Paraclinus nigripinnis*)
- **Checkered blenny** (*Starksia ocellata*)
- **Key blenny** (*Starksia starcki*)

Background

Labrisomid blennies have scaled, tapered bodies; a lateral line; tufts of cirri above the eyes, on the nape (behind the eyes), and (often) on the nostrils; and a long, continuous dorsal fin. Males and females are often differentially colored. Several species from the genera *Nemaclinus*, *Paraclinus*, and *Starksia* are small (≤2 in), secretive, and rarely observed. Members of the genera *Labrisomus* and *Malacoctenus* are more conspicuous and are found on reef and hardbottom habitats.

In recent years, DNA analyses have revealed a number cryptic species that closely resemble each other but differ in their molecular fingerprints. Many of these are blennies and gobies. Hairy blennies, one of the most-common shallow-water labrisomid blennies on Florida's east coast, are thought to be composed of several very similar sister-species, all formerly lumped together under the name *Labrisomus nuchipinnis*. The three species found in Florida are hairy blennies, mock blennies, and masquerader hairy blennies. Although some very subtle characters seen in live individuals may help distinguish these species, it is more practical to treat them here as the "hairy blenny complex." In all three, the females have dark-green to gray-brown bodies, with lighter mottling. Breeding males seasonally develop red-orange heads and

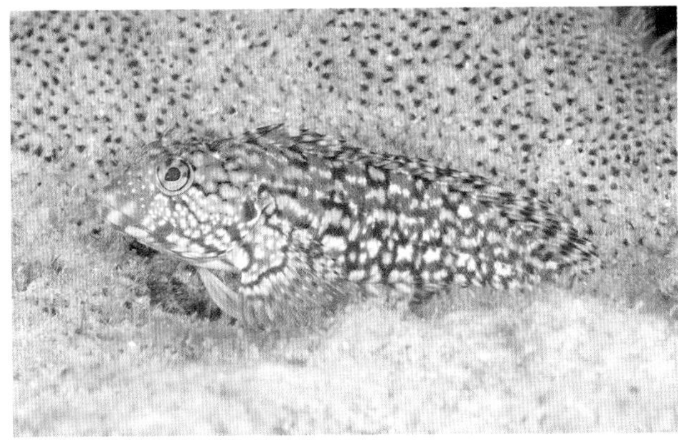

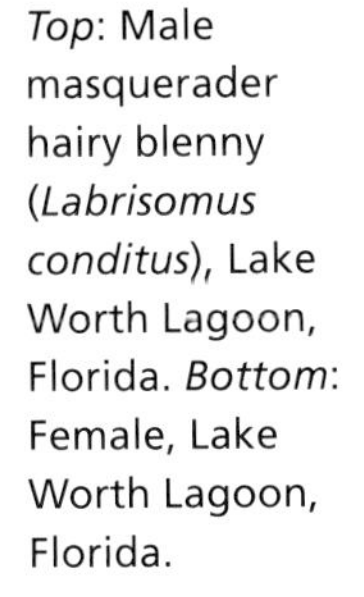

Top: Male masquerader hairy blenny (*Labrisomus conditus*), Lake Worth Lagoon, Florida. *Bottom*: Female, Lake Worth Lagoon, Florida.

Left: Male hairy blenny (*Labrisomus nuchipinnis*), Lake Worth Lagoon, Florida. *Right*: Female hairy blenny, Lake Worth Lagoon, Florida.

bellies and have gray bodies, with dark bars. Nonbreeding males tend to be a more-uniform brown to maroon, often with faint bars, and both sexes have a dark opercular spot. One of the subtle characters said to distinguish these species is the nature of this large, obvious opercular spot: completely white-edged (ocellated) in hairy blennies; only partially circled in white or pale-orange in masquerader hairy blennies; and completely covered in white in mock blennies. In mock blennies, the small pale- or light-blue spots are restricted to the head. These three species are the largest (9 in) of the labrisomid blennies in Florida. Our observations suggest that masquerader hairy blennies are the most common of the three on nearshore hardbottom habitats off southeastern Florida. The intricacies of the hairy blenny complex should keep ichthyologists off the streets for some time into the future.

Palehead blenny (*Labrisomus gobio*)

Two additional species, spotcheek and mimic blennies, also have prominent, ocellated opercular spots. Spotcheek blennies can be distinguished by their smaller (3 in) size and 8–9 well-defined body bars that fully extend to the margin of the dorsal fin. Males are yellow to tan, with reddish-orange bars; females are tan to white, with dark-green to brown or black bars. Mimic blennies (4.5 in) have a shorter snout than spotcheeks and members of the hairy blenny complex. Mimic blennies are brown to grayish, with irregular dark bars that barely extend onto the dorsal fin. The opercular spot ranges from pale-gray to black, and cirri on the nape may be white.

Like mimic blennies, puffcheek, palehead, downy, and longfin blennies have shorter snouts and bars that nudge onto the dorsal fin, but they differ in lacking opercular spots. Downy blennies (3 in) are the easiest of this group to identify. The head is a dull–red-brown,

Downy blenny (*Labrisomus kalisherae*), Palm Beach, Florida

Goldline blenny (*Malacoctenus aurolineatus*)

with a white saddle extending from behind the eyes to the opercular margin, including the first 3 dorsal spines and the base of the pectoral fin. The body is lavender to brownish-green, with 7 well-defined, alternating dark and lighter bars that extend onto the dorsal fin. The first dorsal-fin spine is longer than the subsequent (third to fifth) spines, giving the front of the fin a downhill look, a trait shared with puffcheek and longfin blennies. In contrast, the first dorsal spine is shorter than the fourth and fifth spines In the smaller (2.5 in) palehead blennies, making the front of the fin level. Palehead blennies are gray-brown, with 4–5 dark bars that are darker above the midline than below, and have large eyes relative to their body size. Puffcheek (3.5 in) and longfin (3 in) blennies are tan to brown, with a dark bar under the eye and 3 light spots on the base of the caudal fin (the latter are also present in downy blennies). Puffcheek blennies have 4–5 dark flank bars and 2 dark spots behind the eyes. Longfin blennies have 6–7 irregular dark bars and longer pelvic fins than others in their genus.

The three *Malacoctenus* species have longer and more-pointed snouts than those in *Labrisomus*. All are small (2–2.5 in long). Goldline blennies have golden-yellow fins and dark bars on the flanks. The bars immediately behind the pectoral fin merge into an H pattern; there are dark commas on the base of the pectoral fin; and the underside of the head has several thin, dark bars. Female rosy blennies are tan, with fine spots forming diffuse miniblotches. Males are tan or grey to brown, with numerous dark bars; are darker above than below the midline; have wine-red spots and lines on the head; and have a wine-red tinge to the belly. Saddled blennies have 3–5 triangular, V-shaped saddles along the back. Females have a white background color, sometimes with small red dots. In some individuals, the saddles merge to form a single dark region, with orange spots from the midline to the tail. Breeding males develop a solid red-orange background, with a dark–charcoal-gray head and black saddles.

The rarely seen species of *Paraclinus* are small (most are 1–2 in, but one reaches 4 in) and more slender than the *Malacoctenus* species. Coral blennies have 5 brown bars that

Left: Male rosy blenny (*Malacoctenus macropus*), Lake Worth Lagoon, Florida. *Right*: Female, Lake Worth Lagoon, Florida.

Left: Male saddled blenny (*Malacoctenus triangulatus*), Lake Worth Lagoon, Florida. *Right*: Female, Palm Beach, Florida.

extend onto the dorsal and anal fins: the first 3 are dark, and the rear 2 are much lighter and more diffuse. The dorsal and anal fins lack ocellated spots. Banded blennies are tan to brown to brownish-green, with mottling and ill-defined bars on the flanks. The caudal peduncle has a broad dark bar, and usually there are 1–2 (ranging from 0 to 4) ocellated blue spots on the rear of the dorsal fin (none on the anal fin). Marbled blennies have a very long snout, and the first 3 dorsal-fin spines are long and mostly separated from the rest of the dorsal fin. The base coloration is tan to greenish-brown, and the flanks can be barred (usually 6 bars) or mottled. Individuals have from none to 3 ocellated spots, with blue centers, on the rear of the dorsal fin and 1–2 on the rear of the anal fin. Horned blennies are brown, with long, pointed snouts and distinctive long cirri (reaching the dorsal fin) on the eyes. They do not have ocellated spots on the dorsal or anal fins. Bald blennies have relatively long snouts; are reddish-brown, with short cirri (not reaching the dorsal fin) on the eye; and have two ocellated spots on the rear dorsal fin. Blackfin blennies also have long snouts, and the front of the dorsal fin is imperfectly separated from the rest of the fin. The body and dorsal fin have 7 dark bands, with the rear 5 bands extending onto the anal fin. A single, ocellated spot appears on the rear of the dorsal fin. Threadfin blennies are very elongated, with low and level dorsal and anal fins, and look almost like a tiny (<1.5 in) wrasse. The pectoral fins are quite long, and the body coloration is reddish to brown.

The best way to identify checkered blennies is to look for the orange to gold spots on the head and pectoral-fin bases; body coloration varies from a brown checkerboard pattern to solid brown. Key blennies are pale-orange to tan, with 2 dark spots beneath the eye. On the flanks they have 3 complete

Female masquerader hairy blenny (*Labrisomus conditus*) deposits eggs on the substrate while the male looks on, Lake Worth Lagoon, Florida

straight dark bars, followed by 5 or 6 broken dark bars.

Distribution and Habitat

With few exceptions, labrisomid blennies inhabit shallow (≤15 ft) hardbottom, reefs, and seagrass meadows. The outliers are saddle and longfin blennies (which live as deep as 80 ft) and threadfin blennies (the deepest dwellers, at about 100–800 ft). Members of the hairy blenny complex only occur along Florida's east coast, where they are common on nearshore hardbottom, jetties, and other hard structures, at 2–30 ft depths. Palehead, mimic, spotcheek, goldline, saddled, bald, horned, Key, puffcheek, coral, marbled, and blackfin blennies occur on the southern tip of the peninsula and in the Florida Keys. Some species, including saddled and palehead blennies, nestle in sea anemones. Longfin blennies are known only from the southwest coast and Florida Keys, in 6–80 ft depths. Downy and rosy blennies are distributed from Tampa Bay and Cape Canaveral southward, in depths of 1–30 ft. Banded and marbled blennies are found from the Big Bend ecoregion in the Gulf to Cape Canaveral, in waters 1–30 ft deep. Banded blennies prefer seagrass meadows and algae-covered hardbottom, even inhabiting clumps of unattached algae rolling along the bottom. The widespread checkered and marbled blennies have an affinity for hardbottom and sponges.

Natural History

Labrisomid blennies are substrate spawners that deposit eggs on hardbottom, usually in small areas defended by the male. Mating of members of the hairy blenny complex begins with a drab, gray-green female approaching a brightly colored male's territory. After some nudging and nipping of fins by the male, the female releases eggs onto the hard substrate, usually covered with low-growing turf algae, and moves away. The male then moves over the site and shudders as he fertilizes the eggs. The female promptly returns to the site and, after some additional nudging and nipping from the male, releases a few more eggs for the male to fertilize. This process goes on cyclically for some time, with each mating bout separated by less than a minute. On occasion, more than one female, in turn, may deposit eggs at the nest. Once mating ceases, the male steadfastly guards the fertilized eggs, chasing away all intruders, including the maternal female(s). When snorkeling on nearshore reefs, it is not uncommon to see mature male members of the hairy blenny complex in aggressive lip locks. We have observed spawning by members of the hairy blenny complex on numerous occasions in southeast Florida, on both artificial and natural reefs, usually from January to April.

Rosy blenny males also defend territories in which females deposit eggs to be fertilized. These red-orange breeding males are easy to spot in summer months. Marbled blennies deposit eggs within tubular sponges, but details of their mating are unknown. At least one species of *Starksia* gives birth to live young.

Unlike the algae-eating combtooth blennies, the carnivorous labrisomid blennies feed on small fishes and mobile invertebrates. With the exception of breeding males, most individuals blend in well with the background, which promotes a classic lie-in-wait feeding mode.

Tube Blennies
Family Chaenopsidae

- **Roughhead blenny** (*Acanthemblemaria aspera*)
- **Papillose blenny** (*Acanthemblemaria chaplini*)
- **Spinyhead blenny** (*Acanthemblemaria spinosa*)
- **Yellowface pikeblenny** (*Chaenopsis limbaughi*)
- **Bluethroat pikeblenny** (*Chaenopsis ocellata*)
- **Flecked pikeblenny** (*Chaenopsis roseola*)
- **Banner blenny** (*Emblemaria atlantica*)
- **Sailfin blenny** (*Emblemaria pandionis*)
- **Pirate blenny** (*Emblemaria piratula*)
- **Blackhead blenny** (*Emblemariopsis bahamensis*)
- **Glass blenny** (*Emblemariopsis diaphana*)
- **Wrasse blenny** (*Hemiemblemaria similis*)
- **Blackbelly blenny** (*Stathmonotus hemphillii*)
- **Eelgrass blenny** (*Stathmonotus stahli*)

Background

Members of the tube blenny family are small, elongate fishes that reside in holes and cylinders drilled in limestone rock by something else (worms, clams, etc.). The three *Acanthemblemaria* species, collectively referred to as "spiny-heads," are very similar in outward appearance: usually just the head is visible, with close-set but independently rotating eyes peering out from the home tube. These species can be distinguished by color and by the nature of the fleshy cirri on their heads. Roughhead blennies have bushy cirri above the eyes and on the snout; their body color varies from yellow to reddish-brown to pale-gray. Spinyhead blennies are similarly colored, but their color can vary from pale-gray to yellow; the cirri are short and not extensively branched. Papillose blennies are superficially similar but usually are grayish. They have

Roughhead blenny (*Acanthemblemaria aspera*), Palm Beach, Florida

Papillose blenny (*Acanthemblemaria chaplini*), Hillsboro, Florida

Yellowface pikeblenny (*Chaenopsis limbaughi*), Lake Worth Lagoon, Florida

blue on the branchiostegal rays (below the operculum) and on the membranes between these rays, while yellowface pikeblennies have pale-blue coloration only on the rays. Bluethroats grow to 5 in, as opposed to 3.5 in for yellowfaces. Flecked pikeblennies are easy to distinguish from the former two species by their short snouts and coloration—8 regularly spaced, dark blotches along the body, which is also flecked with rusty or pinkish spots.

Wrasse blennies are yellow, with a thick, black, lateral stripe and a red iris (these individuals look like young blueheads, one of the wrasse species). The young are transparent around the dark lateral stripe; as they grow, the transparent areas turn yellow. Three members in the genus *Emblemaria* are found in Florida waters: banner blennies, sailfin blennies, and pirate blennies. Sailfin blennies have a blunt snout and a broad dorsal fin (which is broader in males). Males are black to cobalt-blue, with small white flecks on the lower jaw and face. Females are light-brown, with faint bars, and have a smaller dorsal fin than males. Banner blennies are similarly shaped but are grayish-brown, with small white dots on the face and lower jaw. Female banner blennies are light-brown, with more-defined, dark-brown bars on the body. Pirate blennies are smaller than the other two species, reaching about 1 in. Males are black on the head and the anterior third of the body. A distinct patch of yellow-orange is located at the base of the black dorsal fin. Female pirate blennies are drab-gray.

Two species in the genus *Emblemariopsis*—glass blennies and blackhead blennies—are found in Florida but are rarely encountered. The state's two *Stathmonotus* species are small, eel-like fishes. Male blackbelly blennies can be white, black, orange, or greenish, with short, stubby cirri that grow along a single plane above the eyes, rather than the elongate, branching cirri found in the other two species.

Pikeblennies (*Chaenopsis*) have long, blunt-ended snouts and sail-like dorsal fins that run continuously around the tail and into the anal fin. Bluethroat and yellowface pikeblennies are the most common in Florida waters. They can be confusing to separate, as bluethroat males can have yellow faces, and some yellowface individuals may have blue on the throat. Both species have an orange spot, rimmed with white and black, on the leading edge of the dorsal fin. Bluethroat pikeblennies have

Bluethroat pikeblenny (*Chaenopsis ocellata*), Lake Worth Lagoon, Florida

Male sailfin blenny (*Emblemaria pandionis*), Lake Worth Lagoon, Florida

Female sailfin blenny (*Emblemaria pandionis*), Lake Worth Lagoon, Florida

a black-edged white spot behind the eye and a series of white marks on the dorsal and anal fins. Females are pale, with dark lines radiating from the eye. Neither males nor females have cirri on their heads. Eelgrass blennies are greenish to brown, with a reddish blotch on the cheek and a single cirrus above the eye.

Distribution and Habitat

For most tube blenny species, home is a hole drilled in rock or coral by something else, usually a rock-boring clam or worm. Roughhead, papillose, and spinyhead blennies tend to occupy holes in larger, stationary rock formations, where wave energy is high. You will see just their tiny heads protruding, with eyes that rotate independently, in a comical way. These *Acanthemblemaria* species are found only offshore of southeast Florida and the Florida Keys, in water depths of up to 50 ft. Pikeblennies prefer a sandy or rubbly bottom, where they inhabit available holes in the substrate. These holes frequently are abandoned worm tubes, including the parchment type constructed by the polychaete worm genus *Chaetopterus*. Yellowface and bluethroat pikeblennies are most common off southeast Florida and the Florida Keys, to about 12 ft depths for bluethroats and 70 ft for yellowfaces. Flecked pikeblennies were first discovered off the Florida panhandle, in rubble–shell-hash bottom that provides excellent habitat for a host of small fishes.

Sailfin blennies prefer shallow, flat, limestone bottom but will

Wrasse blenny (*Hemiemblemaria similis*)

make do with almost any suitable hole. Pirate blennies appear to occupy holes in loose rubble or shells on low-relief hardbottom, to at least 250 ft depths, on the west Florida shelf. Banner blennies are also known from offshore reefs (to 150 ft deep) on the west Florida shelf and northeast Florida shelf, but not from the southern portion of the peninsula. Blackhead and glass blennies are only known from shallow reefs of the Dry Tortugas. Wrasse blennies occur in depths less than 100 ft, offshore of southeast Florida and the Florida Keys. Blackbelly and eelgrass blennies occur mostly on shallow reefs in the Florida Keys and Dry Tortugas (but we have seen pictures of blackbelly blennies from off of Ft. Lauderdale).

Natural History

Tube blennies reproduce by laying demersal eggs in the male's home tube. The male guards the eggs until hatching. Single females may spawn with multiple males, and males will accept eggs from several females. Details of mating vary with the species. A roughhead blenny male signals females by bobbing in and out of his hole. Females will visit and presumably lay eggs in multiple tubes. Male pikeblennies will confront rival males or display to passing females. Although details are not well known, it appears that following some courtship by male pikeblennies, receptive females deposit a clutch of eggs into the male's burrow and then move on. The male guards the eggs until they hatch.

Sailfin blenny males are chosen by females, following some fairly dramatic courting efforts. Males display by rapidly flexing the dorsal fin, a behavior known as flagging. We have seen males flagging from discarded beer bottles, barnacle shells, and holes in wooden pilings. Although flagging may be observed throughout the day, it is most prevalent in the afternoon. Females initiate the ritual by leaving their holes and approaching a male. He will flag repeatedly and intensively—occasionally rising over 1 ft above the tube, with all fins tautly spread. Once the female is impressed by his display, she will enter the male's tube, deposit her eggs, and exit. The male will then fertilize and guard the eggs until they hatch. Courting or spawning behavior has not been described for banner or pirate blennies, but it seems likely that similar processes takes place.

There is no information on reproduction for eelgrass and blackbelly blennies. These two species do not appear to be obligate hole-dwellers, but they are very secretive and seldom seen. Eelgrass blennies are misnamed, as they actually live on reefs and hardbottom areas, not necessarily seagrass meadows.

With the exception of the pikeblennies and wrasse blennies, other members of this family can be seen watching the passing feast from their holes. Pikeblennies are larger, with bigger mouths, enabling them to feed on larger, bottom-dwelling invertebrates; these fishes will leave their hole or worm tube to hunt for prey. Wrasse blennies are thought to mimic initial-phase blueheads, using wrasses or schools of wrasses as cover for their predatory exploits. Wrasse blennies maintain a home tube but spend much of their time outside of the hole.

Clingfishes
Family Gobiesocidae

- **Emerald clingfish** (*Acyrtops beryllinus*)
- **Stippled clingfish** (*Gobiesox punctulatus*)
- **Skilletfish** (*Gobiesox strumosus*)

Background

Clingfishes are small (<3 in), flattened fishes that are equipped with a remarkable suction disc, formed from the pelvic fin and part of the pectoral fins. The disc is covered with tiny protuberances (papillae) that help the fish cling to hard surfaces in wave-swept environments. Emerald clingfishes grow to about 1 in long and range in color from a striking emerald-green to drab-brown. In some individuals, the body is covered in small white spots; in others, there are only flecks of white or gold against the emerald-green background. Stippled clingfishes (about 3 in) are light-brown to pale-gray, with numerous small, dark spots. Some have faint dark bands across the body (usually between the pectoral fins and in front of dorsal fin). Skilletfishes are light–olive-brown to dark-brown, often with 5 dark body bands, the last a distinct dark bar that is always present at the base of the tail.

Distribution and Habitat

Emerald clingfishes live in shallow (<2 ft) waters, off southeast Florida and the Florida Keys, in seagrass meadows having a high current flow and relatively clean grass blades. Skilletfishes characteristically inhabit oyster bars and reefs, shell rubble, sponges, tide pools, and pilings; they are also found in seagrass beds that may be interspersed with oysters or shells. Stippled clingfishes are tropical reef-dwellers, known in Florida from a single Keys specimen, but the clingfishes we have seen attached to loggerhead sponges, in shelf waters offshore of southwest Florida, may be this species.

Natural History

Emerald clingfishes cling to blades of turtle grass (and occasionally manatee grass), usually attaching themselves to the underside of the blade. Females lay their adhesive eggs, 10–25 eggs at a time, on the surface of turtle grass blades. Embryos develop into larvae in about 10 days. Emerald clingfishes feed on tiny crustaceans (copepods, ostracods, amphipods, and isopods) that also live on the surface of the seagrass blades. While feeding, emerald clingfishes scour individual grass blades, moving from blade to blade as currents or waves bring adjacent blades into contact with one another.

From April to August (most commonly in June–July), Chesapeake Bay skilletfishes lay 300–2500 adhesive eggs on the concave inner surfaces of uninhabited oyster shells. Spawning does not take place until the water temperature reaches 65°F, so reproduction in

Skilletfish (*Gobiesox strumosus*), viewed from below, showing the suction disc, Singer Island, Florida

Florida waters probably occurs during our cooler-water periods (aquarium spawning of Florida-caught skilletfishes was observed in October–December). The embryos develop for about 5–7 days before hatching at about 0.1 in long. The male guards and aerates the eggs for the entire incubation period. The distinctive pelvic sucking disc is developed by the time the young grow to a length of 0.2 in, and maturity is reached at 1.5 in. Skilletfishes feed on amphipods, isopods, and small worms.

Dragonets
Family Callionymidae

- **Spotted dragonet** (*Diplogrammus pauciradiatus*)
- **Spotfin dragonet** (*Foetorepus agassizii*)
- **Palefin dragonet** (*Foetorepus goodenbeani*)
- **Lancer dragonet** (*Paradiplogrammus bairdi*)

Background

Dragonets are small bottom dwellers, with flattened heads, beak-like mouths, and large eyes. Tiny (2.5 in) spotted dragonets are light-tan in coloration, with small dark brown spots, and have a hook-shaped spine protruding from the gill cover. The 4 very tall first dorsal-fin spines are obviously separated, and these fishes have a low ridge on the lower part of the body, between the anal and caudal fins. The somewhat larger (4 in) lancer dragonets have a hook-shaped spine and are light-brown, with various dark-brown to red markings that are rimmed with electric-blue lines. The spinous dorsal fin in males is high (with membranes found between the spines) and variably colored, with wavy lines; in females the spinous dorsal fin is shorter and jet-black. Spotfin dragonets (9 in) have a lanceolate tail and an orange-pink background color, with yellow spots on the dorsal and

Male spotted dragonet (*Diplogrammus pauciradiatus*), Lake Worth Lagoon, Florida

Female spotted dragonet (*Diplogrammus pauciradiatus*), Lake Worth Lagoon, Florida

Male spotted dragonets (*Diplogrammus pauciradiatus*) displaying, Lake Worth Lagoon, Florida

caudal fins. The rear of the first dorsal fin has a prominent, ocellated black spot. Palefin dragonets, the largest of the group (12 in), are orange-red, with very small spots on the median fins. The first dorsal-fin spine is highly elongated.

Distribution and Habitat

Spotted dragonets occur around the state, in inshore areas less than 30 ft deep, usually in or near seagrass meadows. Lancer dragonets are found around the state, in water depths ranging from 1 to 300 ft, where the bottom is composed of shell hash or coarse sand. Spotfin and palefin dragonets occur around the state in much deeper waters, ranging from 300 to 2100 ft and 160 to 1200 ft, respectively. Both are found on soft sediment.

Natural History

Lancer dragonets engage in a social hierarchy, with dominant males holding large territories that contain varying number of females. Dominant males patrol territories that range from 75 to 120 sq ft, while females hold smaller territories (2–15 sq ft) within these areas. Subdominant males sneak within and among territories of the dominant males, seeking mating opportunities. When dominant and subordinate males cross paths, mouths gape and dorsal fins flag in impressively aggressive displays. At dusk, mating pairs rush into the water column to spawn.

The shallow-water-dwelling dragonets blend in very well with the bottom. If you find one while diving, keep an eye on it; usually another will appear. They feed on small, bottom-dwelling invertebrates. Spotted dragonets search out tiny crustaceans (known as harpacticoid copepods), using their protrusible mouths to pluck individual morsels from the sand surface. Estimates show that individual spotted dragonets can process as many as 1000 of these copepods per day.

Sleepers
Family Eleotridae

- **Fat sleeper** (*Dormitator maculatus*)
- **Largescaled spinycheek sleeper** (*Eleotris amblyopsis*)
- **Smallscaled spinycheek sleeper** (*Eleotris perniger*)
- **Emerald sleeper** (*Erotelis smaragdus*)
- **Bigmouth sleeper** (*Gobiomorus dormitor*)
- **Guavina** (*Guavina guavina*)

Background

Sleepers, which are close relatives of the gobies, inhabit marine and estuarine waters of Florida. They differ from gobies in having pelvic-fin rays that are not connected (in gobies they are connected). Fat sleepers reach almost 1 ft in length and have a stocky body and small mouth. They are light-brown to blue-gray in color, with 6 dark, slightly diagonal bars (particu-

Fat sleeper (*Dormitator maculatus*), Vero Beach, Florida

Bigmouth sleeper (*Gobiomorus dormitor*), Loxahatchee River, Florida

larly in juveniles) and a bright-blue blotch above the origin of the pectoral fin. Dark lines radiate posteriorly from the eye, and the dorsal, anal, and broadly rounded caudal fins have dark bands, characteristics they share with largescaled spinycheek sleepers (12 in). The latter are coffee-colored on the upper back and darker along the flanks, with dark blotches and spots; they lack a spot above the pectoral fin. This species is easily confused with smallscaled spinycheek sleepers (12 in): both have a downward-pointing spine on the cheek (preopercle) and large, oblique mouths. Smallscaled spinycheeks are gray-brown, with a dark spot above the pectoral fin, and have smaller scales than their sibling species. Emerald sleepers (9 in) are uniformly dark-brown or gray, with a dark spot above the base of the pectoral fin. The tail is pointed; the oblique mouth is large; and the median fins are not banded. Bigmouth sleepers grow to over 2 ft long; have a flattened head; lack cheek spines; and have large mouths, with protruding lower jaws. They are dark-brown above and lighter–yellow-brown below; the dorsal, pectoral, and caudal fins are banded. Young are lighter colored, with a thick, dark, lateral stripe running through the eye to the tail. Guavinas have a robust body, with a blunt snout and protruding lower jaw. They are dark-brown, with 3 dark lines radiating downward from the eye; the median fins are edged in yellow.

Distribution and Habitat

All the sleeper species live in freshwater or brackish-water reaches of coastal rivers. Fat sleepers and largescaled spinycheek sleepers are reported from around the state. Smallscaled spinycheek sleepers are found from Cape Canaveral to Tampa Bay, and emerald sleepers occur from Cape Canaveral to the panhandle. Bigmouth sleepers are found from eastcentral Florida to the panhandle, but they probably reach peak abundance in eastcentral and southeastern Florida. Only one guavina has been reported from Florida (the sole specimen was collected from a land-crab burrow).

Natural History

Sleepers spawn in brackish or freshwater reaches, but fertilized eggs are carried downstream to the ocean, where larvae live in the pelagic environment for a short period of time. As the larvae grow, they move back into the estuary and upstream to the adult habitat. This life-history strategy is known as amphidromy. The best-studied species is the bigmouth sleeper. Females lay 4000–6000 demersal eggs in a 12 in long mass. Eggs reportedly hatch at night, to reduce the risk of the larvae being eaten by predators. In Central American streams, the larval stages of sleepers move back into the estuaries, along with masses of larvae of other fishes and shrimp, in a migration that is known to locals as "tismiche." Influxes of fish and shrimp young of this magnitude have not been documented in Florida rivers, but it is likely that a scaled-down facsimile may occur.

Fat sleepers feed on detritus and some invertebrates, while smallscaled and largescaled sleepers eat a combination of shrimps, crabs, insects, and fishes. Bigmouth sleepers are lie-in-wait predators, reported to bury themselves in leaf litter or sand. There have been reports of bigmouth sleepers feeding at night on fishes, shrimps, and insects (even leaving the water to catch the latter!). Their diets shift with growth: subadults feed on insects and small fishes, and larger bigmouths consume shrimps, crabs, and fishes. Many an angler throwing a lure in east coast rivers for snooks or tarpons has been puzzled when one of these guys hits it.

Gobies
Family Gobiidae

- **River goby** (*Awaous banana*)
- **Bearded goby** (*Barbulifer ceuthoecus*)
- **Notchtongue goby** (*Bathygobius curacao*)
- **Twinspotted frillfin** (*Bathygobius geminatus*)
- **Checkerboard frillfin** (*Bathygobius lacertus*)
- **Island frillfin** (*Bathygobius mystacium*)
- **Frillfin goby** (*Bathygobius soporator*)
- **White-eye goby** (*Bollmannia boqueronensis*)
- **Ragged goby** (*Bollmannia communis*)
- **Shelf goby** (*Bollmannia eigenmanni*)
- **Barfin goby** (*Coryphopterus alloides*)
- **Sand-canyon goby** (*Coryphopterus bol*)
- **Colon goby** (*Coryphopterus dicrus*)
- **Pallid goby** (*Coryphopterus eidolon*)
- **Bridled goby** (*Coryphopterus glaucofraenum*)
- **Glass goby** (*Coryphopterus hyalinus*)
- **Kuna goby** (*Coryphopterus kuna*)
- **Peppermint goby** (*Coryphopterus lipernes*)
- **Masked goby** (*Coryphopterus personatus*)
- **Spotted goby** (*Coryphopterus punctipectophorus*)
- **Bartail goby** (*Coryphopterus thrix*)
- **Sand goby** (*Coryphopterus tortugae*)
- **Darter goby** (*Ctenogobius boleosoma*)
- **Sashcheek goby** (*Ctenogobius pseudofasciatus*)
- **Dash goby** (*Ctenogobius saepepallens*)
- **Freshwater goby** (*Ctenogobius shufeldti*)
- **Emerald goby** (*Ctenogobius smaragdus*)
- **Marked goby** (*Ctenogobius stigmaticus*)
- **Spottail goby** (*Ctenogobius stigmaturus*)
- **Yellowline goby** (*Elacatinus horsti*)
- **Neon goby** (*Elacatinus oceanops*)
- **Yellowprow goby** (*Elacatinus xanthiprora*)
- **Sponge goby** (*Evermannichthys spongicola*)
- **Lyre goby** (*Evorthodus lyricus*)
- **Goldspot goby** (*Gnatholepis thompsoni*)
- **Violet goby** (*Gobioides broussonetii*)
- **Highfin goby** (*Gobionellus oceanicus*)
- **Naked goby** (*Gobiosoma bosc*)
- **Seaboard goby** (*Gobiosoma ginsburgi*)
- **Rockcut goby** (*Gobiosoma grosvenori*)
- **Twoscale goby** (*Gobiosoma longipala*)
- **Code goby** (*Gobiosoma robustum*)
- **Paleback goby** (*Gobulus myersi*)
- **Crested goby** (*Lophogobius cyprinoides*)
- **Dwarf goby** (*Lythrypnus elasson*)
- **Island goby** (*Lythrypnus nesiotes*)
- **Convict goby** (*Lythrypnus phorellus*)
- **Bluegold goby** (*Lythrypnus spilus*)
- **Seminole goby** (*Microgobius carri*)
- **Clown goby** (*Microgobius gulosus*)
- **Banner goby** (*Microgobius microlepis*)
- **Green goby** (*Microgobius thalassinus*)
- **Orangespotted goby** (*Nes longus*)
- **Spotfin goby** (*Oxyurichthys stigmalophius*)
- **Mauve goby** (*Palatogobius paradoxus*)
- **Rusty goby** (*Priolepis hipoliti*)
- **Scaleless goby** (*Psilotris alepis*)
- **Toadfish goby** (*Psilotris batrachodes*)
- **Highspine goby** (*Psilotris celsus*)
- **Tusked goby** (*Risor ruber*)
- **Tiger goby** (*Tigrigobius macrodon*)
- **Leopard goby** (*Tigrigobius saucrus*)
- **Orangebelly goby** (*Varicus marilynae*)

Background

Gobies are the most diverse family of marine fishes in Florida—and the world, for that matter. We do not have the space (or time) to discuss each of the many species in detail, so we will focus on the more-common species, particularly those most likely to be seen while netting, wading, snorkeling, or scuba diving in Florida. The diversity of gobies reflects the continual splitting of species into clusters of sister-species, distinct genetic lineages that can only be separated using minute pigment characters and, ultimately, DNA sequences.

The largest and arguably the most distinctive of the gobies are violet gobies (called "dragonfishes" by the aquarium trade), which have

River goby (*Awaous banana*), Loxahatchee River, Florida

Frillfin goby (*Bathygobius soporator*), Lake Worth Lagoon, Florida

Crested goby (*Lophogobius cyprinoides*), Loxahatchee River, Florida

a single, continuous dorsal fin; a slender, eel-like body; and grow to 2 ft long. River gobies (12 in) are tan to pale-gray on the back, with white undersides. They have broken diagonal marks on the body; thin, dark bands on the dorsal and tail fins; a black spot above the pectoral fin; and an upper jaw that extends past the lower jaw.

The genus *Bathygobius* forms one of those closely knit clusters of species, represented by frillfins: notchtongue gobies, twinspotted frillfins, checkerboard frillfins, and island frillfins. All are medium-sized (2–5 in), with stout bodies and blunt snouts. They are dark-brown to gray on the back, with mottled flanks, and the upper pectoral rays are free (hence the name "frillfin"). Notchtongue gobies have very few dark markings on the flanks, and no paired rows of spots. Checkerboard frillfins have 2 rows of offset spots. Twinspotted frillfins have 6–7 sets of paired spots along the flanks. Island frillfins have dark, rectangular blotches. Frillfin gobies have a distinctive black bar on the first dorsal fin and a dark bar on the body, under the dorsal fin. The body type of crested gobies resembles that of frillfins, but the former species differs by the presence of a fleshy crest on top of the head; the lack of frilly, separated pectoral rays; and dark-gray to green, rectangular markings on the sides. In addition, males have a black blotch, with a bright-orange spot on the first dorsal fin.

Members of the genus *Coryphopterus*, sometimes called "hyaline gobies" because of their translucent bodies, also present a chal-

Top left: Colon goby (*Coryphopterus dicrus*), Lake Worth Lagoon, Florida. *Top right*: Pallid goby (*Coryphopterus eidolon*), Lake Worth Lagoon, Florida. *Bottom left*: Bridled goby (*Coryphopterus glaucofraenum*), Lake Worth Lagoon, Florida. *Bottom right*: Sand-channel goby (*Coryphopterus bol*), Jupiter, Florida.

lenge to the eyeball taxonomist. These species can be distinguished by subtle markings, but separating the closely related species in this genus will test one's patience. Barfin gobies are distinctive, with a dark bar beneath the dorsal fin and (sometimes) red on the cheeks. Bartail gobies have a large dark spot at the base of the pectoral fin. Glass and masked gobies are the only members of the genus *Coryphopterus* to hover in the water column, and they look very much alike. Both species are orange anteriorly, with black eyes; a white stripe begins on the top of the eyes and extends, in broken segments, above the midline to the base of the tail. A distinguishing characteristic between these two species is the position of the anus, relative to a black ring on the undersides. On masked gobies, the anus is near the center of this ring, while the anus on glass gobies is farther forward,

Spotted goby (*Coryphopterus punctipectophorus*), Tampa, Florida

but still within the black ring. Colon gobies have dark-brown markings on the flanks, and 2 distinct spots (resembling the punctuation mark) just forward of the base of the pectoral fin.

Bridled gobies, the most common member of the *Coryphopterus* group, appear to form a species-complex with sand-canyon, sand, pallid, and spotted gobies. Some markings and color patterns vary with habitat. Bridled gobies usually have 2 spots on the base of the caudal fin and 2 white stripes behind the eye. Sand-canyon gobies have spotted eyes and broken yellow stripes on the body. Sand gobies look very much like sand-canyon gobies, but with subtle differences in markings. Pallid gobies are very similar to the other members of the bridled goby complex, but they have 2 yellow-gold stripes behind the eye and a black bar at the base of the tail. Spotted gobies have reddish-brown blotches along the flanks and a single spot at the base of the pectoral fin.

The genus *Gobiosoma* includes five small (<2 in), blunt-nosed species: code, seaboard, naked, rockcut, and twoscale gobies. Naked gobies are pale, with 9–11 wide, dark-gray to reddish bars that extend into the dorsal fins. The other four species have a tan background color, with darker bars. Code gobies have 10–12 dark bars; rockcut gobies have 9 bars; seaboard gobies have 8 dark bars; and twoscale gobies have 9 dark bars, plus 2 scales at the base of the tail.

The *Elacatinus* species differ from other gobies in their color patterns. The upper body of neon gobies is black, with an electric-blue stripe running through the eye to the base of the tail; the underside is white. Yellowprow and yellowline gobies look very similar to neon gobies, but the stripe running through the eye is yellow instead of electric-blue. Yellowprows have white undersides and a small yellow dash on the snout. Yellowline gobies are dark-brown

Left: Glass goby/masked goby (*Coryphopterus hyalinus/personatus*), Jupiter, Florida. *Right*: Peppermint goby (*Coryphopterus lipernes*).

Left: Naked goby (*Gobiosoma bosc*), Crystal River, Florida. *Right*: Code goby (*Gobiosoma robustum*), Crystal River, Florida.

Tiger goby (*Tigrigobius macrodon*), Lake Worth Lagoon, Florida

all over (no white underside). Two of the Florida *Elacatinus* species—tiger and leopard gobies—were recently placed in the genus *Tigrigobius*. Tiger gobies are distinctive: light-gray, with 12 thin, black bars along the body. Leopard gobies are translucent, with pale–orange-gold spots covering the body.

Members of the genus *Microgobius* have large mouths and pointed tails. Seminole gobies are white to tan, with a blue stripe behind the eye and a distinctive yellow-orange lateral stripe, bounded by electric-blue, that extends into a yellowish tail. The dorsal-fin spines are elongated in males and may have off-white tips. Banner gobies are gray to olive dorsally, with silvery flanks; the median fins and tail are pale-lavender, with a red band along the base of the dorsal fin; a thin yellow strip runs the length of the dorsal fin and onto the tail. This species has 3 iridescent marks under the eye. Coloration in males intensifies during the mating season. Clown gobies have dark–gray-brown blotches on the flanks; a thin, white, vertical bar under the origin of the dorsal fin; a large mouth; and elongate dorsal spines. Males have larger mouths than females; the tips of their elongated dorsal spines are orange-yellow; the soft dorsal fin has a series of electric-blue spots, with black centers, just above a parallel orange stripe; and the margin of the tail is banded with yellow-orange and black. Green gobies are pale-green and have 2 white bars on the head, behind the eyes. Males also exhibit 3 white bars under the spinous dorsal fin.

Gobies in the genera *Ctenogobius* and *Gobionellus* are closely related and very similar-appearing. Darter gobies are tan colored, with dark, V-shaped markings on the flanks under the second dorsal fin, and the tail is pointed. Slashcheek,

Top, Neon goby (*Elacatinus oceanops*), Jupiter, Florida. *Bottom*: Yellowprow goby (*Elacatinus xanthiprora*), Florida Middle Grounds.

Seminole goby (*Microgobius carri*), Lake Worth Lagoon, Florida

freshwater, emerald, marked, dash, blotchcheek, and spottail gobies look very similar to darter gobies but can separated by the subtle differences in their color patterns. Dash gobies are the most distinctive, with a white background color; 5 dark dashes along the midline; a vertical black line through the eye and down to the lower jaw; and a dark, triangular mark on the gill cover. Slashcheek gobies have dark markings on the side of the head and a dark spot on the caudal peduncle. Freshwater gobies are tan, with a dark line from the corner of the mouth to

Top: Clown goby (*Microgobius gulosus*), St. Johns River, Florida. *Bottom*: Male, Crystal River, Florida.

Top: Banner goby (*Microgobius microlepis*), Lake Worth Lagoon, Florida. *Bottom*: Males, Lake Worth Lagoon, Florida.

Dash goby (*Ctenogobius saepepallens*), Lake Worth Lagoon, Florida

the upper edge of the gill cover, 4 dark marks along the midline, and a spot at the base of the tail. Emerald gobies look like darter gobies, but the former have pale-green spots on the head and flanks, and a dark shoulder spot just above and behind the gill opening. Marked gobies have 3 short bars under the eye, and the flanks have 10 or more thin, pale bars. Blotchcheek gobies have a single spot below the base of the pectoral fin. Spottail gobies are grayish, with a V-shaped dark mark below the eye, 4 dark marks along the midline, and a small spot at the base of the tail. Highfin gobies reach a length of 9 in and may be distinguished by a large, dark spot above the pectoral fin and elongate extensions on the dorsal spines.

Lyre gobies are similar to *Ctenogobius* species: light-brown, with 5–6 thin bars, 2 dark marks under the eye, and a double spot on the base of the tail. Males develop higher dorsal fins than females. Goldspot gobies have a blunt head; 6 dark smudges along the body; and a gold spot, rimmed in black, above and behind the gill opening. Orangespotted gobies are gray, with pale-orange spots on the head and body, and have 5–7 dark blotches on the flanks. Spotfin gobies are pale, with 5 golden blotches on the flanks and a distinct black spot on the back of the dorsal fin.

Distribution and Habitat

Gobies are bottom dwellers that occur from the freshwater reaches of coastal rivers to the edge of the continental shelf. Different species live on sedimentary bottom, often

Spottail goby (*Ctenogobius stigmaturus*), Lake Worth Lagoon, Florida

Goldspot goby (*Gnatholepis thompsoni*), Lake Worth Lagoon, Florida

Orangespotted goby (*Nes longus*), with a sand snapping shrimp (*Alpheus floridanus*), Lake Worth Lagoon, Florida

adjacent to reefs, seagrass meadows, saltmarshes, or mangroves. Violet gobies are found in dark, brackish water and in saltmarshes, and they are sometimes reported from the intake screens of coastal power plants. Members of the bridled goby complex are found along the east coast, from Cape Canaveral through the Florida Keys, and to about Tampa on the west coast, mostly in shallow-water seagrass, sand, and rubble habitats, where salinity is high.

Some species intimately associate with live corals or sponges. Neon gobies are often found lying on living brain or star corals, and peppermint gobies are almost always found on living stony coral colonies. Yellowprow gobies exclusively lie on live sponges. At least two species not discussed elsewhere in this chapter, because of their secretive habits—tusked and sponge gobies—are small, cylindrical fishes, adapted to live within the labyrinthine canals that permeate sponges.

Natural History

Gobies are demersal spawners. All species studied to date lay eggs on the substrate, in nests, or in burrows, where, after a brief period, they develop into pelagic larvae. The number of eggs per clutch may be small, depending on the species. The mating habitats of only a few species have been studied in any detail. Male clown gobies mate with multiple females; eggs are deposited inside the burrow, and the male then guards the eggs until they hatch. Male clown gobies have large, elongated dorsal-fin elements that they flare at passing fishes, including other males.

Although most gobies rest directly on the seafloor, a few—banner, Seminole, glass, masked, shelf, and white-eye gobies—hover in the water column, usually above a home burrow, picking at plankton as it passes by their burrows.

Many species or species-groups—including *Microgobius*, some *Ctenogobius*, and some *Coryphopterus*—live in burrows constructed by the individual or, in some cases, another organism, such as a snapping shrimp. Orangespotted gobies are strongly linked with snapping shrimps (*Alpheus floridanus*). Other gobies—most notably dash gobies—will live with snapping shrimps, but the association is not as well developed as with orangespotted gobies. A snapping shrimp has poor eyesight and relies on the goby to stand watch

Spotfin goby (*Oxyurichthys stigmalophius*), Jupiter, Florida

for approaching predators while it bulldozes sediment from inside the burrow and out the entrance. On sandy areas with lots of shrimp burrows, you can snorkel over the bottom and only glimpse puffs of sediment, indicating where the gobies wheeled around and dove head-first into the burrow. A snapping shrimp has long, thin antennae that touch the goby's back as the shrimp emerges from the burrow with a load of sand. If the goby flicks its fins against the antennae, the shrimp will retreat. Individual gobies have their own tolerance level for the approach of divers. If you find a tolerant goby, the burrowing process can be observed for a long time. Other gobies—most notably bridled, notchtongue, and spotfin gobies—have been reported to associate with snapping shrimps, but of these three species, we have only seen spotfin gobies with snapping shrimps in Florida.

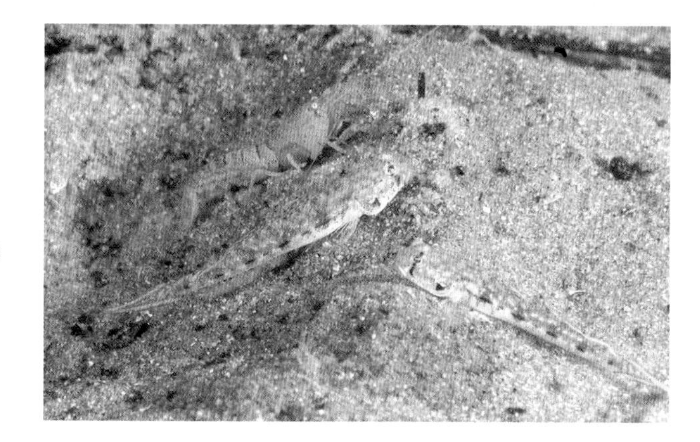

Dash gobies, with a sand snapping shrimp (*Alpheus floridanus*), Lake Worth Lagoon, Florida

Wormfishes
Family Microdesmidae

- **Pugjaw wormfish** (*Cerdale floridana*)
- **Pink wormfish** (*Microdesmus longipinnis*)

Background

Wormfishes look more like eels than gobies, being elongated, slender, and moderately laterally compressed little fishes, with tiny eyes and short, blunt snouts. The oblique mouth is small, with the lower jaw extending below the upper jaw. They lack lateral lines. The low, long-based dorsal and anal fins are continuous with the short, rounded caudal fin. Pugjaw wormfishes (3.5 in) are a translucent white, with tiny black peppering. They are not as large or as long and thin as pink wormfishes (11 in), which are pinkish-, orangish-, or reddish-tan, with scattered brown speckles on the back. An iridescent–dark-blue spot is found on the rear of the operculum.

Distribution and Habitat

Wormfishes burrow in muddy or sandy bottoms in shallow water. Pugjaw wormfishes occur in southeast Florida, from Jupiter through the Keys, in less than 50 ft of water; they are usually associated with coral reefs and open coastal habitats. Pink wormfishes normally are found in the Gulf, in shallow nearshore waters, to a depth of 125 ft, although waifs do occur on the Atlantic coast.

Pugjaw wormfish (*Cerdale floridana*), Lake Worth Lagoon, Florida

Natural History

Pink wormfishes are known to occupy mud-shrimp and jawfish burrows. They are eaten by croakers and probably by other bottom-dwelling fishes. Crustacean eggs were observed in the stomach of a pugjaw wormfish. Little is known about reproduction in wormfishes, but their larvae are well known and readily distinguished.

Dartfishes
Family Ptereleotridae

- **Blue dartfish** (*Ptereleotris calliura*)
- **Hovering dartfish** (*Ptereleotris helenae*)

Background

Dartfishes are close relatives of the gobies and are often called "hovering gobies," due to their habitat of maintaining a station above the home burrow. Blue dartfishes, which reach 7 in long, usually are white but sometimes have a light–blue-green cast. The dorsal fin usually has a black margin, which can be red in some individuals. Adults have an electric-blue line that runs along the dorsal midline, from the head to the origin of the dorsal fin. The snout is blunt, with an upturned mouth, and the tail is long and pointed. Hovering dartfishes look very similar, but the tail is rounded instead of pointed. The young of blue dartfishes also have rounded tails and thus may be confused with hovering dartfishes.

Distribution and Habitat

Both species inhabit sand- and shell-rubble–bottom near reefs. Blue dartfishes are found around the state, in nearshore and shelf waters, as deep as 200 ft. They are abundant in some areas of the west Florida shelf. Hovering dartfishes are primarily an insular species, found off southeastern Florida and the Florida Keys.

Natural History

Dartfishes inhabit U-shaped burrows and, when frightened, zip head-first into their self-made homes. Individuals hover several inches to over a foot above their burrows, head-down, at a slight angle. Reproductive characteristics are unknown, but it is likely that demersal eggs are deposited in the burrow. Blue dartfishes feed on plankton as they hovers above their burrows. Both species co-occur with yellowhead jawfishes and sand tilefishes in sandy, near-reef areas.

Blue dartfish (*Ptereleotris calliura*), Tampa, Florida

Spadefishes
Family Ephippidae

- **Atlantic spadefish** (*Chaetodipterus faber*)

Background

Atlantic spadefishes are the only member of this family occurring in the western Atlantic. They grow to over 2 ft long and resemble angelfishes in having a deep and laterally compressed body, with long extensions on the median fins. There are usually 6 black bars on the body, over a silvery background, but sometimes the bars are not visible and the body is uniformly grey. Small juveniles have a higher profile than adults and are uniformly dark-brown to almost coppery, with flecks of white inside black flecks; their soft dorsal, anal, and tail fins are transparent.

Distribution and Habitat

Atlantic spadefishes occur throughout Florida, usually associated with structured habitat, in depths less than 200 ft. Some adults and juveniles enter the estuarine segments of coastal rivers or lagoons.

Natural History

Observations of spawning in Atlantic spadefishes indicate that adults assemble in groups above structured habitat. Individual pairs undergo a courting ritual, which results in a sort of liplock prior to rushing toward the surface, presumably to release eggs and sperm. On some glassy-calm days, adult Atlantic spadefishes can be seen swimming at the water surface, with their dorsal fins protruding—prompting cries of "Shark!" from onlookers. This behavior may be related to courtship or spawning. Following fertilization, the pelagic eggs transform into larvae that spend an average of 20–30 days in the plankton before settling to the seafloor. These dark-colored juveniles are thought to mimic decaying plant debris, particularly mangrove seed-pods. The mimicry is enhanced by their behavior—

Atlantic spadefishes (*Chaetodipterus faber*), Jupiter Island, Florida

Juvenile Atlantic spadefish (*Chaetodipterus faber*), Lake Worth Lagoon, Florida

gently rolling around in shallow littoral waters, much like the adjacent debris—thereby hiding in plain sight from predatory fishes and birds. Juveniles also can be spotted in the nearshore wash-zone in the surf.

Although adults often are seen by themselves, they generally school in the water column, around structured habitat. Atlantic spadefishes feed in the water column, consuming jellyfishes and other gelatinous plankters.

Surgeonfishes
Family Acanthuridae

- **Doctorfish** (*Acanthurus chirurgus*)
- **Blue tang** (*Acanthurus coeruleus*)
- **Ocean surgeon** (*Acanthurus tractus*)

Background

The family Acanthuridae has three representatives in Florida waters: ocean surgeons, blue tangs, and doctorfishes. Yet another regional form—Gulf surgeons—recently have been shown to be the same species as ocean surgeons. The name "surgeonfish" or "doctorfish" is derived from the presence of a scalpel-like spine protruding from the caudal peduncle on all members of the family. The three

Doctorfish (*Acanthurus chirurgus*), Lake Worth Lagoon, Florida

Left: Blue tang (*Acanthurus coeruleus*), Looe Key, Florida. *Right*: Juvenile, Palm Beach, Florida.

Transitional juvenile blue tang (*Acanthurus coeruleus*), Juno, Florida

Ocean surgeon (*Acanthurus tractus*), Lake Worth Lagoon, Florida

species occurring in Florida waters are all similarly disc-shaped and are best separated by color pattern. Blue tangs are the most distinctive, having a royal- to cobalt-blue body, with thin, wavy, blue stripes. The median fins and tail are a lighter, brighter-colored blue. Juvenile blue tangs are uniformly bright-yellow and gradually transform to the adults' blue coloration as they grow. Individuals of varying sizes are often observed in transitional coloration, usually partly yellow, with blue fins. Ocean surgeons vary from light-brown to dark-brown, with a white bar around the caudal peduncle and a white caudal-fin margin. Colorations in this species and in doctorfishes are variable and can be changed. Doctorfishes resemble ocean surgeons at small sizes, but the former are distinguished by looking at the margin of the tail: ocean surgeons have a more-curved tail profile, while doctorfishes have a straight-cut tail margin. Doctorfishes also usually have 8–12 thin, vertical bars on the body. Newly settled members of the genus are transparent and are referred to as acronurus larvae.

Distribution and Habitat

All three species reach peak abundances in shallow, fully saline waters of southeastern Florida and the Florida Keys. Surgeonfishes are uncommon outside of this area; their occasional occurrence elsewhere is the result of larval dispersal.

Natural History

Surgeonfishes spawn in the water column (in March for ocean surgeons). Male blue tangs change color when courting: the face becomes pale, and the posterior half of the body remains the normal blue. Doctorfishes are more ecologically flexible than the other species, and they can also be found in deeper waters and multiple habitats. Ocean surgeons aggregate during winter to spawn. Juveniles of all species settle in shallow water.

All species are herbivores, commonly observed grazing on algae, either in groups or as single individuals. Each species exhibits individual preferences for brown, green, or red algae. Ocean surgeons and doctorfishes ingest sand while foraging, which may help with the mechanical digestion of algae and other plant matter. Adult blue tangs typically form schools that rove around reefs, foraging on algae.

Barracudas
Family Sphyraenidae

- **Great barracuda** (*Sphyraena barracuda*)
- **Sennet** (*Sphyraena borealis*)
- **Guaguanche** (*Sphyraena guachancho*)
- **Southern sennet** (*Sphyraena picudilla*)

Background

Barracudas are elongated, silvery fishes, with a protruding lower jaw and prominent crooked, snaggled teeth. Great barracudas are the largest and most commonly encountered of the barracudas, growing to about 6 ft (stories about 8 ft barracudas dismembering potential trophy fishes not withstanding!). Adults have inky-black smudges on their sides; small juveniles have a series of wavy dark bars along the flank that gradually fade with growth. Sennets and southern sennets, which may represent a single widespread species, are smaller, not exceeding 2 ft long. They can be distinguished from the other barracudas by having a short upper jaw that does not reach the eye; an adult body without black smudges or spots; and a wash of yellow along the midline in adults and larger juveniles. Guaguanches are equally small, and adults also lack black pigmentation, but the upper jaw extends past the front of the eye, and the body has a yellow lateral stripe. There are also differences in fin placement, and, when they are young, sennets and southern sennets have a distinctive, fleshy black tab extending from the lower jaw that is lacking in guaguanches.

Distribution and Habitat

Great barracudas are found around the state (and the world), in coastal and offshore habitats. Young great barracudas settle in inshore waters, around mangrove shorelines or seagrass meadows; adults occur from inshore habitats (but not estuaries) out to oceanic waters of the Gulf Stream Current. They are common on reefs and (pelagically) in oceanic waters. Guaguanches, sennets, and southern sennets also are found—uncommonly—throughout Florida, in inshore waters. Their young most often occur in seagrass beds, and the adults in a variety of habitats, including reefs, to depths of 100 ft. Guaguanches are migratory along Florida's southern coasts.

Adult great barracuda (*Sphyraena barracuda*), Lake Worth Lagoon, Florida

Natural History

Little is known about the spawning behavior of great barracudas, but they undergo classic developmental migration: from shallow inshore waters to deeper offshore reefs and the open sea. Spawning occurs in outer-shelf waters, where eggs are broadcast into the water column. Larvae settle in shallow waters, usually around seagrasses or mangroves, and some settle into drifting sargassum. Young are seen in the shallows of southern Florida during the spring and summer. Individuals that are 2 in or smaller hang in the water column, with their snouts pointing upward, often among red mangrove prop roots or black mangrove pneumatophores (aerial roots), where they appear to mimic drifting plant debris. Larger (4–16 in) juveniles are found in shallow water, near mangrove shorelines or over sand- and seagrass bottoms.

Great barracudas are second only to sharks as fish-catching and eating machines. The jaw teeth are plenty imposing, but farther back, on the roof of the mouth,

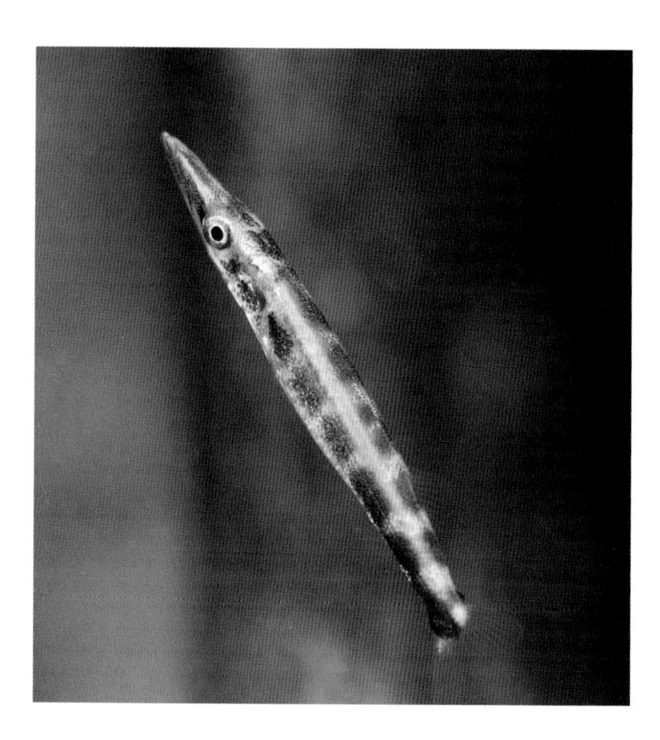

Juvenile great barracuda (*Sphyraena barracuda*), Key West, Florida

there is a row of tightly seated, serrated teeth capable of shearing a relatively large fish cleanly in two. Great barracudas use their snaggled teeth to grip a fleeing fish that is often too large to swallow whole; they then methodically cut this catch with their shearing teeth and gobble down the pieces. Much to the chagrin of anglers, great barracudas frequently attack hooked fishes, shearing off the rear half of the body with a single, bandsaw-like bite. Juveniles frequently consume pilchards and silversides; adults eat almost any species of fish, including morays, pilchards, needlefishes, jacks, grunts, snappers, surgeonfishes, and porcupinefishes.

Great barracuda are known to sequester ciguatoxin in their tissues. This toxin is produced by a tiny planktonic alga known as a dinoflagellate (another dinoflagellate species causes red tide). Human consumers of ciguatoxic fishes experience gastrointestinal distress, as well as neurological effects that vary in severity and can last for years. Reports of ciguatera poisoning in Florida are not common, and many cases are ultimately traced to barracudas that originated in the Bahamas or the Caribbean Sea.

Sennets and guaguanches tend to settle in groups around structured hardbottom, both on the shelf and in nearshore waters. Young sennets are seen schooling with herrings or scads around artificial and natural hardbottom habitat. Guaguanches seem to prefer turbid water over mudbottom. They often are caught, along with migrating bluefishes and Spanish mackerels, off Florida's east coast. Southern sennets consume fishes and squids.

Snake Mackerels
Family Gempylidae

- **Snake mackerel** (*Gempylus serpens*)
- **Escolar** (*Lepidocybium flavobrunneum*)
- **Black snake mackerel** (*Nealotus tripes*)
- **American sackfish** (*Neoepinnula americana*)
- **Black gemfish** (*Nesiarchus nasutus*)
- **Roudi escolar** (*Promethichthys prometheus*)
- **Oilfish** (*Ruvettus pretiosus*)

Background

Snake mackerels are a variable group of fishes that have projecting lower jaws and prominent, sharp teeth. Pelvic fins often are missing or are rudimentary. They have 2 dorsal fins and all but one species (American sackfishes) have finlets following the second dorsal and anal fins. The silvery American sackfishes, the smallest (12 in) of the group, have a more laterally compressed and deeper body than most other species in the Gempylidae. The pelvic fins are well developed, and the second dorsal fin originates over the almost equal-sized anal fin. Escolars and oilfishes, which also have these traditional pelvic fins, are the largest members of the family (6.75 ft and 10 ft, respectively; >130 lb). Both are dark-brown to black, streamlined species, with large eyes. Escolars have 4–6 dorsal finlets and a wandering, dippsy-doo lateral line, while oilfishes have 2 finlets and a relatively straight lateral line. The extremely elongated snake mackerels (3.5 ft) and black gemfishes (4.5 ft) have tiny pelvic fins; lower jaws with pointed tips; and very long, low first dorsal fins. The color and number of dorsal and anal finlets separate these two species: silvery-gray, with 5–7 finlets in snake mackerels; dark-brown to violet, with a pair of finlets, in gemfishes. Black snake mackerels (11.5 in) and roudi escolars (3.5 ft), are moderately elongated species. They both lack pelvic fins but have nearly symmetrical second dorsal and anal fins and a pair of dorsal and anal finlets. Black snake mackerels have a straight lateral line, while this line is highly arched anteriorly in roudi escolars.

Oilfish (*Ruvettus pretiosus*), Ft. Pierce, Florida. *Photo by George H. Burgess.*

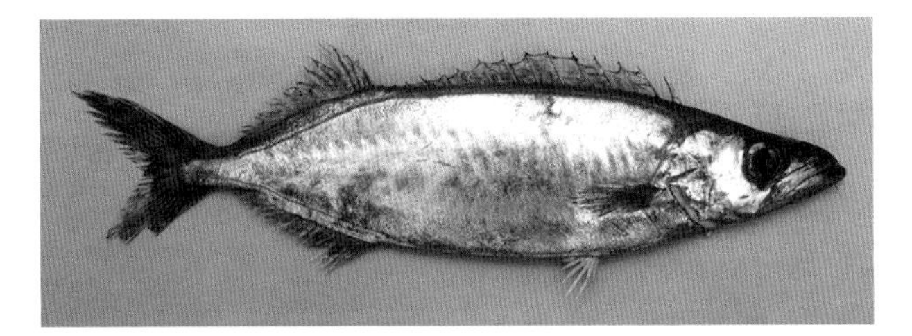

American sackfish (*Neoepinnula americana*), Ft. Pierce, Florida. *Photo by George H. Burgess.*

Distribution and Habitat

All species in this family occur in appropriately deep water around Florida. Oilfishes, black snake mackerels, and snake mackerels are oceanic, continental-slope species that occasionally wander over outer continental-shelf waters. In one study, oilfishes were observed by day, hovering, often in pairs, from 4 in to 23 ft over even bottom, at a depth of 3805 ft. At night, they vertically moved to epipelagic depths of 320 ft. The other species are found over outer shelf and upper continental-slope waters. American sackfishes and oilfishes are largely benthopelagic, living, respectively, near the bottom at depths of 600–1500 ft and from the surface to 26,250 ft deep. Benthopelagic oilfishes also consume meso- and epibenthic prey items, which suggests marked vertical movements. Other benthopelagic species include escolars (to 650 ft deep) and roudi escolars (320–2500 ft). Male roudi escolars live at greater depths than females. Black snake mackerels and snake mackerels are mesopelagic (living in midwater) and epipelagic (living in the upper water column) fishes that occur as deep as 650 ft. Snake mackerels frequently are encountered at the water's surface at night, where they are attracted to lights on ships, even jumping onto such vessels. Black gemfishes are bentho- and mesopelagic residents of 650–3900 ft depths.

Natural History

Snake mackerels, as a family, are vertical migrators that live deep by day and move upward in the water column by night, feeding on a variety of pelagic fishes, cephalopods, and crustaceans, in varying proportions. For example, oilfishes primarily eat fishes and cephalopods at a 2:1 ratio, in one study focusing on flyingfishes, roudi escolars, snake mackerels, dolphinfishes, rainbow runners, and black jacks, and on hakes, congers, cutlassfishes, grenadiers, and shark eggs in another. Escolars eat tunas, dolphinfishes, pomfrets, and ribbonfishes. Snake mackerels dine on flyingfishes, tunas, and lanternfishes, and they have been captured by recreational fishers using surface-trolled lures. Reproduction appears to be year round for many species. Black snake mackerels reach sexual maturity at about 7 in, and roudi escolars at about 20 in. Snake mackerels mature at 1.5–2 ft and produce 300,000–1,000,000 eggs. Escolars appear to be early summer to midsummer spawners; one female had 36 million eggs. Snake mackerel larvae undergo vertical migration, but in the opposite direction of the adults: up by day and down by night. Oilfish eggs and larvae probably are pelagic, and black gemfish larvae are epi- and mesopelagic. Roudi escolars live at least 11 years. Young snake mackerels are regularly eaten by two seabird species—black noddies and red-footed boobies—in the Hawaiian Islands, and they probably fall prey to oceanic seabirds in our area. Oilfishes may be intentionally ignored as prey, because their flesh has strong purgative properties. A 9.5 ft, short-finned pilot whale died after the 3 ft oilfish it had consumed bit through its stomach wall (best to avoid the fresh sashimi!). Oilfishes maintain neutral buoyancy by storing low-density oil in the skull and under the skin.

Cutlassfishes
Family Trichiuridae

- **Atlantic cutlassfish** (*Trichiurus lepturus*)

Background

This family has several deepwater members that occur in water depths greater than 600 ft, but Atlantic cutlassfishes (sometimes called "ribbonfishes" or "hairtails" by anglers) are the only species that also is found in shallow water. This fish is unmistakable—an elongated, laterally compressed, silvery-blue body; a lower jaw that juts out farther than the upper jaw; and a mouth full of large, fang-like teeth. The body tapers to filament (no caudal fin), and the dorsal fin spans the entire body. Adults reach almost 5 ft long. A long, silvery fillet strip cut from an Atlantic cutlassfish is highly coveted as a trolling bait, particularly for king mackerels.

Distribution and Habitat

Atlantic cutlassfishes are a worldwide species, and they occur around the state in shelf, coastal, and inshore waters, to depths of at least 400 ft, usually over soft-bottom.

Natural History

Atlantic cutlassfishes mature at 2.5–3 ft and apparently spawn in coastal and inshore waters of Florida. Limited information indicates gonadal maturity is highest during winter months in southeastern Florida. Females produce an estimated 3400 eggs per spawning batch. Atlantic cutlassfishes feed on shrimps, squids, and especially on fishes, such as herrings, anchovies, drums, butterfishes, jacks, and marine catfishes. They frequently eat each other, as well. We have observed Atlantic cutlassfishes feeding on small fishes attracted to the lights of remotely operated vehicles in the Gulf of Mexico, in waters depths of 200–300 ft. Individuals hang vertically in the water column, with their heads up, darting rapidly upward to grab prey items. Atlantic cutlassfishes fall prey to porpoises and predatory fishes, including billfishes. They reach a maximum age of 7 or 8 years. Young are found in deeper water than the nearshore-inhabiting adults, indicating an offshore-to-inshore recruitment pattern. Atlantic cutlassfishes are prone to the thickening of certain bones (hyperossification), forming skeletal stones.

Atlantic cutlassfish (*Trichiurus lepturus*). *Photo by Jay Fleming Photography.*

Mackerels
Family Scombridae

- **Wahoo** (*Acanthocybium solandri*)
- **Bullet mackerel** (*Auxis rochei*)
- **Frigate mackerel** (*Auxis thazard*)
- **Little tunny** (*Euthynnus alletteratus*)
- **Skipjack tuna** (*Katsuwonus pelamis*)
- **Atlantic bonito** (*Sarda sarda*)
- **Atlantic chub mackerel** (*Scomber colias*)
- **King mackerel** (*Scomberomorus cavalla*)
- **Spanish mackerel** (*Scomberomorus maculatus*)
- **Cero** (*Scomberomorus regalis*)
- **Albacore** (*Thunnus alalunga*)
- **Yellowfin tuna** (*Thunnus albacares*)
- **Blackfin tuna** (*Thunnus atlanticus*)
- **Bigeye tuna** (*Thunnus obesus*)
- **Bluefin tuna** (*Thunnus thynnus*)

Background

Mackerels and tunas are sleek, tireless swimmers and the staples of marine sport and commercial fishing in Florida. These species are built for speed, with stiff, forked tails; narrow caudal peduncles; large flank muscles; and tiny scales. In general, mackerels are slender and tunas are rotund. Species within these two groups may be easily recognized from external characteristics. Starting with the mackerels, wahoos have a long snout, as long as the rest of the head; a sail-like dorsal fin; a steely-blue body, with dark bars along the flanks; and conical teeth. Wahoos can grow to 7 ft long and weigh up to 180 lb. Ceros and Spanish, king, and Atlantic mackerel have snouts that are shorter than the rest of the head. Ceros and king and Spanish mackerels have flat (not conical) teeth in the jaws. King mackerels are gray dorsally, with silver flanks and a white underside. The lateral line curves sharply downward under the soft dorsal fin. King mackerels grow to over 5 ft long and weigh 100 lb. Small (<12 in) king mackerels are often confused with Spanish mackerels; again, a quick look at the lateral line is a reliable way to separate the two. Spanish mackerels are blue-green to grayish dorsally, with silvery sides and a white belly; the lateral line gradually curves toward the tail; the flanks have yellow or golden spots; and the anterior portion of the spinous dorsal fin is jet-black. Spanish mackerels reach almost 3 ft long and weigh 12 lb. Ceros look very much like Spanish mackerels but, along the midline of the body, their golden spots coalesce to form a single, long line of variable length. Ceros grow to 3 ft and weigh 18 lb. Juveniles (<8 in) can be hard to distinguish

Spanish mackerels (*Scomberomorus maculatus*), Pecks Lake, Florida

King mackerel (*Scomberomorus cavalla*). *Photo by George H. Burgess.*

without counting things, but the lateral-line pattern will separate king mackerels from Spanish mackerels and ceros. Atlantic chub mackerels are small (total length 20 in), more slender, and rounded in cross-section, compared with the *Scomberomorus* mackerel species. Atlantic chub mackerels are blue-green, with black, wavy lines on the back and faint spots on the silvery flanks.

Tunas are represented by the genera *Auxis*, *Euthynnus*, *Katsuwonus*, *Sarda*, and *Thunnus*. These are the football-shaped water rockets. Atlantic bonitos are the most distinctive, with conical teeth in a jaw that extends posterior to the orbit. They are dark–steel-blue dorsally, with 9–12 obliquely angled, black stripes above the midline. The other tunas do not have conical teeth or a long upper jaw like Atlantic bonitos. Frigate mackerels and bullet mackerels (still considered tunas, despite their vernacular names) look very much alike: dark-blue backs, with black, wavy lines and silvery sides. The dorsal fin is widely separated, and there are no black spots on the flanks. Frigate and bullet mackerels are often mistaken for juvenile little tunnies. Little tunnies and skipjack tuna have very similar shapes and fin placement but differ in color patterns. Little tunnies are dark-blue to blue-green dorsally, with wavy black lines on the back. The flanks are silver, with scattered black spots between the pectoral and pelvic fins. Skipjack tunas have blue backs, silver sides, and 3–5 longitudinal stripes on the belly, but no black spots.

Tunas in the genus *Thunnus* include albacores and blackfin, yellowfin, bigeye, and bluefin tunas. These fishes have a similar body shape but differ slightly in color, fin size, fin placement, and gill-raker counts (yes, we broke our promise of no counting). At similar sizes, *Thunnus* species can be confusing to identify, at least based on their external appearances. Albacores are easiest to separate from the group, by the white margin on the tail and the extremely long pectoral fins that reach past the origin of the anal fin. Pectoral-fin length in the remaining tuna species is much shorter. Bigeye tunas can weigh up to 400 lb and measure over 7 ft long. The pectoral fins reach the second dorsal fin, and there are between 23 and 31 gill rakers. Bluefin tunas are the gargantuans of the group, growing to over 1000 lb and reaching more than 7 ft long. Bluefins have between 33 and 43 gill rakers, and their pectoral fin is short, not reaching the second dorsal fin. Blackfin tunas grow to over 3 ft and weigh about 40 lb. The body is dark-blue to black on the back, with a wide band of golden-yellow just above the midline, as well as a series of thin, pale bars and vertically arranged spots on

Cero (*Scomberomorus regalis*), Palm Beach, Florida

Little tunny (*Euthynnus alletteratus*), Jupiter, Florida

Skipjack tuna (*Katsuwonus pelamis*), Jupiter, Florida

Blackfin tuna (*Thunnus atlanticus*), Jupiter, Florida

the sides. Total gill-raker counts are 25 or less, and the dorsal and anal finlets are dusky, with a trace of bronze color. Yellowfin tunas grow to over 6 ft long and weigh as much as 400 lb. Their gill-raker counts range from 26 to 35. The dorsal and anal fins in adults can be very long, and the finlets are bright-yellow, with a thin black edge.

Distribution and Habitat

Mackerels and tunas are wide-ranging migrators, found in both coastal and oceanic habitats. Wahoos are found around the state, in oceanic waters from the shelf-break and beyond. The *Scomberomorus* mackerels are restricted coastal and shelf waters around the state. Spanish and king mackerels form subpopulations, broadly distinguished as Atlantic and Gulf of Mexico groups (from their origins). The Gulf of Mexico subpopulation of Spanish mackerels extends from the panhandle to Florida Bay. The Atlantic subpopulation moves between the Florida Keys and northeastern Florida and beyond (to Cape Hatteras, North Carolina). Within the subpopulations, individual fishes tend to segregate into size-related groups, consisting of a few dozen to many thousands. Spanish mackerels will venture inside inlets, into lagoons and estuaries, in places like the St. Johns River, Indian River Lagoon, Lake Worth Lagoon, Biscayne Bay, Tampa Bay, St. Andrews Bay, and Pensacola Bay. Ceros are an insular species, generally restricted to southeast Florida and the Florida Keys, but on occasion they will occur outside these areas. Solitary individuals are a common sight on structured habitats in southeast Florida. King mackerels also have recognized Gulf and Atlantic subpopulations; these two subpopulations intermix to an unknown extent off southeastern Florida, creating chronic headaches for fishery managers from the Gulf and Atlantic regions. Atlantic chub mackerels occur around the state, in shelf and upper slope waters.

Tunas are highly migratory, generally preferring blue oceanic water beyond the shelf break. They tend to segregate by size and will travel in mixed-species associations. Atlantic bonitos and albacores are not common anywhere off Florida, and when they do appear, it is usually off the northeastern coast, near the western edge of the Gulf Stream. Little tunnies are very common and occur in shelf waters year round, migrating inshore in the spring and summer, and moving offshore during winter months. Skipjack tunas are rarely found inshore of the shelf edge, but they are common in the Gulf Stream and Loop Currents. Blackfin tunas do not migrate to the extent that other tunas do. They form populations that remain within broad areas, rather than migrating on a larger geographical scale. Blackfin tunas regularly appear in places such as the Hump (a seamount) off the Florida Keys, the shelf edge offshore of Palm Beach, De Soto Canyon, and off of Cape Canaveral and Jacksonville. Bluefin, yellowfin, and bigeye tunas are wide ranging, preferring the blue oceanic waters of the Gulf Stream and Loop Currents; they are rarely found inside the shelf edge. Large adult bluefin tunas are occasionally observed in shallow shelf waters off southeast Florida. These larger tunas are relatively uncommon in Florida; only fishers with boats having a long-range capacity will see them. Off Florida's east coast, migrating yellowfin tunas usually stay on the eastern side of the Gulf Stream—essentially in Bahamian waters—but on occasion schools will pass near the coast. Some combination of water temperature, current flow, and the presence of forage fishes is needed to hold migrating schools in particular areas. A wave buoy moored over 100 mi off Cape Canaveral attracts a regular complement of yellowfin, blackfin, and skipjack tunas, as well as wahoos. Bullet mackerels occur mostly offshore of northwestern Florida, over the outer shelf. Frigate mackerels are found in shelf and oceanic waters around the state.

Natural History

Mackerels are water-column spawners. Spanish and king mackerels spawn during spring and summer, over the shelf. It is not clear if there are precise locations where spawning takes place. Although not documented, spawning appears to follow a lunar pattern in king mackerels. Female mackerels produce thousands to millions of eggs that are released in batches over the spawning season. The young are pelagic for a short time before settling in shelf and coastal waters. Juvenile king and Spanish mackerels often school together and will associate with herring and anchovy schools. Wahoos spawn in summer months, producing between 500,000 and 4.5 million eggs, depending on the size of the female.

All mackerels are rapacious predators. Wahoos feed on little tunnies, dolphinfishes, rainbow runners, blue runners, and blackfin tunas. Spanish mackerels readily slash through schools of anchovies, Spanish sardines, or other small, silvery fishes at the water surface. During these feeding episodes, it is common to see individuals jumping out of the water. King mackerels consume fishes, squids, and shrimps. Fishes are the most-important food, with Spanish sardines, round scads, Atlantic thread herrings, Atlantic cutlassfishes, and flyingfishes being the favored items. When feeding, king mackerels can launch themselves like a missile, in a tight trajectory, rising more than 30 ft in the air (a phenomenon known as skyrocketing).

Tunas spawn in the water column, with most species producing millions of eggs. Yellowfin tunas release as many as 8 million eggs. Larger species, such as bluefin tunas, migrate around the Atlantic, from New England to the Gulf of Mexico. Evidence (based on the collection of very young larvae) indicates that the Loop Current in the eastern Gulf of Mexico is a spawning area. Blackfin tunas probably spawn in several areas around the state, including southeast Florida, the Straits of Florida (west of Key West), and the De Soto Canyon area.

Feeding tuna schools roam the blue waters off both coasts of Florida. The best way to find them is to locate birds that follow the same baitfish schools targeted by the tunas. Look for terns and other seabirds, especially sooty terns, royal terns, petrels, fulmers, and—the grand-master tuna locators—frigate birds. Feeding tunas will break the surface and skyrocket when chasing baitfishes at the surface. The sight of multiple large yellowfin tunas in the air is unforgettable.

Swordfishes
Family Xiphiidae

- **Swordfish** (*Xiphias gladius*)

Background

The swordfish is the sole member of this family. They are sometimes called "broadbills" or "broadbill swordfishes," because the bill is flattened in cross-section, as opposed to the cylindrical profile of the bill in the related billfish family (Istiophoridae). Unlike marlins and sailfishes, swordfishes lack pelvic fins, and the dorsal fin has a short base. Swordfishes are brownish to blue on the back, and white on the undersides. They grow to about 15 ft and weigh more than 1000 lb. Swordfishes are an important commercial and recreational fishery species in Florida.

Distribution and Habitat

Swordfishes are distributed in oceanic waters around the world. Concentrations occur offshore of southeastern Florida and the Florida Keys, but the species is found wherever the Gulf Stream and Loop Currents hug the shelf edge, and especially in 64°F–72°F waters.

Natural History

Swordfishes are fast-moving and highly migratory fishes that summer in temperate waters and overwinter in tropical waters; more females make the northerly trek than do males. They spawn year round, in warm waters (68°F–72°F) of the central Gulf and off eastern Florida, with peak activity in spring in the Gulf and summer in the Straits of Florida (north to Cape Canaveral). Females release between 1 and 4.2 million eggs in a batch. The pelagic larvae look like little crocodiles (long jaws, with lots of pointy teeth). The young grow quickly—at 1–3 in long they begin increasing by about 1.5 in per week—and gorge on copepods and fish larvae near the water surface. Adults live about 10 years; a fish weighing 250 lb is between 3 and 5 years old. Since males have shorter life spans and grow more slowly, swordfishes larger than 7.5 ft are females.

In New England, swordfishes are seen basking at the water surface (and historically were captured there with harpoons), but in seeking their preferred temperature range, they occupy progressively deeper waters as they move southward. Swordfishes are attracted to underwater lights; fishers effectively pairing chemical lightsticks with baits has contributed to overfishing of the species. They are an aggressive species, and there are numerous accounts of swordfishes attacking boats, lighted research submersibles, and remotely operated vehicles. Numerous swordfishes have become entangled in the rigging of submarine vessels or have left parts of their bills in hulls. A 300 lb swordfish even killed a Massachusetts harpooner by piercing his fishing dory and stabbing him.

Because swordfishes are vertical migrators, in deeper water (to about 2000 ft) by day and shallower water by night, an individual may experience temperature changes of 34°F within a 2-hour period. Swordfishes have large eyes, and they feed very actively on moonlit nights, suggesting that they can see well under those conditions. As fishers from southeastern Florida and the Florida Keys have discovered, swordfishes also feed during

Swordfish (*Xiphias gladius*). *Photo © Franco Banfi/SeaPics.com.*

daylight hours, but do so near the bottom (ca. 1200–1800 ft). Swordfishes off southeast Florida heavily consume squids, to a lesser degree eat a variety of fishes, and occasionally munch on shrimps. The broad bill is used to stun their prey, making them easy grabs. Some swordfishes have pinkish-orange, salmon-like flesh and are called "pumpkins" by fishers. The cause of this condition in unknown, but it may be related to feeding on shrimps or other organisms containing carotenoid (yellow to red) pigments.

Billfishes
Family Istiophoridae

- **Sailfish** (*Istiophorus platypterus*)
- **White marlin** (*Kajikia albida*)
- **Blue marlin** (*Makaira nigricans*)
- **Roundscale spearfish** (*Tetrapturus georgii*)
- **Longbill spearfish** (*Tetrapturus pfluegeri*)

Background

Billfishes do not have any commercial market value as edible fishes in Florida, but they are the gold standard for bluewater gamefishes, their economic importance being attributable to the aerial gymnastic displays they provide for happy recreational fishers. The distance from shore to blue water is the primary barrier for anglers seeking billfishes, and this is why fishers in southeast Florida and the Keys boast about their bounty of sailfishes and marlins.

Sailfishes are officially designated as the state's marine fish, and nowhere is this species more important than in Stuart, the self-proclaimed sailfish capital of the world. The bill, formed from elongated skull bones, is rounded in cross-section in all species, which differs from the horizontally flattened broadbill of the related swordfishes. Sailfishes are dark-blue to bronzy-brown above, with a bronze to yellow margin above a white belly. The fins are dark-blue, but their unique distinguishing feature is the broad, high dorsal fin (the sail), which is much higher than the body is deep. The sail is dark-blue and covered with small black spots. When feeding or otherwise excited, individuals will display electric-blue vertical bars and spots along the flanks. The pelvic fins are very long and slender, twice or more the length of the pectoral fins.

The body and dorsal fin of white marlins are colored like those of sailfishes (minus the body spots), but only the front portion of the dorsal fin is elevated. The tips of the dorsal, pectoral, and anal fins are distinctively rounded; the pelvic fins are shorter, not much longer than the pectoral fins. Both sailfishes and white marlins reach lengths of about 9 ft and weights of 200 lb, although most are smaller. Roundscale spearfishes look very much like white marlins but have pointed pectoral fins and lack dorsal-fin spotting; in addition, the anus is not located as close to the anal fin as in marlins. Roundscale spearfishes also are smaller, only reaching 6.6 ft, and their scales have rounded leading edges. Individuals referred to as "hatchet marlins"—a rarely encountered morph bearing truncated dorsal and anal fins—probably can be allocated either to this species or to white marlins.

Blue marlins, achieving a size of over 14 ft and a weight of about 1000 lb, are the largest of the bony fishes. Atlantic blue marlins do not get as large as their genetically distinct Indo-Pacific mates, which reach weights of about 1500 lb. Blue marlins are colored like white marlins, but the anterior portion of the dorsal fin is not as high as the body is deep, and it and the anal fin are pointed. Longbill spearfishes are skinny billfishes, with a very long dorsal fin that originates at a point about above the preopercle. They have pointed dorsal-, pectoral- and anal-fin tips; lack dorsal-fin spotting; and have a shorter bill than other billfishes. Longbill spearfishes reach a length of a little over 8 ft.

Distribution and Habitat

All billfishes are wide-ranging inhabitants of either the epipelagic or bluewater realms of Florida, usually staying within the upper 300 ft of the water column. Marlins are usually found offshore of the shelf break, and spearfishes appear

Sailfish (*Istiophorus platypterus*). *Photo by James R. Abernethy.*

to be even more oceanic in nature, but sailfishes venture into shallow nearshore waters. We have seen sailfishes in 10 ft of water off both coasts of Florida, and they have been hooked off fishing piers and jetties.

Natural History

Billfishes are large, fast-swimming (up to 80 mph) predators. They possess effective adaptations that help them in their daily lives, such as having color vision and being able to regulate their body temperatures. Their bills are used to stun prey and as offensive weapons (portions of bills have been found buried in the bodies of fishes, including their own species). A white marlin recently repeatedly stabbed an unlucky Brazilian.

All species are broadcast spawners, capable of producing huge numbers of eggs. Sailfishes reproduce during the summer, off Florida's east and west coasts; individuals may spawn more than once during the season. Based on the collection of early-stage larvae, sailfishes appear to spawn near the leading edges of small eddies (4–10 mi across) formed in the Gulf Stream. During spawning season, a lone female may be pursued by multiple males. A typical mature female produces about 2 million eggs and releases between 750,000 and 1.6 million during an individual spawning event. Larvae reside in surface waters, where they grow rapidly into juveniles, which usually associate with drifting sargassum. Adults live to at least 10 years old. Sailfishes are primarily fish eaters, feeding heavily on flyingfishes, herrings, jacks, needlefishes, and squids. Feeding sailfishes encircle schools of round scads, round herrings, or Spanish sardines and use their bills to dispatch individuals from the schools.

As with most other billfish species, blue marlins spawn in the open ocean during summer, peaking in July. Larvae and young grow very quickly, at 0.5 in per day. Blue marlins are formidable predators that feed on tunas, dolphinfishes, rainbow runners, and little tunnies.

Watching an adult blue marlin (midnight-blue body, with electric-blue markings aglow on the flanks) pursue an equally colorful dolphinfish is a memorable sight. Blue marlins prefer water temperatures above 66°F. Although blue marlins migrate across vast expanses of the open ocean, adults remain within areas of favorable temperature and productivity for entire seasons. Larval occurrence indicates that blue marlins spawn in the Straits of Florida.

White marlins are not common in Florida; they aggregate instead in certain other areas, such as the outer shelf off the mid-Atlantic states and Venezuela. Females produce 4–11 million eggs and spawn from March to June. Larvae develop in surface waters. White marlins feed on fishes: jacks, mackerels, dolphinfishes, and flyingfishes. Little is known about the natural history of roundscale spearfishes, because of their long-term confusion with white marlins, with whom they apparently share a similar abundance. There are indications of some spatiotemporal differences of abundance, with greater numbers of roundscale spearfishes found off Florida and in the Sargasso Sea during winter months, and apparent movement northward as summer water temperatures rise.

Limited tagging reveals that longbill spearfishes occupy the top 500 ft of the water column, with the majority of their time spent in less than about 80 ft depths. There they feed on such oceanic fishes as snake mackerels, mackerels, and flyingfishes, as well as squids. Longbill spearfishes spawn well away from land, probably to the south in Florida and in the Caribbean Sea. Unlike the other billfishes, they may spawn in winter months (although they are summer spawners off Venezuela). They probably do not live much longer than 3–4 years.

Medusafishes
Family Centrolophidae

- **Black driftfish** (*Hyperoglyphe bythites*)
- **Barrelfish** (*Hyperoglyphe perciformis*)

Background

Medusafishes have a single, long dorsal fin and relatively larger jaws than the related driftfishes, ariommatids, squaretails and butterfishes. Their jaws have a single row of conical teeth, and the pelvic fins fold into a groove in the body. Black driftfishes and barrelfishes are brownish-gray above and whitish below. They have broadly rounded snouts, and tails that are less forked than those of their relatives. The front portion of the dorsal fin is much shorter than the rear portion and bears short, stout spines. The dorsal spine and ray counts are said to differentiate the two species: 7–8 spines + 22–25 rays in black driftfishes and 8–9 spines + 19–21 rays in barrelfishes. Barrelfishes are larger (to 3 ft) than black driftfishes (2 ft).

Distribution and Habitat

Young are found at the water surface, where they associate with flotsam and Portuguese men o' war. Adults are found near the bottom, over the outer continental shelf and upper slope. Ranges in Florida are uncertain, but black driftfishes are found in the Gulf, and barrelfishes occur off the southeast coast.

Natural History

Barrelfishes exhibit typical deep-water life-history attributes: long lives (to 85 years), slow growth, and late sexual maturity. Since adults are taken by deep-fishing recreational and commercial fishers, there is concern about the long-term sustainability of fisheries for this species. Adults mature at 24–27 in, and spawning appears to occur year round, perhaps with a winter peak. Barrelfishes feed primarily on salps (gelatinous planktonic invertebrates); lesser amounts of small fishes and squids also enter the diet.

Driftfishes
Family Nomeidae

- **Man-of-war fish** (*Nomeus gronovii*)
- **Freckled driftfish** (*Psenes cyanophrys*)
- **Silver driftfish** (*Psenes pellucidus*)

Background

Driftfishes inhabit near-surface oceanic waters, in close company with jellyfishes, sargassum, or other flotsam. The most distinctive species in the group is man-of-war fishes, sometimes called "blue bottle fishes" or "shepherd fishes." Juveniles have a dark-blue back and silvery flanks, with dark-blue and black blotches. The fan-like pelvic fins are mostly black; the tip of the forked tail lobes are black; and royal-blue spots, circled in black, occur at the bases of the tail lobes. This color pattern is seen in fishes at lengths of 8 in or less. Adults grow to about 16 in and are uniformly dark-brown, with longer pectoral fins and shorter pelvic fins than juveniles. Juveniles swim mostly with their pectoral fins. Freckled driftfishes grow to about 8 in and have a deep body, small head, and forked tail. Adults and subadults are pale-yellow, with linear rows of small black spots or dashes along the flanks. Small juveniles have irregular black spots covering the body. Silver driftfishes reach a length of 30 in; adults are dark-brown and more slender than the silvery juveniles. Juveniles have a more-rounded profile than adults; the median fins are broad; and their background color is uniformly silver, with some irregular black spots.

Distribution and Habitat

Adult man-of-war fishes apparently live deep in the water column (>600 ft), but little is known other than that. The juvenile stage, however, is well known. Juveniles prefer offshore blue waters, where they associate with the jellyfish-like Portuguese men o' war (*Physalia*

Juvenile man-of-war fish (*Nomeus gronovii*), with a Portuguese man o' war (*Physalia physalia*), Jupiter, Florida

Freckled driftfish (*Psenes cyanophrys*), Jupiter, Florida

physalia). Portuguese men o' war are seasonal visitors off Florida, with peak numbers occurring during April–May. When onshore winds prevail, man-of-war fishes follow their hosts into beach, tide-pool, and estuarine habitats, where they have little chance of surviving. Freckled driftfishes are fairly common under sargassum mats, jellyfishes, and other flotsam in offshore waters around the state. They, too, will end up in difficult shore or estuary situations, or in other unfavorable areas, if their home mats of sargassum are blown ashore. Adult silver driftfishes live in deeper portions of the offshore water column around the state. Juveniles associate with jellyfishes, particularly moon jellyfishes (*Aurelia aurelia*), in near-surface waters. Moon jellyfishes are most common in summer and fall off eastern Florida. When swimming inside the bell of the jellyfish, silver driftfishes can easily be mistaken for young jacks, particularly Atlantic bumpers (see chapter 77).

Natural History

Man-of-war fishes are obligate associates of Portuguese men o' war. This association is usually described as commensal (one species benefits while the other is unaffected); the fish gains protection from potential predators by hiding among the host's powerful stinging tentacles. It could be argued, however, that the presence of man-of-war fishes attracts other fishes into the siphonophore's deadly net, providing a nutritional benefit for Portuguese men o' war. Contrary to popular belief, man-of-war fishes are not immune to the Portuguese man o' war's potent sting, nor do they acclimate to stinging tentacles, like Indo-Pacific clownfishes do with their host, sea anemones. Apparently individuals simply avoid the tentacles by being very agile swimmers. How long the juveniles stay with the Portuguese men o' war is unknown. The fishes gain an additional benefit from the association—they nip off nonstinging portions of the Portuguese men o' war and eat them.

If you snorkel under a big sargassum patch in blue water, you may see freckled driftfishes, usually in groups but occasionally as singletons. Little is known about their biology, aside from the fact that they are often found (not surprisingly) in the stomachs of dolphinfishes. Nor is much known about the biology of silver driftfishes, except that they spend their early life history in the upper layer of the open ocean, prior to moving deeper into the water column as they mature.

Ariommatids
Family Ariommatidae

- **Silver-rag** (*Ariomma bondi*)
- **Brown driftfish** (*Ariomma melanum*)
- **Spotted driftfish** (*Ariomma regulus*)

Background

Ariommatids resemble driftfishes in possessing large eyes, small mouths, blunt snouts, first dorsal and pelvic fins that fit into grooves in the body, and forked tails. They differ in having a pair of poorly developed, fleshy keels on the narrow caudal peduncle. Spotted driftfishes are the most distinctive, having black spots on the upper half of the oval, silvery body. The first dorsal and pelvic fins are black. Silver-rags and brown driftfishes are more streamlined and are uniformly silver to brownish. Separation of the two species requires looking carefully at details of scalation.

Distribution and Habitat

All three species in this family are found statewide, but they are most commonly encountered in

Brown driftfish (*Ariomma melanum*), Dry Tortugas, Florida. *Photo by George H. Burgess.*

the Straits of Florida and Gulf of Mexico. Silver-rags are primarily an outer-continental-shelf species, found at depths of 250–600 ft. They are replaced in waters deeper than 600 ft on the continental slope by brown driftfishes. Spotted driftfishes are more erratically encountered over a wide range of depths (60–1400 ft), especially around southern Florida, where they occur in shelf waters as adult waifs and as larval drift.

Natural History

The larvae and young are of all three species are pelagic, and adults are epibenthic, over deep water. Off Brazil, silver-rags spawn from autumn to spring. They eat benthic invertebrates and are consumed by sailfishes and other pelagic predators.

Squaretails
Family Tetragonuridae

- **Bigeye squaretail** (*Tetragonurus atlanticus*)

Background

The rarely encountered bigeye squaretails are not likely to be confused with any other fishes in our area. They are uniformly dark-brown or black in coloration, and reach about 2 ft long. The elongate, slender body is round in cross-section. The long caudal peduncle is squarish in cross section and supports a pair of low keels, formed by specialized scales. Adherent body scales are placed in a geometric pattern; the scales bear hefty lateral keels, resulting in skin that is very rough to the touch. The head and eyes are large; the snout is blunt; and the mouth is box-like. The lower jaw, when closed, fits completely within the upper jaw. Bigeye squaretails have laterally flattened, knife-like teeth. The long, low first dorsal fin fits into a groove when depressed; the taller second dorsal and anal fins are pretty much equal in size and shape.

Distribution and Habitat

Bigeye squaretails are an oceanic, pelagic species that occasionally wander into continental-shelf waters as adults. The larvae are more commonly found in these waters. Young are epipelagic (near the water surface) and adults presumably are mesopelagic (midwater).

Natural History

The young often are commensal with colonial tunicates (marine invertebrates), living inside the branchial chambers of salps and the cloacal chambers of pyrosomes. The teeth of bigeye squaretails are well designed for grabbing jellyfishes, comb jellies, and salps, the chief prey items. Larger plankton also are consumed. Bigeye squaretails fall prey to swordfishes and probably to other larger, pelagic predators. This species is thought to spawn in the spring and summer, in the North Atlantic.

Butterfishes
Family Stromateidae

- **Gulf butterfish** (*Peprilus burti*)
- **Harvestfish** (*Peprilus paru*)
- **Butterfish** (*Peprilus triacanthus*)

Background

Members of the butterfish family might be confused with jacks (and often are found in association with them), since they have silvery, laterally compressed, deep bodies and deeply forked caudal fins. Butterfishes differ, however, in lacking pelvic fins and spiny first dorsal fins, and in having small mouths and fatty, adipose (transparent) eyelids. Their snouts are short and blunt. The scaled, single dorsal and anal fins are symmetrical (long and low, except for elongated front portions). Harvestfishes have the highest such anterior rays and the deepest body; they look like a stealth fighter flying on its side. The dorsal, anal, and yellow caudal fins are tipped in black. Butterfishes and Gulf butterfishes are less deep bodied; have shorter fin elongations; and bear a long, horizontal line of pores, mostly under the base of the dorsal fin. These two species are best distinguished by their distributions, but Gulf butterfishes are a bit deeper bodied and lack the diffuse, grey, lateral spotting found in butterfishes. Harvestfishes and butterfishes reach about 1 ft in length; Gulf butterfishes grow to 10 in.

Distribution and Habitat

Gulf butterfishes and butterfishes largely segregate geographically, with butterfishes found along the northeast Florida coast and Gulf butterfishes present in the Gulf of Mexico, primarily off the panhandle. Gulf butterfishes do occur fairly occasionally off the east coast of Florida, however, leading to identification difficulty in that area. From Cape Hatteras, North Carolina, south to Florida there are two populations of butterfishes (one offshore and the other inshore) that are morphologically distinguishable. Harvestfishes are found along both coasts. All three species are largely continental-shelf forms, and none of them are common south of Cape Canaveral and Tampa. Gulf butterfishes are most abundant in outer-shelf waters, at temperature of 59°F–66°F; butterfishes and harvestfishes tend to occupy shallower waters.

Natural History

These are schooling species that have supported small fisheries. All three species are facultative symbiotants (occasional, opportunistic co-inhabitants) with jellyfishes. Some young settle under the bells of jellyfishes and are initially engaged in a commensal relationship, in which the young fishes receive some protection from their gelatinous buddies. As they grow, these fishes begin to engage in a bit of selective nipping of their host (ectoparasitism); eventually, as subadults and adults, this evolves into outright predation. It is not fully understood whether these species are immune from the nematocyst stings of the jellyfishes or if they simply avoid their near-proximity to death, although it has been suggested that the slime of Gulf butterfishes might make them able to fend off the sticky tentacles.

Gulf butterfishes spend daylight hours near the bottom and migrate vertically at night. They reach sexual maturity at about 6 in, around 1 year of age. Spawning occurs offshore, in the northcentral Gulf, in the late fall and late winter–early spring, coinciding with longshore currents. Developing young are swept inshore, to 15–90 ft depths that serve as nursery areas; they move to midshelf waters (120–320 ft) at 9–12 months of age. The maximum age for Gulf butterfishes is about 2.5 years. Butterfishes

Harvestfish (*Peprilus paru*). Photo by George H. Burgess.

Butterfish (*Peprilus triacanthus*). Photo by George H. Burgess.

reach sexual maturity in their second year of life, at a length of about 8 in, and live 3–6 years. They spawn from late winter to early summer. The young eat jellyfishes, and then switch to small fishes, squids, crustaceans, worms, and other invertebrates as adults. Harvestfishes are summer spawners.

Boarfishes
Family Caproidae

- **Deepbody boarfish** (*Antigonia capros*)
- **Shortspine boarfish** (*Antigonia combatia*)

Background

Boarfishes are odd, deep-bodied, highly laterally compressed fishes, with short snouts and concave foreheads. Their dorsal and anal spines are stout, and their small jaws are upturned and protrusible. The two boarfish species exhibit changes in body shape and coloration related to size / age. Both are pink to red in base coloration, darker above than below. At all sizes, deepbody boarfishes (8.5 in) have a deeper body, which is taller than long. In contrast, the body of shortspine boarfishes (6 in) is relatively less deep and is longer than tall. The anterior dorsal spines of deepbody boarfishes are relatively longer than those of shortspine boarfishes.

Distribution and Habitat

Boarfishes are epibenthic species that are found at appropriate depths, usually near hardbottom. In Florida, those depths are about 200–1100 ft for deepbody boarfishes and around 400–1500 ft for shortfin boarfishes. Both occur from off of Palm Beach and Miami (in the Straits of Florida) and in the Gulf of Mexico southward. The rarity of these species off much of Florida's east coast is enigmatic, as boarfishes have been taken off the Carolinas and as far north as Block Island, Massachusetts.

Natural History

Little is known about the biology of these two species. Juveniles are pelagic, and young are found epibenthically in schools or aggregations. Small invertebrates are the primary food items for boarfishes.

Deepbody boarfish (*Antigonia capros*). *Photo by George H. Burgess.*

Turbots
Family Scophthalmidae

- **Windowpane** (*Scophthalmus aquosus*)

Background
Windowpanes are the only member of their flatfish family in the western Atlantic. The very wide body and separate dorsal-fin spines on the head distinguish this species from all other flounders bearing eyes on their left sides. In addition, the body is largely translucent and very thin, hence the vernacular name.

Distribution and Habitat
Windowpanes are a temperate species that distributionally extend southward only to northeastern and eastcentral Florida. They are inshore fishes that frequently enter estuaries, including the lower St. Johns River.

Natural History
Adult windowpanes reach sexual maturity in their fourth year of life, at lengths a bit over 8 in. They spawn in the water column, over continental-shelf waters, and produce pelagic eggs and larvae. The young settle in estuaries and move offshore as they grow larger. They prefer sandbottom over mudbottom. Growth is rapid; the spring-spawned young grow 0.02–0.024 in per day, and they reach sizes of 3.1–5.3 in prior to the onset of cooler winter temperatures. Windowpanes feed on mysid shrimps, shrimps, amphipods, other crustaceans, and small fishes. They fall prey to great and double-crested cormorants.

Windowpane (*Scophthalmus aquosus*). Photo by George H. Burgess.

Sand Flounders
Family Paralichthyidae

- **Three-eye flounder** (*Ancylopsetta dilecta*)
- **Ocellated flounder** (*Ancylopsetta quadrocellata*)
- **Gulf Stream flounder** (*Citharichthys arctifrons*)
- **Sand whiff** (*Citharichthys arenaceus*)
- **Horned whiff** (*Citharichthys cornutus*)
- **Spined whiff** (*Citharichthys dinoceros*)
- **Anglefin whiff** (*Citharichthys gymnorhinus*)
- **Spotted whiff** (*Citharichthys macrops*)
- **Bay whiff** (*Citharichthys spilopterus*)
- **Mexican flounder** (*Cyclopsetta chittendeni*)
- **Spotfin flounder** (*Cyclopsetta fimbriata*)
- **Fringed flounder** (*Etropus crossotus*)
- **Shelf flounder** (*Etropus cyclosquamus*)
- **Smallmouth flounder** (*Etropus microstomus*)
- **Gray flounder** (*Etropus rimosus*)
- **Shrimp flounder** (*Gastropsetta frontalis*)
- **Gulf flounder** (*Paralichthys albigutta*)
- **Summer flounder** (*Paralichthys dentatus*)
- **Southern flounder** (*Paralichthys lethostigma*)

- **Fourspot flounder** (*Paralichthys oblongus*)
- **Broad flounder** (*Paralichthys squamilentus*)
- **Shoal flounder** (*Syacium gunteri*)
- **Channel flounder** (*Syacium micrurum*)
- **Dusky flounder** (*Syacium papillosum*)

Background

Most of the sand flounders do not exceed 1 ft in length, and several are potato chip–sized as adults, but a few notable members of this family grow larger and are highly sought-after for the table. Sand flounders tend to look alike to the casual observer: eyes on the left side of the greatly flattened body; the eyed side often remarkably well colored, to match the surrounding seafloor; and the blind side white. The more-astute viewer will note that these species usually have some teeth. Species in this family can be tough to separate, but there are some characters that divide them into smaller groups, notably the size of the mouth (small or large), the shape of the lateral line (arched or straight), and the presence/absence of ocellated spots.

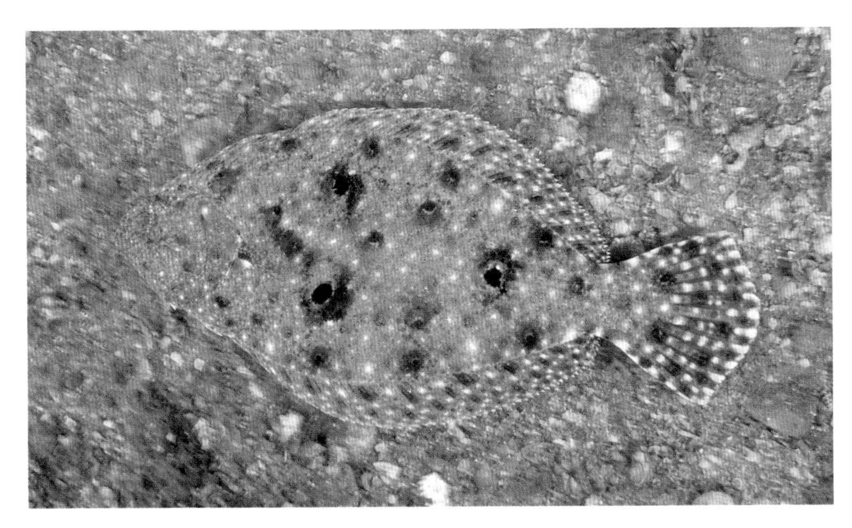

Gulf flounder (*Paralichthys albigutta*), Lake Worth Lagoon, Florida

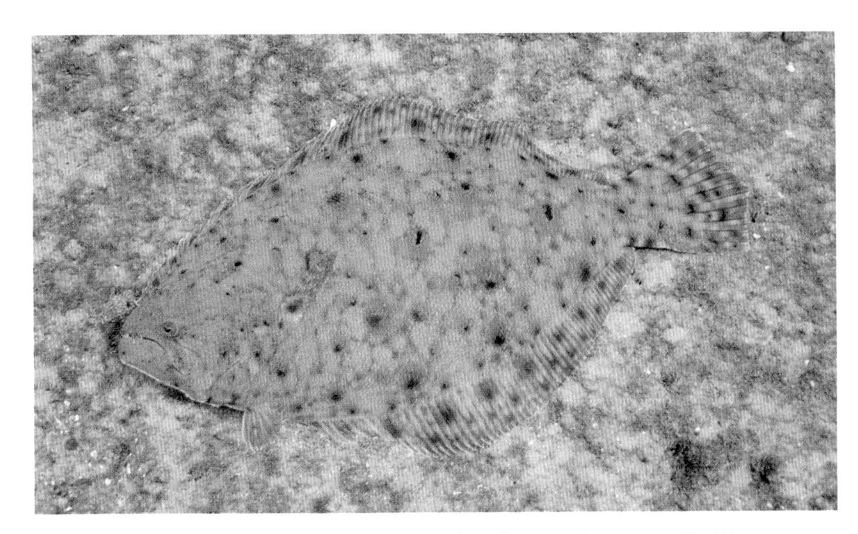

Southern flounder (*Paralichthys lethostigma*), Lake Worth Lagoon, Florida

The *Etropus* species are small flounders with small mouths, straight lateral lines, close-set eyes, relatively large scales, and a dull (tan to brownish) coloration. Unfortunately, these species are best distinguished by looking at scale-distribution patterns and morphology. All *Etropus* species (except fringed flounders) have scales on the snout and body scales that contain secondary scales (essentially scales on the scales). Fringed flounders (8 in) often have a dark margin to the caudal fin and 2–3 dark blotches above the lateral line. In smallmouth flounders (6 in), the 2 mandibles of the lower jaw are relatively symmetrical in size, and the secondary scales cover only the front half or less of a scale. Gray flounders (5 in) and similarly sized shelf flounders have asymmetric mandibles, the left (eyed side) being smaller than the right (look at the critter head on!), and secondary scalation covers three-quarters or more of each scale. Gray flounders have more-extensive snout scalation than shelf flounders but lack the 4–6 small dark circles often seen on the eyed side of the latter species.

All other sand flounders have large mouths. Members of the genera *Ancylopsetta*, *Gastropsetta*, and *Paralichthys* have lateral lines that arch over the pectoral fins. Most of these species have 3–5 prominent ocellated spots, but two *Paralichthys* species don't: southern (2.5 ft) and broad (11 in) flounders.

Both usually have non-ocellated dark and light spots or mottling, but broad flounders are deeper bodied, and these two species are separated ecologically. The related Gulf, summer, and fourspot flounders are separated by the number and placement of their ocellated spots. Gulf flounders (2.3 ft) have 3 conspicuous, ocellated spots that form a triangle: 2 of the spots are placed above one another, just past the pectoral fin; and the third is centered posteriorly, on the lateral line. The larger (3 ft) summer flounders have numerous ocellated spots, but posteriorly, 5 of them form a 2-1-2 pattern: 2 just past midbody, near the dorsal and anal fins; 2 posteriorly, near those fins; and 1 in between, on the lateral line. The appropriately named fourspot flounders (18 in) are more slender than the other *Paralichthys* species and have 4 ocellated spots: 2 located just past the middle of the body, near the dorsal and anal fins; and 2 posteriorly, near the rear bases of those fins.

Three-eye flounders (10 in) have a pattern of 3 ocellated spots, similar to that seen in Gulf flounders, but the spots are much larger, and the triangle is pushed a little farther back on the body. The anterior dorsal-fin rays and pelvic fins are elongated. Ocellated flounders (10 in) have the same pattern as three-eye flounders but add a fourth, smaller ocellated spot just over the pectoral fin and the arch of the lateral line. The fronts of the dorsal and pelvic fins are not especially elongated. Shrimp flounders (10 in) have a broadly rounded head (the head is more pointed in other species), and the anterior dorsal-fin and pelvic rays are elongated. There are 3 large ocellated spots: 2 at midbody, near the dorsal and anal fins, and 1 over the pectoral fin.

Species in the genus *Syacium* have large mouths, straight lateral lines, and are brown or tan in coloration, without any ocellated spots. Their eyes are relatively widely separated from each other. Dusky flounders (12 in) are brownish, with some blotches on the flanks and median fins; significantly, the pectoral fin has 3 dark, crossing bands. Male dusky flounders have elongated upper pectoral-fin rays, even more–widely separated eyes, and turquoise markings around the mouth and eyes. The similar-looking channel flounders (12 in) have a clear pectoral fin, and the distance between the eyes is less pronounced. There are small, dusky spots along the base of the dorsal fin, and larger blotches are located on the lateral line: the larger one is placed under the clear pectoral fin, and other between the rear bases of the dorsal and anal fins. Shoal flounders (8 in) have a deeper body than either dusky or channel flounders and are otherwise unspectacularly marked.

Spotfin flounders, in the genus *Cyclopsetta*, have a large spot on the pectoral fin and tail and 2 ocellated spots on the dorsal and anal fins. Mexican flounders look very similar to spotfin flounders but have 3 large spots along the margin of their caudal fin.

That leaves the seven species of *Citharichthys*: small (<10 in) lookalikes, with large mouths, straight lateral lines, close-set eyes (in all but one species) and tan or brownish coloration. Two species—spotted and anglefin whiffs—can be readily separated by color patterns. Spotted whiffs (8 in) are light–gray-brown, with numerous small, dark and white spots uniformly distributed over the body and fins, giving these fishes a salt-and-pepper look. Tiny (3 in) anglefin whiffs have large black spots on the pointed tips of their sharply angled dorsal and anal fins. Adult males also have a series of horny projections

Channel flounder (*Syacium micrurum*), Lake Worth Lagoon, Florida

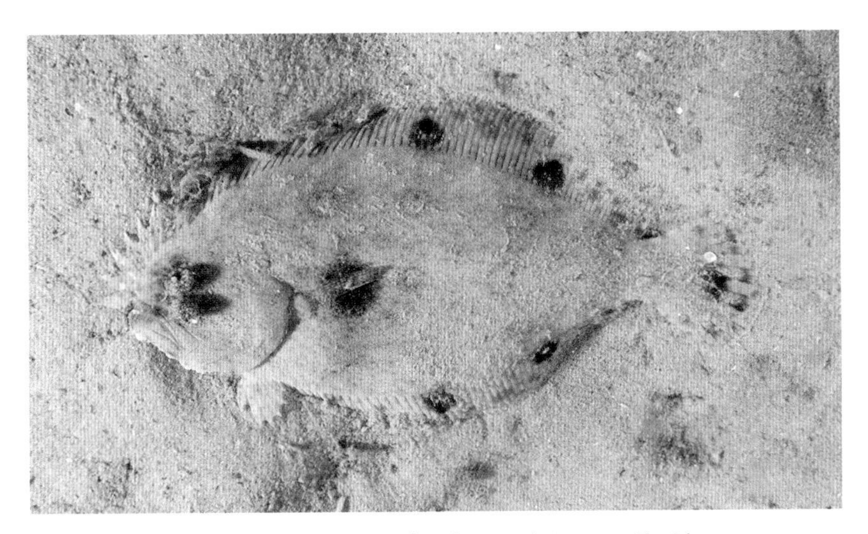

Spotfin flounder (*Cyclopsetta fimbriata*), Lake Worth Lagoon, Florida

coming off their heads. Two other species have similar projections. Gulf Stream flounders (7 in) have a single, stout projection (3 in) on the tip of the snout, a mini Cyrano de Bergerac! Adult male horned whiffs (4 in) have a single, thin snout spine; both sexes have a dark spot at the base of the pectoral fin and relatively wide-set eyes. The less deep-bodied bay whiffs (8 in) are drab-brown or tan, with small, white and dark body blotches and a diffuse spot at the base of the tail. Spined whiffs have a series of small spots on both the dorsal and anal fins. Sand whiffs (8 in) are brownish, with many tiny, dark spots on the body.

Distribution and Habitat

Ocellated and three-eye flounders occur statewide, on sand- or mudbottom, from nearshore to 600 ft and from 200 to 1200 ft, respectively. *Citharichthys* species are distributed in shallow and deep-water groups. Horned, spined, and anglefin whiffs and Gulf Stream flounders reside in deeper water (depths >150 ft) and seldom are seen. Spotted and bay whiffs are common in inshore and inner-shelf waters around Florida. Sand whiffs occur only along the southeast coast, in water depths of 0–50 ft.

Gulf flounders are distributed around the state, from nearshore to 500 ft depths, but they are most common in waters less than 100 ft deep. Summer flounders occur only in northeast Florida, where they are found out to 500 ft water depths. Summer and southern flounders will enter estuarine systems, but southern flounders do not visit areas with very low salinities; instead, they ascend into freshwater reaches in the St. Johns River. Individuals may be found in the sand periphery around reefs or wrecks in shelf waters along the east coast. Southern, summer, Gulf, and broad flounders migrate along the coast during fall and winter. Fourspot flounders are a northern species, found in water depths of 600 ft or more off eastern Florida. Spotfin flounders reside in shelf and coastal waters around the state. Mexican flounders occur only in shelf waters of the western Panhandle ecoregion. Fringed, gray, shelf, and smallmouth flounders may be beyond the 100-fathom curve (and thus outside our depth limit for fishes in Florida).

Bay whiff (*Citharichthys spilopterus*), Jupiter, Florida

Natural History

Sand flounders are water-column spawners. The larvae are born with eyes on separate sides of the head, like other fishes. During early

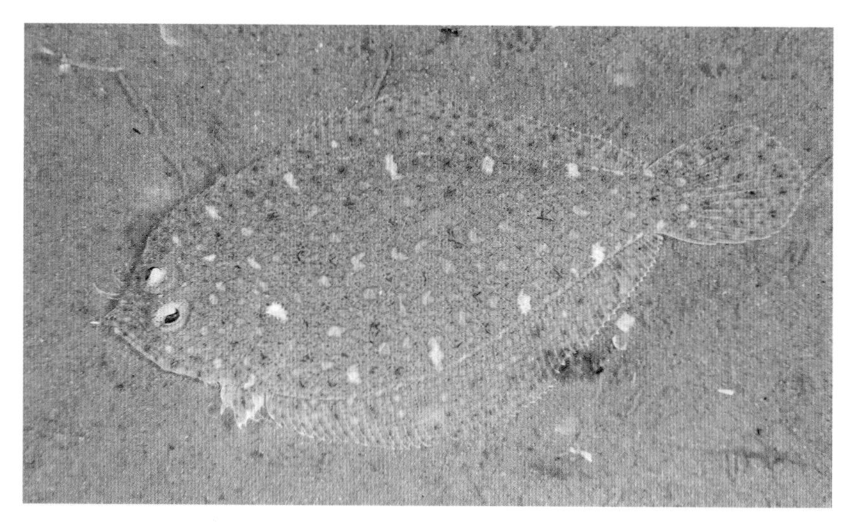

Spotted whiff (*Citharichthys macrops*), Jupiter, Florida

growth, the eyes migrate to the left side of the body and the young begin to look like little flounders. Not much is known about mating or spawning in sand flounders. Some species migrate to the shelf edge to spawn.

Sand flounders are supremely adapted for their lie-in-wait feeding mode. Individuals rest on the bottom, either matching the substrate or buried under a veneer of sediment. Some species seem to be more active at night. The favored types of prey generally follow the body size of the fish. Smaller species from the genera *Citharichthys* and *Etropus* feed on small, bottom-dwelling invertebrates, including worms, clams, crabs, shrimps, and amphipods. Intermediate-sized species, such as dusky flounders, dine on shrimps, crabs, worms, snails, clams, and small fishes. Large Gulf, summer, southern, fourspot, and broad flounders are primarily fish eaters but will also consume larger shrimps and crabs. We have seen Gulf flounders feeding on anchovies and mullets at the waters surface. Live topminnows or mullets are favored baits for anglers seeking catch big summer, Gulf, and southern flounders (large individuals of all three species are called "doormats" by anglers).

Lefteye Flounders
Family Bothidae

- **Peacock flounder** (*Bothus lunatus*)
- **Eyed flounder** (*Bothus ocellatus*)
- **Twospot flounder** (*Bothus robinsi*)
- **Spiny flounder** (*Engyophrys senta*)
- **Slim flounder** (*Monolene antillarum*)
- **Deepwater flounder** (*Monolene sessilicauda*)
- **Sash flounder** (*Trichopsetta ventralis*)

Background

Lefteye flounders, as their name implies, have both eyes on their left side and lay their right side on the bottom. They may be distinguished from other flounders by their wider, stouter body shape and by having a lateral line with a distinct upward hump over the pectoral fin, rather than a gradual curve from tail to head. As with other flatfishes, they exhibit plain to intricate, eyed-side color patterns overlaying various shades of gray-brown backgrounds. Peacock, eyed, and twospot flounders have round or nearly round bodies, with their eyes widely separated from each other. Peacock flounders (18 in) are the most colorful of the group, with a pattern of dark spots against a pale background, accented with tiny, black-rimmed, electric-blue spots and larger, black-rimmed blue rings. On occasion, individuals will turn very dark—almost black. Males have elongated pectoral rays. Eyed flounders are smaller (<10 in), with multiple rings, spots, and blotches overlapping a gray–light-brown background. A triangular black spot is present on the lateral line, about two-thirds of the way past the head; 2 black spots on the tail are arranged vertical to the long axis of the body (<>:). The space

Peacock flounder (*Bothus lunatus*)

between the eyes on mature males exceeds that of females. Twospot flounders look very much like eyed flounders, with one subtle but important distinction: the 2 black spots on the tail are aligned with the long axis of the body (<>××). Spiny flounders have more–closely spaced eyes, long eyelashes (cirri), and linear blotches on the midbody. Sash flounders are more slender than the others; have close-set, relatively large eyes; and are gray-brown, with few obvious markings, although a dark spot just posterior to the pectoral fin is usually present. Deepwater flounders have an elongated body, with a pointed head and large eyes. Slim flounders look very similar to and are closely related to deepwater flounders; some ichthyologists considered them to be the same species.

Distribution and Habitat

Lefteye flounders spend their lives on sedimentary bottoms, from inshore waters to beyond the shelf break. The larval stage drifts with plankton for months, contributing to some of the distributional patterns. Eyed and twospot flounders reside around the entire state, from inland waters across the shelf to water depths of 300 ft. Peacock flounders occur in the Big Bend ecoregion on the west coast and Cape Canaveral on the east coast, although they are most common in the Florida Keys and southeastern Florida, in water depths ranging from nearshore to about 200 ft. Spiny flounders are found around the state, in water depths ranging from 100 to 600 ft, in places where the bottom is sandy. Deepwater and slim flounders occur over softbottom around the state, in water depths from 200 to 2000 ft. Sash flounders are only known from Cape San Blas westward, on sandbottom, in 350 ft to over 1400 ft depths.

Natural History

Lefteye flounders spawn in the water column and release pelagic eggs. Eyed flounders spawn late in the day, usually after some animated courtship. Their spawning exploits are readily seen by patient observers, wherever water clarity allows. Mature males, easily recognized by their more widely spaced

Eyed flounder (*Bothus ocellatus*), Lake Worth Lagoon, Florida

eyes, patrol spawning territories that accommodate harems of up to six females. Males mate with the females by sliding underneath the female; the pair then rise in the water column and release their gametes. Males spawn multiple times during the day within the harem. Eyed flounders feed by extracting small shrimps, snails, clams, and worms that live on or in the sediments. Little is known about the life history of the other lefteye flounder species.

American Soles
Family Achiridae

- **Lined sole** (*Achirus lineatus*)
- **Naked sole** (*Gymnachirus melas*)
- **Fringed sole** (*Gymnachirus texae*)
- **Scrawled sole** (*Trinectes inscriptus*)
- **Hogchoker** (*Trinectes maculatus*)

Background

American soles, as a group, are the smallest flatfishes known from Florida. They are distinguished from other flatfishes by their close-set eyes, placed on the right side of the fish; rounded bodies and tails; small, rudimentary pectoral fins (or none at all); and small, turned-down mouths. The two *Gymnachirus* species are readily separated from the others by their lack of scales and by the presence of prominent bars, on the eyed side, that extend out onto the dorsal, anal, and caudal fins. The larger (to 8.5 in) naked soles (sometimes called "zebra soles") have 15–32 (usually 20–30) thin black bars, offset by equally wide or slightly wider pale bars. In contrast, the smaller (to 5.5 in) fringed soles usually have more than 30 (range = 25–49) dark bars, offset by wider pale bars (usually twice the width of the dark bars). Juveniles of both species have a dark-brown background, with the dark bars not readily visible, but present. Lined soles, the largest species (10 in) in

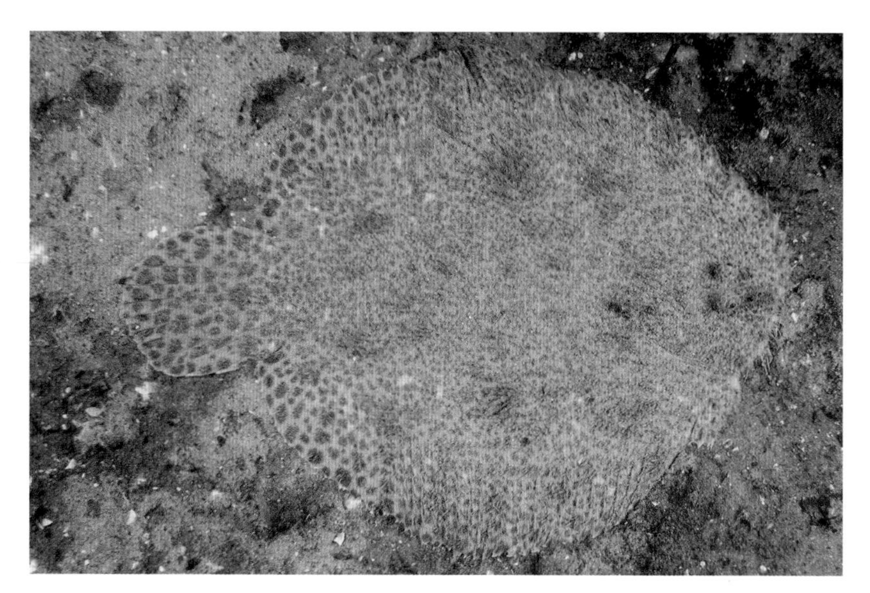

Lined sole (*Achirus lineatus*), Indian River Lagoon, Florida

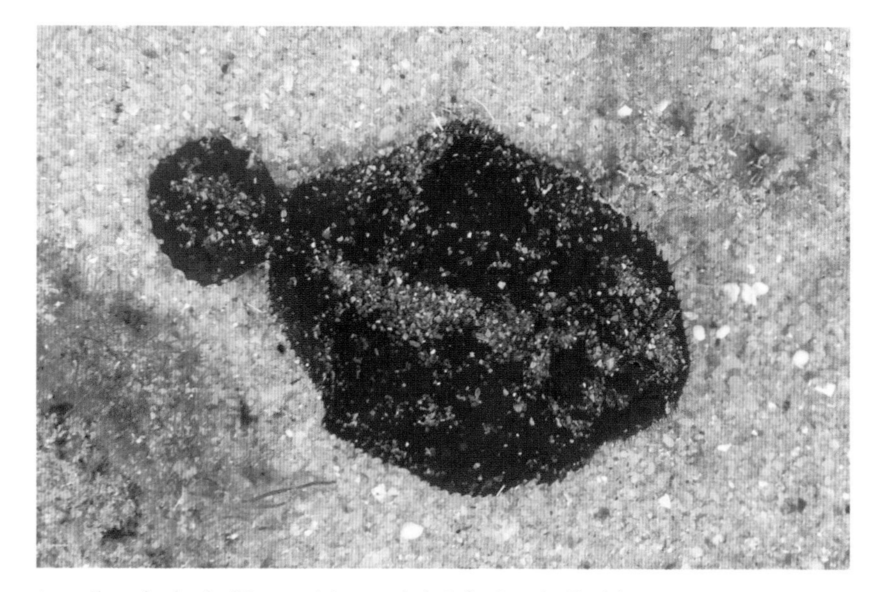

Juvenile naked sole (*Gymnachirus melas*), Palm Beach, Florida

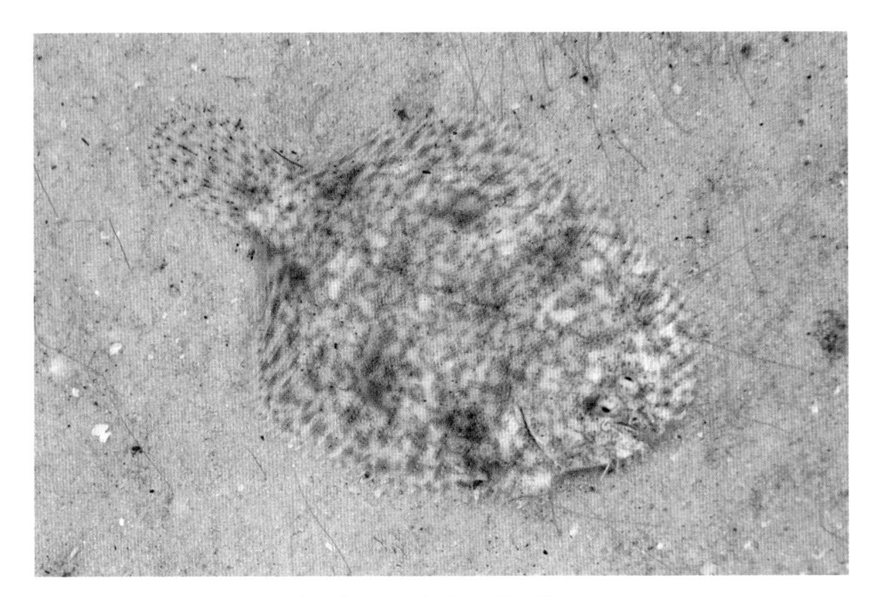

Hogchoker (*Trinectes maculatus*), Crystal River, Florida

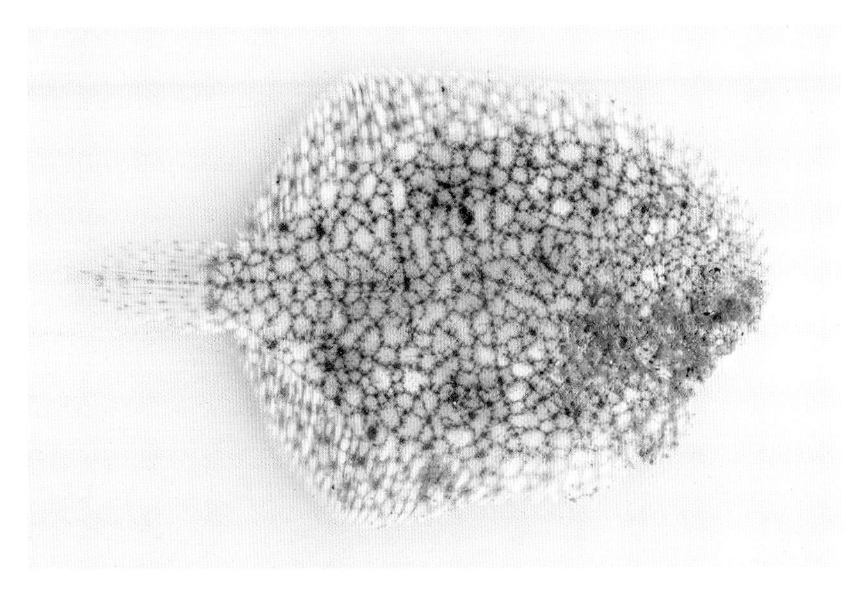

Scrawled sole (*Trinectes insrciptus*), Jupiter, Florida

the group, have 8 faint, thin bars that cross the gray-brown body but do not extend onto the fins. The fins and body are covered with dark spots of varying sizes, the largest being reserved for the body. Hogchokers, the most common Florida sole, reach 8 in long and are brown to gray, with wavy lines and variable mottling. They are the only species in this family to lack pectoral fins, a character that readily separates them from lined soles. Scrawled soles are orange-brown, with a few small, dark blotches and a distinctive, reticulating network of dark lines covering the entire body. They reach a maximum size of 6 in.

Distribution and Habitat

Lined soles commonly occur in salinities higher than 15 psu, in inshore, coastal, and shelf waters, at depths less than 200 ft. Naked soles are uncommon in inshore waters, but they inhabit all Florida coastal and shelf waters, to about 600 ft deep. Fringed soles distributionally replace naked soles from the panhandle westward in the Gulf, so these two species may co-occur there. Scrawled soles are a tropical insular species, periodically found in southeast Florida and the Florida Keys. Hogchokers occur in abundance in the state's estuaries and in the lower reaches of coastal rivers. Young, quarter-sized hogchokers are regularly seen in many of Florida's freshwater springs.

Natural History

American soles spawn in the water column, just above the substrate. Eggs and larvae are pelagic, but specific details on reproductive behavior are unknown. The spawning season for naked soles extends from May to November, and from April to November for lined soles. Hogchokers spawn from May to September in coastal and shelf waters, but they may spawn year round in the St. Johns River.

The members of this family feed on tiny, bottom-dwelling invertebrates. All American soles bury themselves in the substrate and are difficult to see. Adult lined soles are very striking when observed uncovered. Individuals seem to move around more at night, and this may be the best time to see one. Young hogchokers inhabit freshwater and move downstream into higher salinities as they grow. This is a remarkable feat, considering that they have underdeveloped (aglomerular) kidneys, which must have to work overtime balancing internal and external salt concentrations in a changing environment. We have seen small (<4 in) individuals in many of Florida's freshwater springs, including those around the Crystal, Ichetucknee, and St. Johns Rivers, many miles from the ocean.

Tonguefishes
Family Cynoglossidae

- **Caribbean tonguefish** (*Symphurus arawak*)
- **Chocolatebanded tonguefish** (*Symphurus billykrietei*)
- **Offshore tonguefish** (*Symphurus civitatium*)
- **Spottedfin tonguefish** (*Symphurus diomedeanus*)
- **Margined tonguefish** (*Symphurus marginatus*)
- **Largescale tonguefish** (*Symphurus minor*)
- **Pygmy tonguefish** (*Symphurus parvus*)
- **Longtail tonguefish** (*Symphurus pelicanus*)
- **Deepwater tonguefish** (*Symphurus piger*)
- **Blackcheek tonguefish** (*Symphurus plagiusa*)
- **Blotchfin tonguefish** (*Symphurus stigmosus*)
- **Spottail tonguefish** (*Symphurus urospilus*)

Background

Unless you are a shrimper or like to dive at night over bare, sandy bottoms, you may never see a tonguefish, even though twelve species occur in Florida waters. All tonguefishes have a similar, teardrop shape; small, close-set eyes on the left side of the body; a small, twisted mouth; pigmented eyed sides and pale blind sides; and continuous dorsal and anal fins, running around the body, that merge with the tail fin. Individual species can be distinguished by subtle differences in the coloration of the eyed side and the fins (although, if you're hardcore, you can use dorsal-fin ray counts, most in the 80s and 90s!). We restrict our treatment here to the three largest (7–9 in) and most commonly observed species, which (happily) are easily identified. Spottedfin tonguefishes are the only species with 1–5 (usually 2) prominent black spots on the rear portions of both the dorsal and anal fins. The color of the eyed side is tan to brown, without obvious markings, although some indistinct bars occasionally are present. Spottail tonguefishes have a single, conspicuous, ocellated black spot on the caudal-fin rays. The eyed side is light-brown or tan, transversed by 4–11 (mostly 6–11) dark-brown bars that are alternatively narrow and wide. Blackcheek tonguefishes are light-brown or tan, with faint to well-marked bars, formed by small blotches; usually there is a black spot on the gill cover. Except for deepwater tonguefishes, which reach 6 in long, the other species in our area are small (<4 in) and are all drably colored, with small differences in mottling and striping.

Spottedfin tonguefish (*Symphurus diomedeanus*), Lake Worth Lagoon, Florida

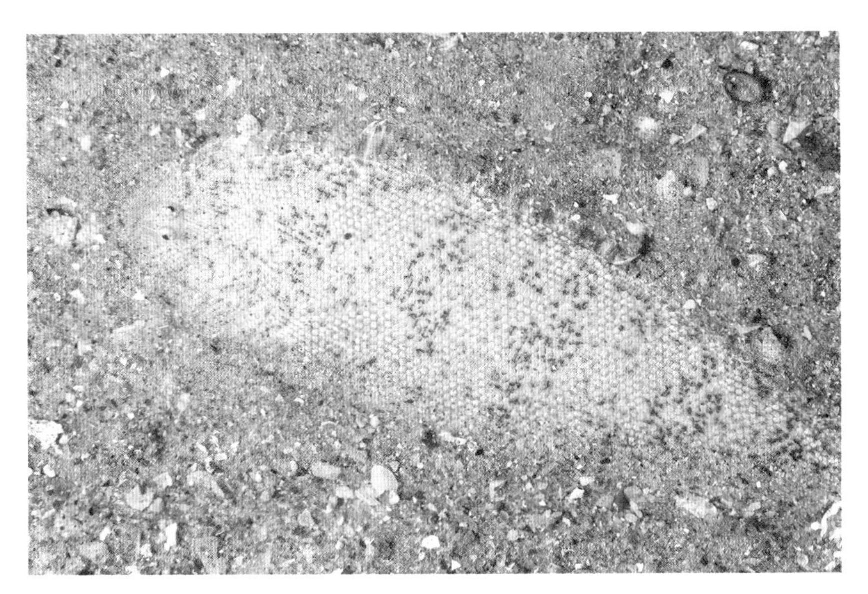

Blackcheek tonguefish (*Symphurus plagiusa*), Lake Worth Lagoon, Florida

Distribution and Habitat

Although the three highlighted species have similar statewide ranges, they are segregated by substrate type and water depth. Blackcheek tonguefishes are the most common species, and they are the only tonguefishes that regularly occur in softbottom habitats in lagoons, bays, and riverine estuaries. Adults also inhabit inner-shelf waters, but they rarely are found deeper than 100 ft. Spottail tonguefishes live at depths of 20–130 ft, usually on carbonate sediments near low-relief hardbottom. Spottedfin tonguefishes rarely enter inshore waters, preferring the open shelf, with shell-hash and carbonate-sand bottoms, at depths of 60–600 ft.

Natural History

Little information is available on tonguefish reproduction, beyond knowing that they spawn in the water column. Blackcheek tonguefishes spawn from February to March on the west Florida shelf; spottedfins spawn from March to August; and spottails spawn from August to November. Blackcheek tonguefishes live for only about 5 years, growing quickly in the first year and then more slowly once sexual maturity is reached in the second year.

Tonguefishes feed on bottom-dwelling invertebrates living in the sand, and the majority of species appear to be most active at night. Juvenile blackcheeks eat bottom-dwelling copepods, and adults choose small molluscs and crustaceans. Spottedfins consume crabs, shrimps, worms, and small snails.

Spikefishes
Family Triacanthodidae

- **Spotted spikefish** (*Hollardia meadi*)
- **Jambeau** (*Parahollardia lineata*)

Background

Spikefishes resemble the related triggerfishes and filefishes in having a deep, laterally compressed body; small mouth; moderately elongated snout; and short, slit-like gill openings, located just before the base of the pectoral fin. They differ in having a first dorsal fin with 6 hefty spines; a pair of pelvic fins with a single, long spine and 1–2 rays; and large eyes. Their scales are rough, like sandpaper. Spotted spikefishes and jambeaus are diamond shaped, about as long as they are high, with the greatest depth occurring at the tips of the stout first dorsal and anal spines, near the midbody. The small caudal fin is broadly rounded. Jambeaus are attractively colored in pastels: the head and body are light-pink to yellow, with 5–10 moderately wide, horizontal, reddish-brown lines. Each jaw has at least 2 teeth in the inner row. In contrast, the teeth in spotted spikefishes are in a single row. The base head and body coloration is pink to red, overlaid by 2 yellow stripes on the side (1 from the eye to the upper base of the caudal fin, and 1 running obliquely from the eye to the base of the anal fin) and rows of brown spots on the upper body. Jambeaus achieve a length of 8 in, twice the size of spotted spikefishes.

Distribution and Habitat

Both spikefish species are deep-water forms that wander in the lower reaches of continental-shelf waters but are more common over the edge, in the slope waters. Jambeaus range from depths of 278 to 7879 ft, and the spotted spikefishes occur in depths of 180–1485 ft. A third species, *Hollardia hollardi*, doesn't inhabit areas shallower than 900 ft and thus is beyond the scope of this volume.

Natural History

Nothing is known about the life history of spikefishes.

Triggerfishes
Family Balistidae

- **Gray triggerfish** (*Balistes capriscus*)
- **Queen triggerfish** (*Balistes vetula*)
- **Rough triggerfish** (*Canthidermis maculata*)
- **Ocean triggerfish** (*Canthidermis sufflamen*)
- **Black durgon** (*Melichthys niger*)
- **Sargassum triggerfish** (*Xanthichthys ringens*)

Background

Triggerfishes are deep-bodied, small-mouthed fishes lacking pelvic fins. They have tough skin, with rough, diamond-shaped scales, and possess a unique spiny dorsal fin that can be locked in the upright position. The lock mechanism is released by pushing down the second dorsal spine (the trigger). Florida species can be distinguished primarily by body shape and color pattern, which varies from gaudy to dull.

Gray and queen triggerfish are about the same size (16–18 in) and have similar body shapes, but they differ considerably in coloration. Gray triggerfishes are blue-gray, with pale undersides. They have wavy blue lines on the median fins and tail, and a dark saddle on the back, between the spiny and soft dorsal fins. Juveniles are light-green to pale-yellow, with irregular, electric-blue spots covering the body; the dark saddle is also present. In adult queen triggerfishes, the face below the eye and the area under the pectoral fin are bright-orange. Electric-blue encircles the mouth, and a broad, electric-blue stripe angles from above the mouth toward the pectoral fin, where it abruptly turns downward. Dark-blue lines radiate from the eye. The body varies from steel-gray to olive-green. The leading dorsal-fin ray and upper and lower filaments of the caudal fin are elongated. Juveniles resemble adults, but with more-muted colors.

The similarly shaped ocean triggerfishes, the largest member of the family (reaching 24 in), are a deep-bodied species, with tall dorsal and anal fins. They are gray-brown laterally, with a whitish belly and an obvious, large dark spot on the base of the pectoral fin. The somewhat slimmer and smaller (to 19 in) black durgons are uniformly black, with orange coloration on the cheek below the eye. A distinct white line runs along the base of the median fins.

Queen triggerfish (*Balistes vetula*), Jupiter, Florida

Top: Gray triggerfish (*Balistes capriscus*), Ft. Lauderdale, Florida. *Bottom*: Juvenile, Jupiter, Florida.

Rough triggerfishes also are a bit slimmer and rarely exceed 20 in. The dark-gray to cobalt-blue body coloration is punctuated by white or pale-blue spots that are smaller than the diameter of the eye. There is a groove in front of the eyes, and the scales have keels, making this species particularly rough to the touch. Sargassum triggerfishes are the most slender and smallest (to about 10 in) of the triggerfishes. They have a blue to lavender body, with black spots or dashes that form horizontal stripes. There are 3 dark lines on the cheeks, associated with grooves, and the tail margin is bright–reddish-brown.

Distribution and Habitat

Most triggerfish species are reef and hardbottom dwellers as adults; as juveniles, they associate with sargassum and flotsam. Gray triggerfishes, a continental species, are the most widespread, occurring statewide, in shelf waters. Queen triggerfishes, a tropical species, are rare outside of southeastern Florida and the Keys. Black durgons and ocean and sargassum triggerfishes also prefer the clear waters of southern Florida. The latter two species are found mostly on deeper reefs (100–300 ft). Rough triggerfishes are the sole exception to the above adult/juvenile habitat-choice pattern. An epipelagic bluewater species, they usually live their entire life in oceanic waters, in association with flotsam (although occasional adults are found on deep reefs).

Natural History

Reproductive behavior is best known for gray triggerfishes. During summer (June and July in southeast Florida), males select sandy areas adjacent to artificial or natural structures, in water depths ranging from 30 to 100 ft, and build one to several shallow, circular, 2 ft wide depressions in the sand that serve as nests. During courtship, the males turn dark–gray-blue; females have a whitish head and electric-blue markings on the median fins. There may be a harem social structure, with one male building several nests in an area for multiple females. A receptive female is led to the nest by the male and deposits a sticky, doughnut-sized demersal egg mass in the depression, which she then guards. The male patrols the perimeter, aggressively chasing away potential predators. Females maintain the nest by moving larger sedimentary fragments with their mouths and bath the eggs with water jetted out of the mouth, an act that also blows away finer sediment. Eggs hatch in about 2 days.

Queen and ocean triggerfishes also excavate nests with their body and fins and use water jetting. Little is known about their life histories. We have seen ocean

Left: Juvenile ocean triggerfish (*Canthidermis sufflamen*), Jupiter, Florida. *Right*: Rough triggerfish (*Canthidermis maculata*), Jupiter, Florida.

triggerfishes guarding nests in sand patches, near hardbottom habitats, off southeastern Florida. Observations concerning the details of nesting by queen triggerfishes are limited. The demersal eggs hatch into pelagic larvae that drift in the water column until settling around flotsam and sargassum in offshore waters. Triggerfishes appear to remain with the cover provided by floating debris or plants until suitable benthic settlement sites are encountered. Black durgons, sargassum triggerfishes, and rough triggerfishes have not been studied, and we have not observed spawning or reproductive behavior by these species in Florida. Gray triggerfishes are relatively long lived, reaching 17 years of age, while queen triggerfishes last for only 7–11 years.

Triggerfishes are vocal fishes, making noises by clicking their tooth plates and drumming their pectoral fins. They swim by undulating their dorsal and anal fins and will forage, primarily by day, up to 100 ft away from cover. Historically, queen triggerfishes fed almost exclusively on long-spined sea urchins (*Diadema*), but with the latter's marked decline since the 1980s, they have switched to crabs, brittle stars, and chitons. Black durgons and sargassum triggerfishes mostly feed on plankton but will scavenge on larger food items if available; black durgons consume a variety of invertebrates and limited amounts of algae. Gray and ocean triggerfishes eat various motile and attached invertebrates, especially hard-shelled critters like chitons, clams, amphipods, crabs, brittle stars, sand dollars and sea urchins. It is not uncommon to see triggers blowing jets of water out of the mouth to dislodge invertebrates from their hiding places

Black durgon (*Melichthys niger*)

Gray triggerfish (*Balistes capriscus*) guarding an egg mass, Jupiter, Florida

under rocks or in shallow sediments. Jetting is often used by gray triggerfishes to expose buried sand dollars or to flip over sea urchins or slipper lobsters to get to their vulnerable ventral sides.

Gray triggerfishes are the scourge of bottom-fishing anglers. Their staccato bites and small mouths make them expert bait stealers. Triggerfishes in general have strong jaw muscles and teeth capable of extracting a plug of flesh from an unwary diver or angler. Gray triggerfishes can be very curious and have been known to attack divers, especially when guarding nests. We have had gray triggerfishes bite at the ports of underwater camera housings and other inanimate objects, and one took a chunk out one of the authors' hands when he overconfidently reached into an on-deck pile of trawl catch to retrieve a prized specimen!

Filefishes
Family Monacanthidae

- **Dotterel filefish** (*Aluterus heudelotii*)
- **Unicorn filefish** (*Aluterus monoceros*)
- **Orange filefish** (*Aluterus schoepfii*)
- **Scrawled filefish** (*Aluterus scriptus*)
- **Whitespotted filefish** (*Cantherhines macrocerus*)
- **Orangespotted filefish** (*Cantherhines pullus*)
- **Fringed filefish** (*Monacanthus ciliatus*)
- **Slender filefish** (*Monacanthus tuckeri*)
- **Planehead filefish** (*Stephanolepis hispidus*)
- **Pygmy filefish** (*Stephanolepis setifer*)

Background

Filefishes, relatives of the triggerfishes, similarly have small mouths and gill openings; have rough, sandpaper-like scales; and lack pelvic fins. They differ, however, in being unable to lock their dorsal fins in the up position. Color patterns offer the best clues for the identification of the multiple filefish species, but these may vary considerably within a species; juveniles tend to differ in coloration from the adults.

Species in the genus *Aluterus* have long and relatively delicate first dorsal spines and differ slightly, but significantly, in body shape and coloration, facilitating identification. Dotterel and scrawled filefishes have bright-blue spots and wavy stripes against a gray to light-brown background. Dotterel filefishes only reach 1 ft in length; they have a steeper head profile and a deeper body than the other members of the genus. Scrawled filefishes are the largest (to 3 ft) in this group and have a long, sloped forehead. Juvenile scrawled filefishes are colored gray, dark-brown, or yellowish, with black spots and white-rimmed, light-blue spots. Juveniles in these two species look a lot alike, however, necessitating counting dorsal-fin rays (scrawled filefishes have 43 or more) to confirm identification.

Dotterel filefish (*Aluterus heudelotii*), Lake Worth Lagoon, Florida

Orange and unicorn filefishes are similarly sized (to 2 ft) and shaped, being more streamlined than dotterels but deeper bodied than scrawleds. They differ notably in coloration, however. Unicorn filefishes are dull-gray to brown, often with a reticulated pattern of interconnected dark blotches. The more-attractive orange filefishes are off-white, with contrasting dark-brown patches, or are bright–silvery-orange; both base colorations are punctuated by small, bright-orange or yellow spots. Unicorn filefishes have a truncate caudal fin, with pointed tips, while orange filefishes have a broadly rounded tail.

The other Florida filefishes have short, stout dorsal spines; are smaller in size; and, with one exception (slender filefishes) are diamond shaped, with deep bodies. The most colorful of these are whitespotted and orangespotted filefishes. Juvenile whitespotted filefishes are dark-brown to golden and covered in white spots. Adults are variable in coloration and can change patterns quickly. Generally they exhibit one of two color morphs, which may represent the two sexes. In one, the base coloration ranges from dull-white to dark-brown anteriorly, grading into rusty-orange on the rear third of the body. In the other common color pattern, the body is uniformly bright-orange, with a pale-yellow to white saddle extending anteriorly from the dorsal-fin origin. In both color patterns, the body is covered in white spots, which are larger in diameter than the eye, and the tail is dark-brown. On either side of the caudal peduncle there are 2–3 pairs of large, yellow-orange, recurved spines. The coloration of orangespotted filefishes is also highly variable, ranging from uniformly dark-brown, with small orange spots, to light–yellow-brown anteriorly, with 5 broad stripes. Usually there are 1 or 2 obvious white spots on the caudal peduncle. Juveniles are dark-brown to golden, with small orange spots.

Four additional smallish and generally drably colored (brown to green) filefishes are found in our waters. All but one are more abundant than the previously

Scrawled filefish (*Aluterus scriptus*), Singer Island, Florida

Juvenile scrawled filefish (*Aluterus scriptus*), Lake Worth Lagoon, Florida

Orange filefish (*Aluterus schoepfii*), Lake Worth Lagoon, Florida

discussed species, the exception being the diminutive (to 3.5 in) slender filefishes. They might be confused with elongated juvenile dotterel or scrawled filefishes if not for their short, stout dorsal spine and coloration. Slender filefishes are variably colored, often presenting a reticulated pattern, with an irregular, white, lateral stripe. The larger (to 8 in) fringed filefishes have a relatively deep body and exhibit a variety of color patterns, ranging from bright-green to tan, with brown blotches. Adult males grow an enlarged, bright-orange belly flap that no doubt is used like the dewlap of anole lizards.

Planehead and pygmy filefishes pretty much look alike, ranging in color from pale-yellow to dark-brown, with various blotches. Pygmy filefishes, which are smaller (to 7.5 in, but usually seen at 1–2 in) than planeheads (to 10 in), often have a series of light dashes on their sides, but counting dorsal rays (planeheads have 29 or more) is mandatory for identification.

Distribution and Habitat

For the most part, filefishes are widely distributed; in fact, scrawled and unicorn filefishes are found throughout the world, in tropical and subtropical waters. One reason for this broad distribution is that all of the species discussed above are rafters—they associate with pelagic sargassum or flotsam as young. Settlement and survival in different parts of the state vary with the individual species. Planehead, fringed, orange, and pygmy filefishes are the most widespread. Unicorn filefishes are found near

Left: Unicorn filefish (*Aluterus monoceros*), Jupiter, Florida. *Right*: Juvenile, Jupiter, Florida.

Left: Spotted phase whitespotted filefish (*Cantherhines macrocerus*). *Right*: Saddle phase.

the shelf break off southeastern Florida, and they are particularly abundant during spring. Whitespotted, orangespotted, and slender filefishes prefer reefs and hardbottom habitats in southeastern Florida and the Keys. Dotterel filefishes are relatively rare but are also found in hardbottom habitats.

Natural History

The reproductive habits of Atlantic filefishes are not well known. Some general observations are that pair spawning is common and eggs are demersal. Whitespotted filefishes appear to form long-lasting relationships, and often the same pair can be seen on the same reef for years. Whether nests are constructed and guarded, as is the case in triggerfishes, is uncertain. Larvae of most of our filefish species have been collected in plankton nets towed in the open ocean. Early juveniles, juveniles, and subadults of all western Atlantic species associate with sargassum and flotsam. Planehead and pygmy filefishes often numerically dominate net samples from drifting sargassum patches (scoop up clumps of sargassum, shake them over a bucket, and more often than not tiny filefishes will drop out). As a consequence of their association with drifting sargassum, filefishes are important food items for oceanic predators such as dolphinfishes, tunas, wahoos, and billfishes.

Adult filefishes generally associate with reef, hardbottom, or seagrass habitats as adults. Individuals move slowly around

Juvenile whitespotted filefish (*Cantherhines macrocerus*), Jupiter, Florida

these habitats, blending into the background through a combina tion of coloration and behavior. For example, scrawled and dotterel filefishes are often seen drifting head down among seagrasses, sargassum, loose ropes, mooring lines, pilings, or other structures in the water column. Their coloration often matches the background very well: green in seagrasses or algae, and various shades of brown or red in hardbottom habitats. Slender filefishes exhibit a particular affinity for octocorals (sea whips). Careful observation is required to find a slender filefish, and it will usually

Left: Adult orangespotted filefish (*Cantherhines pullus*), Lake Stuart, Florida. *Right*: Juveniles, Jupiter, Florida.

Left: Fringed filefish (*Monacanthus ciliatus*), Venice, Florida. *Right*: Slender filefish (*Monacanthus tuckeri*), Indian River Lagoon, Florida.

Slender filefish (*Monacanthus tuckeri*), Palm Beach, Florida

be oriented head down, aligned with the vertical axis of soft coral branches. Individuals casually maintain their orientation with the branches as they sway in the surge.

Filefishes have relatively small mouths and well-developed teeth that are adapted for consuming small, attached invertebrates, as well as small, free-living animals. For this reason, they fall among the most-skilled bait stealers of all time. Planehead filefishes often feed over sandy bottoms by jetting water out of the mouth to displace small invertebrates living among the sand grains. Young Gulf of Mexico planeheads primarily eat amphipods and then move up to bivalves and turtle grass at larger sizes, while co-occurring orange filefishes start on copepods and then switch to epifaunal worms and turtle grass as they grow. Whitespotted filefishes are one of only a few fish species that target sponges and toxic zoanthids (non-stony corals).

Left: Planehead filefish (*Stephanolepis hispidus*), Lake Worth Lagoon, Florida. *Right*: Pygmy filefish (*Stephanolepis setifer*), Jupiter, Florida.

Boxfishes
Family Ostraciidae

- **Honeycomb cowfish** (*Acanthostracion polygonia*)
- **Scrawled cowfish** (*Acanthostracion quadricornis*)
- **Spotted trunkfish** (*Lactophrys bicaudalis*)
- **Trunkfish** (*Lactophrys trigonus*)
- **Smooth trunkfish** (*Lactophrys triqueter*)

Background

Boxfishes are housed in a bony shell composed of hexagonal (or polygonal) plates, sort of the armored cars of the sea. The cowfish species (genus *Acanthostracion*) look like miniature bulls, with bony plates above the eyes forming forward-pointing horns. The base color of scrawled cowfishes is white, pale-yellow, or greenish, with wavy electric-blue lines on the body and head. Honeycomb cowfishes are basally tan or light-yellow or gray, with dark–charcoal-gray or brown hexagonal markings bordering each bony plate, giving them a honeycomb look. The cheeks have a reticulated pattern.

The genus *Lactophrys*, collectively called "trunkfishes," lack head horns. Spotted trunkfishes have a cream to light-brown background coloration that is punctuated with small, dark-brown or blackish spots; often there are white blotches on the flanks. Small juveniles are cream-colored cubes, with dark spots. In contrast, young of the trunkfish species are bright-green, with small white spots over the body and 2 dark blotches on the flanks. The fins of juveniles are

Top: Scrawled cowfish (*Acanthostracion quadricornis*), Lake Worth Lagoon, Florida. *Bottom*: Juvenile, Lake Worth Lagoon, Florida.

Honeycomb cowfish (*Acanthostracion polygonius*), Palm Beach, Florida

Smooth trunkfish (*Lactophrys triqueter*), Looe Key, Florida

Top: Late juvenile trunkfish (*Lactophrys trigonus*), Lake Worth Lagoon, Florida. *Bottom*: Early juvenile, Lake Worth Lagoon, Florida.

transparent, so underwater they look like tiny bouncing peas. Adults are golden-brown, with small white spots covering the body and 1 or 2 dark blotches on the forward flanks, the one under the pectoral fin forming a short chain pattern. The top of the body looks camel-like, with a pronounced hump. The base color of smooth trunkfishes is blackish or dark-brown; tiny, irregular white spots cover the entire body, and all fins but the caudal fin are yellow to gold in color. The bases of the pectoral and dorsal fins have 2 dark-brown to black blotches. Young smooth trunkfishes are dark-brown balls, with large white or yellow spots and pale fins. Scrawled cowfishes, trunkfishes, and spotted trunkfishes reach about 18 in; honeycomb cowfishes are a bit smaller (16 in); and smooth trunkfishes are the squirt, at about 1 ft.

Distribution and Habitat

All boxfishes except smooth trunkfishes, which do not occur in the Gulf of Mexico, are found around the state, in seagrass and structured hardbottom habitats, including algal spongebottoms and reefs. All species are known from nearshore to 300 ft water depths.

Natural History

Boxfishes generally spawn as pairs, but aggregations have been reported in some Caribbean locales. The larvae are planktonic for about 1 week; individual smooth and spotted trunkfishes settle on structured habitat across the shelf. Young trunkfishes and scrawled cowfishes settle in seagrass meadows, algal beds, or other structured habitats in inshore waters. In the tropics, scrawled cowfishes spawn year round, with distinct summer and winter peaks.

Adult boxfishes prefer structured habitat, where they feed on a variety of algae and invertebrates, including sponges, tunicates, corals and hydroids, crustaceans, mollusks, and polychaete worms. Cowfishes especially favor sponges. Individuals of all species use their mouths to waterjet the shallow sediment, dislodging desirable food items. The appearance of mysterious white rings on endangered *Acropora* coral recently was attributed to foraging honeycomb cowfishes. In addition to providing protection, the boxfishes' carapace helps produce hydrodynamic stability in swimming and, in tandem with fin control, results in superb stability and maneuverability in their structurally complex environment. Boxfishes secrete a powerful toxin (ostracitoxin) from their skin cells, but its value against predators is somewhat debatable, since boxfish species have been found in the stomachs of herons, tiger sharks, Nassau groupers, gags, amberjacks, and cobias. Nonetheless, the toxin is believed to deter predation somewhat, and it can kill invertebrates and fishes. For this reason, you should exercise caution when keeping the cute little cubes in your aquarium.

Trunkfish (*Lactophrys trigonus*), Lake Worth Lagoon, Florida

Spotted trunkfish (*Lactophrys bicaudalis*), Jupiter, Florida

Puffers
Family Tetraodontidae

- **Goldface toby** (*Canthigaster jamestyleri*)
- **Sharpnose puffer** (*Canthigaster rostrata*)
- **Smooth puffer** (*Lagocephalus laevigatus*)
- **Oceanic puffer** (*Lagocephalus lagocephalus*)
- **Marbled puffer** (*Sphoeroides dorsalis*)
- **Northern puffer** (*Sphoeroides maculatus*)
- **Southern puffer** (*Sphoeroides nephelus*)
- **Blunthead puffer** (*Sphoeroides pachygaster*)
- **Least puffer** (*Sphoeroides parvus*)
- **Bandtail puffer** (*Sphoeroides spengleri*)
- **Checkered puffer** (*Sphoeroides testudineus*)

Background

Puffers are so named because of their habit of inflating their extra-large stomachs with water (or air, if the fish is taken out of the water) when threatened. Puffers are the more diverse of the two related, inflation-happy families (the other being porcupinefishes). Puffers have rough, sandpapery skin and teeth that are fused into grinding stones. The beak-like jaws, composed of paired premaxillary and dentary bones, each have a central gap, distinguishing puffers from porcupinefishes, which lack the gaps and also have unique, hard body spines. The teeth allow both groups to break apart shelled prey items.

Oceanic and smooth puffers, sometimes called "rabbitfishes," have lunate tails and are the largest of the puffers, reaching lengths of 2 and 3 ft, respectively. Oceanic puffers are markedly countershaded: white on the belly, silvery laterally, and dark-blue dorsally (perfect coloration for a bluewater species). The pectoral fins also are bicolored—dark above and light below—as opposed to the uniformly light-colored fins of the related smooth puffers. Smooth puffers are also countercolored, but the differentiation between their lighter–green-grey dorsum and silver-white sides and abdomen is not as distinct. They also have a series of dark spots under the pectoral fin.

Traditionally, only a single sharpnose puffer species was thought to be in our area, but recent study reveals the presence of two discrete species. Sharpnose puffers and goldface tobies are similar in body size (3–4 in) and shape, but they differ slightly—yet significantly—in color pattern. Both species have truncate tails; chocolate-colored backs; and white undersides, with blue spots on the sides and belly. Sharpnose puffers have black leading margins on each lobe of the yellow tail fin, as well as a series of thin blue lines on the snout, caudal peduncle, and caudal fin. Goldface tobies lack the dark tail margins and have a gold, web-like pattern on the back; blue spots on the tail and caudal peduncle; and 2 parallel black stripes that extend from the base of the caudal fin anteriorly to the level of the pectoral fin.

The seven *Sphoeroides* species range in size from 6 to 12 in and share a common shape, having large heads attached to smaller, tapering bodies, with truncate or rounded tails (they look a lot like swimming chicken legs!). Identification of the various species within this genus can be resolved by examining the dorsolateral and snout coloration, since all their bellies are white. Blunthead puffers have a broadly rounded snout and are the dullest puffer species, being uniformly grayish-brown. Marbled puffers are a mottled brown to grey above, with a pointed snout; the sides are a blue- and gold-lined network of gold and brown polygons and blotches. There are broad dark bands at the base and outer margin of the truncate caudal fin on this species and on southern puffers, but the latter have a very different head shape and are dark-brown, with white spots forming rings and lines. Bandtail and least puffers also have those 2 tail bands, but both of these species have a narrow, lighter band outside the dark margin; in addition, a bandtail puffer's caudal fin is very rounded (a least puffer's is slightly emarginate). The dorsum of bandtail puffers is covered with small, fleshy tabs (lappets), and this species has a distinctive row of uniform, dark-brown spots, arranged in a line from the lower jaw to the caudal peduncle. Northern puffers also have a line of spots, but these are larger, more diffuse, and don't extend onto the snout. The base color is pale-yellow, and there are

Sharpnose puffer (*Canthigaster rostrata*), Lake Worth Lagoon, Florida

Goldface toby (*Canthigaster jamestyleri*), Hobe Sound, Florida

small, black-pepper spots dorsally. Checkered puffers have dark-brown spots and angular blotches, creating a variegated bull's-eye pattern on the back.

Distribution and Habitat

Sharpnose puffers and goldface tobies prefer structured habitats, including natural and artificial reefs. Goldface tobies generally are restricted to water depths ranging from 100 to 300 ft, and sharpnose puffers occur from shore to about 300 ft deep. Smooth and oceanic puffers are pelagic species, usually found in deep water, and occur around the state, although smooth puffers also venture into shallower waters. The seven *Sphoeroides* species generally are found in shallow inshore (often estuarine) waters. Exceptions are blunthead puffers, found in 180–800 ft depths; marbled puffers, in shallower waters, at 60–300 ft; and bandtail puffers, which range from shore to about 150 ft deep. All species are widespread in Florida waters, except least puffers (only in the northern Gulf), northern puffers (only off the northeast coast), and checkered puffers (east coast and the Keys).

Natural History

Puffers deposit demersal eggs, but details on the reproductive behavior and development of most species are lacking. Northern puffers, the best-studied species, mature early in their second year and spawn in Chesapeake Bay in May, June, and July (the spawning seasons for all *Sphoeroides* species are from late spring to early fall). The left ovary of the female is 1.5–2 times the size of the right; combined, they produce up to half a million eggs (checkered puffers produce even more). Spawning observed in an aquarium revealed a daytime release of eggs, deposited in a circular pattern in the sand. Development from fertilization to the larval stage occurs in 112 hours. In the Loxahatchee River, where checkered puffer juveniles are abundant in the low-salinity waters in winter and spring, we have seen groups of adult checkered puffers migrating downstream, presumably to the ocean, where they are likely to spawn in the fall. We have also observed courting behavior between individual bandtail puffers.

Female sharpnose puffers establish territories in near-reef or rubble habitats, defending them from other females and small males. Males also have territories and form harems with one or more females that reside within their territories. A male makes a house call to a female, and she selects an area within tufts of benthic algae in which to deposit her round, 0.03

Checkered puffer (*Sphoeroides testudineus*), Jupiter, Florida

Bandtail puffer (*Sphoeroides spengleri*), Lake Worth Lagoon, Florida

Southern puffer (*Sphoeroides nephelus*). Jupiter, Florida

in eggs. The male then fertilizes them, and both parents abandon the area. Young, measuring around 0.06 in, hatch in about 96 hours. Mass mortalities of young sharpnose puffers and other *Canthigaster* species have been reported.

Northern puffers live for 5-plus years. Growth occurs fastest during summer and early fall, and females always are larger than equally aged males. Checkered puffers, in contrast, reach smaller sizes and live briefer lives than their northern cousins. Northern puffers consume various mollusks, barnacles, crustaceans, sea anemones, worms, sea urchins, bryozoans, sponges, sea squirts, and algae, and they have even taken a crack at a watermelon seed! Bandtail and checkered puffers concentrate on crabs, bivalves, and snails. Sea whips, crustaceans, and small fishes are the major food items of young smooth puffers in Brazil.

The tissues of puffers are well known for their toxic properties, caused by the presence of salitoxin and tetrodotoxin. Ingesting the skin, liver, gonads, or other viscera from these fishes can result in paralysis or death. While the practice of inflating their bodies and the toxins they contain presumably reduce the threat of predation, these devices are not foolproof: sharpnose puffers are eaten by barracudas, peacock flounders, and seabirds; other puffer species commonly are consumed by billfishes and other predatory fishes.

In southeastern Florida, checkered puffers support a limited but devout following of admirers that covet the flesh of what they call "tasty toadies." These aficionados use cane poles to catch adult puffers as they migrate toward the coast to spawn. The flesh is considered a delicacy, but—as with the famed Asian fugu cuisine—the consumer walks a fine line between a hallucinogenic buzz and death. The entrails of puffers and marine-toad secretions are used in zombification rituals by some Caribbean islanders.

Porcupinefishes
Family Diodontidae

- **Bridled burrfish** (*Chilomycterus antennatus*)
- **Web burrfish** (*Chilomycterus antillarum*)
- **Spotfin burrfish** (*Chilomycterus reticulatus*)
- **Striped burrfish** (*Chilomycterus schoepfii*)
- **Balloonfish** (*Diodon holocanthus*)
- **Porcupinefish** (*Diodon hystrix*)

Striped burrfish (*Chilomycterus shoepfi*), Lake Worth Lagoon, Florida

Web burrfish (*Chilomycterus antillarum*), Lake Worth Lagoon, Florida

Background

Members of the two porcupinefish genera may be separated according to whether the body spines remain rigidly erect or can be alternatively elevated/relaxed. Spines on burrfishes (*Chilomycterus*) are relatively short and fixed in an upright position. In the *Diodon* genus, porcupinefishes and balloonfishes normally keep their long spines folded down against the body but, when threatened, extend them upward as the fish inflates its body with water (or air, if on the deck of a boat), giving it the appearance of a giant sandspur. Inflated, dried, and shellacked *Diodon* were once a staple of Florida curio shops.

Burrfish species can be differentiated by color pattern. Bridled burrfishes are tan, with small spots covering the entire body. They also have a pair of large, white-rimmed brown blotches: one above the pectoral fins, and one at the base

the larger blotches. The upper portion of the body is light-gray or tan and is covered with small black spots. Because of its size and coloration, spotfin burrfishes are easily confused with one of the *Diodon* species, porcupinefishes, which reach 3 ft in length but have movable body spines. The smaller (maximum 18 in) balloonfishes also have flexible spines, but they differ in having large brown blotches on the back.

Bridled burrfish (*Chilomycterus antennata*), Key West, Florida

Spotfin burrfish (*Chilomycterus reticulatus*), Lost Tree, Florida

of the dorsal fin. Web burrfishes are light-tan, with a dark-brown, reticulating pattern over much of the body. There is a dark-brown spot over the base of the dorsal fin, and 2 brown spots just behind the pectoral fin on both sides of the body. Striped burrfishes are yellow to light-brown, with brown wavy lines over the upper half of the body. The underside is pale to golden-yellow. They also have black blotches behind the pectoral fins and on the dorsal- and anal-fin bases. In the pelagic juvenile phase, striped burrfishes are yellow-orange, with small blue circles. Spotfin burrfishes, the largest (to 2.5 ft long) in the genus, differ from the others in lacking

Distribution and Habitat

All porcupinefishes typically associate with floating sargassum and flotsam as juveniles; thus, as natural oceanic rafters, they are widely distributed worldwide in tropical and subtropical waters. A third cosmopolitan species, the pelagic porcupinefish (*Diodon eydouxii*), probably will also be found in Florida's offshore waters. Adult balloonfishes and porcupinefishes associate with high structures, commonly on reefs and other hardbottom habitats. In contrast, burrfishes are obligate inshore species in all life phases, and the Florida species are distributionally restricted to the western Atlantic. Bridled, web, and spotfin burrfishes are tropical species that are confined to southeastern Florida and the Keys. Striped burrfishes are the only temperate species and are found throughout Florida, in seagrass beds, in estuaries, and on reefs and algal spongebottoms, in water depths to 200 ft.

Natural History

As a group, porcupinefishes spawn in pairs, with females releasing pelagic eggs into the water column.

Left: Balloonfish (*Diodon holocanthus*), Palm Beach, Florida. *Right*: Juvenile balloonfish (*Diodon holocanthus*), Jupiter, Florida.

Courtship behavior has been observed in balloonfishes, where pairs interact, usually at dusk. We have seen striped burrfishes engaging in what appears to be courting behavior. But little is known about spawning in the other species in this family. Late-stage burrfish larvae look so different that they were once considered a separate critter, named *Lyosphaera*. Some researchers have speculated that this stage mimics poisonous sea hares (marine gastropod molluscs). On occasion, pelagic juvenile balloonfishes occur in huge numbers, either in floating sargassum or on the sea surface, and are washed ashore en masse.

Porcupinefish (*Diodon hystrix*), Palm Beach, Florida

Burrfishes and porcupinefishes consume hermit crabs, sea urchins, sea biscuits, and other invertebrates. These items are easily dispatched with the fishes' fused crushing plates. Like puffers, their viscera contain the poison tetrodotoxin. Despite their spines and toxin, members of this family (particularly the pelagic young) are eaten by tiger sharks, houndfishes, Atlantic goliath groupers, dolphinfishes, white grunts, great barracudas, and most species of snappers, billfishes, and tunas. Small burrfishes fall prey to wading birds and predatory inshore fishes.

Molas
Family Molidae

- **Sharptail mola** (*Mola lanceolata*)
- **Ocean sunfish** (*Mola mola*)
- **Slender mola** (*Ranzania laevis*)

Background

Molas are large and peculiar-looking—like the front end of what should be a much larger fish. Molas swim by sculling with their high median fins, while the vestige of a tail, known as a clavus, serves as a rudder. Sharptail molas, which grow to about 6.5 ft long, are unique among their kin in having a small, centrally located, pointed tail on the blunt clavus. The egg-shaped body is grayish-brown on the back and the fins, with white spotting, and white on the sides and belly, with dark spotting. The body of similarly colored ocean sunfishes is rounded, and the clavus is curved and scalloped, like the edge of a piecrust. Ocean sunfishes are the largest of the molas, growing close to 12 ft long and weighing almost 5000 lb, thereby garnering the distinction of being the heaviest bony fish in the sea. The much smaller (reaching about 2.5 ft) slender molas are more elongate than other molas, and the long but thin dorsal and anal fins abut the nearly straight and scalloped clavus. Their coloration is dark-blue dorsally, with electric-blue wavy bars against a pale-orange background on the flanks.

Distribution and Habitat

All three mola species are found in pelagic waters worldwide. They occur along the edge of the Gulf Stream off eastern Florida, and in deeper waters of the Gulf. Sharptail molas are much less common than ocean sunfishes, but they may be mistakenly identified as the latter by the uninitiated. Slender molas are so rarely observed that scientists write a paper about every capture of this species.

Natural History

Molas are thought to be water-column spawners, although spawning has never been observed. One female ocean sunfish carried over 300 million eggs in just part of her ovary. The larvae of molas look like little swimming sandspurs. The young grow rapidly during their first year in the pelagic realm.

Ocean sunfishes are well-known consumers of jellyfishes, but they also are reported to feed on some fishes, algae, hydroids, crustaceans, brittle stars, and molluscs. Juvenile sharptail molas also eat some benthic annelids and sponges. Molas fall prey to large sharks, including white sharks, and blue marlins. The flesh of sharptail molas has been reported to be poisonous.

Ocean sunfishes get their name from their habit of sunning (lying on their sides on the sea surface). This behavior may serve to warm up the body before taking forays into deeper, colder waters. Tagging studies demonstrate that ocean sunfishes make regular excursions from shallow to deep water: down by day and up by night. Sharptail molas have been observed from submersibles, at a depth of 2200 ft, and it has been suggested that the relative rarity of capture for this species may reflect the fact that they actually are a midwater fish that takes advantage of concentrations of deep-dwelling comb jellies, medusas, salps, and siphonophores.

Ocean sunfish (*Mola mola*), Key West, Florida. *Photo by Don DeMaria.*

Adult ocean sunfishes move along fronts between water masses where jellyfish aggregate. The edges of the Gulf Stream and Loop Currents off Florida concentrate jellyfish. Aerial surveys for marine mammals have documented considerable numbers of ocean sunfishes (and perhaps some sharptail molas) moving along the U.S. East Coast during winter, to as far south as Cape Canaveral. The presence of ocean sunfishes coincides with that of migrating leatherback turtles, which are also large-bodied jellyfish eaters.

Individual ocean sunfishes are islands unto themselves for tiny external parasites, providing spacious strongholds for multispecies assemblages. Although we have not observed this in Florida waters, ocean sunfishes in Indonesia park themselves above reefs, beckoning small cleaner fishes to reap the bounty of microcrustacean biomass and provide some tactile stimulation at the same time.

Appendix

List of species known from Florida waters from families covered in the book. The order presented is phylogenetic (from oldest to most recent in terms of evolutionary history). Names generally follow Page et al. (2013), except where more recent taxonomic revisions were available.

Family	Species
Lampreys (Petromyzontidae)	sea lamprey (*Petromyzon marinus*)
Nurse sharks (Ginglymostomatidae)	nurse shark (*Ginglymostoma cirratum*)
Whale sharks (Rhincodontidae)	whale shark (*Rhincodon typus*)
Sand tigers (Odontaspididae)	sand tiger (*Carcharias taurus*)
Thresher sharks (Alopiidae)	bigeye thresher (*Alopias superciliosus*) common thresher shark (*Alopias vulpinus*)
Basking sharks (Cetorhinidae)	basking shark (*Cetorhinus maximus*)
Mackerel sharks (Lamnidae)	white shark (*Carcharodon carcharias*) shortfin mako (*Isurus oxyrinchus*) longfin mako (*Isurus paucus*)
Cat sharks (Scyliorhinidae)	marbled cat shark (*Galeus arae*) chain dogfish (*Scyliorhinus retifer*)
Hound sharks (Triakidae)	smooth dogfish (*Mustelus canis*) Florida smoothhound (*Mustelus norrisi*)
Requiem sharks (Carcharhinidae)	blacknose shark (*Carcharhinus acronotus*) bignose shark (*Carcharhinus altimus*) spinner shark (*Carcharhinus brevipinna*) silky shark (*Carcharhinus falciformis*) Galapagos shark (*Carcharhinus galapagensis*) finetooth shark (*Carcharhinus isodon*) bull shark (*Carcharhinus leucas*) blacktip shark (*Carcharhinus limbatus*) oceanic whitetip shark (*Carcharhinus longimanus*) dusky shark (*Carcharhinus obscurus*) reef shark (*Carcharhinus perezii*) sandbar shark (*Carcharhinus plumbeus*) night shark (*Carcharhinus signatus*) tiger shark (*Galeocerdo cuvier*) lemon shark (*Negaprion brevirostris*) blue shark (*Prionace glauca*) Atlantic sharpnose shark (*Rhizoprionodon terraenovae*)
Hammerhead sharks (Sphyrnidae)	scalloped hammerhead (*Sphyrna lewini*) great hammerhead (*Sphyrna mokarran*) bonnethead (*Sphyrna tiburo*) smooth hammerhead (*Sphyrna zygaena*)
Angel sharks (Squatinidae)	Atlantic angel shark (*Squatina dumeril*) disparate angel shark (*Squatina heteroptera*)

Family	Species
Torpedo electric rays (Torpedinidae)	Atlantic torpedo (*Torpedo nobiliana*)
Electric rays (Narcinidae)	lesser electric ray (*Narcine bancroftii*)
Sawfishes (Pristidae)	smalltooth sawfish (*Pristis pectinata*) largetooth sawfish (*Pristis pristis*)
Guitarfishes (Rhinobatidae)	Atlantic guitarfish (*Rhinobatos lentiginosus*)
Skates (Rajidae)	spreadfin skate (*Dipturus olseni*) rosette skate (*Leucoraja garmani*) freckled skate (*Leucoraja lentiginosa*) clearnose skate (*Raja eglanteria*) roundel skate (*Raja texana*)
American round stingrays (Urotrygonidae)	yellow stingray (*Urobatis jamaicensis*)
Whiptail stingrays (Dasyatidae)	southern stingray (*Dasyatis americana*) roughtail stingray (*Dasyatis centroura*) Atlantic stingray (*Dasyatis sabina*) bluntnose stingray (*Dasyatis say*)
Butterfly rays (Gymnuridae)	spiny butterfly ray (*Gymnura altavela*) smooth butterfly ray (*Gymnura micrura*)
Eagle rays (Myliobatidae)	spotted eagle ray (*Aetobatus narinari*) bullnose ray (*Myliobatis freminvillei*) southern eagle ray (*Myliobatis goodei*)
Cownose rays (Rhinopteridae)	cownose ray (*Rhinoptera bonasus*)
Mantas (Mobulidae)	giant manta (*Manta birostris*) devil ray (*Mobula hypostoma*)
Sturgeons (Acipenseridae)	shortnose sturgeon (*Acipenser brevirostris*) Atlantic sturgeon (*Acipenser oxyrinchus*
Gars (Lepisosteidae)	alligator gar (*Atractosteus spatula*) longnose gar (*Lepisosteus osseus*)
Tenpounders (Elopidae)	ladyfish (*Elops saurus*) malacho (*Elops smithi*)
Tarpons (Megalopidae)	tarpon (*Megalops atlanticus*)
Bonefishes (Albulidae)	bonefish (*Albula vulpes*) bigeye bonefish (*Albula* sp.)
Freshwater eels (Anguillidae)	American eel (*Anguilla rostrata*)
Morays (Muraenidae)	pygmy moray (*Anarchias similis*) chain moray (*Echidna catenata*) fangtooth moray (*Enchelycore anatina*) chestnut moray (*Enchelycore carychroa*) viper moray (*Enchelycore nigricans*) green moray (*Gymnothorax funebris*) lichen moray (*Gymnothorax hubbsi*) blacktail moray (*Gymnothorax kolpos*) sharktooth moray (*Gymnothorax maderensis*)

Family	Species
	goldentail moray (*Gymnothorax miliaris*)
	spotted moray (*Gymnothorax moringa*)
	blackedge moray (*Gymnothorax nigromarginatus*)
	polygon moray (*Gymnothorax polygonius*)
	honeycomb moray (*Gymnothorax saxicola*)
	purplemouth moray (*Gymnothorax vicinus*)
	redface moray (*Monopenchelys acuta*)
	reticulate moray (*Muraena retifera*)
	stout moray (*Muraena robusta*)
	marbled moray (*Uropterygius macularius*)
Snake eels (Ophichthidae)	Key worm eel (*Ahlia egmontis*)
	tusky eel (*Aplatophis chauliodus*)
	stripe eel (*Aprognathodon platyventris*)
	academy eel (*Apterichtus ansp*)
	finless eel (*Apterichtus kendalli*)
	sooty eel (*Bascanichthys bascanium*)
	whip eel (*Bascanichthys scuticaris*)
	shorttail snake eel (*Callechelys guineensis*)
	blotched snake eel (*Callechelys muraena*)
	ridgefin eel (*Callechelys springeri*)
	slantlip eel (*Caralophia loxochila*)
	spotted spoon-nose eel (*Echiophis intertinctus*)
	snapper eel (*Echiophis punctifer*)
	irksome eel (*Gordiichthys ergodes*)
	horsehair eel (*Gordiichthys irretitus*)
	string eel (*Gordiichthys leibyi*)
	sailfin eel (*Letharchus velifer*)
	surf eel (*Ichthyapus ophioneus*)
	sharptail eel (*Myrichthys breviceps*)
	goldspotted eel (*Myrichthys ocellatus*)
	broadnose worm eel (*Myrophis platyrhynchus*)
	speckled worm eel (*Myrophis punctatus*)
	margined snake eel (*Ophichthus cruentifer*)
	shrimp eel (*Ophichthus gomesii*)
	blackpored eel (*Ophichthus melanoporus*)
	spotted snake eel (*Ophichthus ophis*)
	palespotted eel (*Ophichthus puncticeps*)
	king snake eel (*Ophichthus rex*)
	diminutive worm eel (*Pseudomyrophis fugesae*)
	blackspotted snake eel (*Quassiremus ascensionis*)
Conger eels (Congridae)	bandtooth conger (*Ariosoma balearicum*)
	bullish conger (*Bathycongrus bullisi*)
	conger eel (*Conger oceanicus*)
	manytooth conger (*Conger triporiceps*)
	blackgut conger (*Gnathophis bathytopos*)
	longeye conger (*Gnathophis bracheatopos*)
	brown garden eel (*Heteroconger longissimus*)
	yellow garden eel (*Heteroconger luteolus*)

Family	Species
	margintail conger (*Paraconger caudilimbatus*)
	splendid conger (*Pseudophichthys splendens*)
	yellow conger (*Rhynchoconger flavus*)
	whiptail conger (*Rhynchoconger gracilior*)
	Guppy's conger (*Rhynchoconger guppyi*)
	threadtail conger (*Uroconger syringinus*)
Anchovies (Engraulidae)	Key anchovy (*Anchoa cayorum*)
	Cuban anchovy (*Anchoa cubana*)
	striped anchovy (*Anchoa hepsetus*)
	bigeye anchovy (*Anchoa lamprotaenia*)
	dusky anchovy (*Anchoa lyolepis*)
	bay anchovy (*Anchoa mitchilli*)
	flat anchovy (*Anchoviella perfasciata*)
	silver anchovy (*Engraulis eurystole*)
Herrings (Clupeidae)	blueback herring (*Alosa aestivalis*)
	Alabama shad (*Alosa alabamae*)
	skipjack shad (*Alosa chrysochloris*)
	hickory shad (*Alosa mediocris*)
	American shad (*Alosa sapidissima*)
	Gulf menhaden (*Brevoortia patronus*)
	yellowfin menhaden (*Brevoortia smithi*)
	Atlantic menhaden (*Brevoortia tyrannus*)
	gizzard shad (*Dorosoma cepedianum*)
	round herring (*Etrumeus teres*)
	false pilchard (*Harengula clupeola*)
	redear sardine (*Harengula humeralis*)
	scaled sardine (*Harengula jaguana*)
	dwarf herring (*Jenkinsia lamprotaenia*)`
	little-eye herring (*Jenkinsia majua*)
	shortband herring (*Jenkinsia stolifera*)
	Atlantic thread herring (*Opisthonema oglinum*)
	Spanish sardine (*Sardinella aurita*)
Sea catfishes (Ariidae)	hardhead catfish (*Ariopsis felis*)
	gafftopsail catfish (*Bagre marinus*)
Lizardfishes (Synodontidae)	largescale lizardfish (*Saurida brasiliensis*)
	smallscale lizardfish (*Saurida caribbaea*)
	shortjaw lizardfish (*Saurida normani*)
	inshore lizardfish (*Synodus foetens*)
	sand diver (*Synodus intermedius*)
	largespot lizardfish (*Synodus macrostigmus*)
	offshore lizardfish (*Synodus poeyi*)
	bluestripe lizardfish (*Synodus saurus*)
	red lizardfish (*Synodus synodus*)
	snakefish (*Trachinocephalus myops*)
Codlets (Bregmacerotidae)	antenna codlet (*Bregmaceros atlanticus*)
	striped codlet (*Bregmaceros cantori*)

Family	Species
	Keys codlet (*Bregmaceros cayorum*) stellate codlet (*Bregmaceros houdei*) spotted codlet (*Bregmaceros mcclellandi*)
Codlings (Moridae)	metallic codling (*Physiculus fulvus*)
Merlucciid hakes (Merlucciidae)	offshore hake (*Merluccius albidus*) luminous hake (*Steindachneria argentea*)
Phycid hakes (Phycidae)	Gulf hake (*Urophycis cirrata*) Carolina hake (*Urophycis earllii*) southern hake (*Urophycis floridana*) spotted hake (*Urophycis regia*)
Pearlfishes (Carapidae)	pearlfish (*Carapus bermudensis*) chain pearlfish (*Echiodon dawsoni*)
Cusk-eels (Ophidiidae)	Atlantic bearded brotula (*Brotula barbata*) blackedge cusk-eel (*Lepophidium brevibarbe*) mottled cusk-eel (*Lepophidium jeannae*) fawn cusk-eel (*Lepophidium profundorum*) twospot brotula (*Neobythites gilli*) stripefin brotula (*Neobythites marginatus*) longnose cusk-eel (*Ophidion antipholus*) shorthead cusk-eel (*Ophidion dromio*) blotched cusk-eel (*Ophidion grayi*) bank cusk-eel (*Ophidion holbrookii*) crested cusk-eel (*Ophidion josephi*) striped cusk-eel (*Ophidion marginatum*) colonial cusk-eel (*Ophidion robinsi*) mooneye cusk-eel (*Ophidion selenops*) sleeper cusk-eel (*Otophidium dormitator*) polka-dot cusk-eel (*Otophidium omostigmum*) dusky cusk-eel (*Parophidion schmidti*) redfin brotula (*Petrotyx sanguineus*)
Viviparous brotulas (Bythitidae)	reef-cave brotula (*Grammonus claudei*) gold brotula (*Gunterichthys longipenis*) Key brotula (*Ogilbia cayorum*) curator brotula (*Ogilbia sabaji*) shy brotula (*Ogilbia suarezae*) black brotula (*Stygnobrotula latebricola*)
Toadfishes (Batrachoididae)	Gulf toadfish (*Opsanus beta*) leopard toadfish (*Opsanus pardus*) oyster toadfish (*Opsanus tau*) Atlantic midshipman (*Porichthys plectrodon*)
Goosefishes (Lophiidae)	reticulate goosefish (*Lophiodes reticulatus*) blackfin goosefish (*Lophius gastrophysus*)
Frogfishes (Antennariidae)	longlure frogfish (*Antennarius multiocellatus*) dwarf frogfish (*Antennarius pauciradiatus*) striated frogfish (*Antennarius striatus*)

Family	Species
	ocellated frogfish (*Fowlerichthys ocellatus*) singlespot frogfish (*Fowlerichthys radiosus*) sargassumfish (*Histrio histrio*)
Batfishes (Ogcocephalidae)	Atlantic batfish (*Dibranchus atlanticus*) pancake batfish (*Halieutichthys aculeatus*) spiny batfish (*Halieutichthys bispinosus*) Gulf batfish (*Halieutichthys intermedius*) longnose batfish (*Ogcocephalus corniger*) polka-dot batfish (*Ogcocephalus cubifrons*) slantbrow batfish (*Ogcocephalus declivirostris*) shortnose batfish (*Ogcocephalus nasutus*) spotted batfish (*Ogcocephalus pantostictus*) roughback batfish (*Ogcocephalus parvus*) palefin batfish (*Ogcocephalus rostellum*) tricorn batfish (*Zalieutes mcgintyi*)
Mullets (Mugilidae)	mountain mullet (*Agonostomus monticola*) striped mullet (*Mugil cephalus*) white mullet (*Mugil curema*) liza (*Mugil liza*) redeye mullet (*Mugil rubrioculus*) fantail mullet (*Mugil trichodon*)
New World silversides (Atherinopsidae)	rough silverside (*Membras martinica*) inland silverside (*Menidia beryllina*) Key silverside (*Menidia conchorum*) Atlantic silverside (*Menidia menidia*) tidewater silverside (*Menidia peninsulae*)
Old World silversides (Atherinidae)	hardhead silverside (*Atherinomorus stipes*) reef silverside (*Hypoatherina harringtonensis*)
Flyingfishes (Exocoetidae)	margined flyingfish (*Cheilopogon cyanopterus*) bandwing flyingfish (*Cheilopogon exsiliens*) spotfin flyingfish (*Cheilopogon furcatus*) Atlantic flyingfish (*Cheilopogon melanurus*) clearwing flyingfish (*Cypselurus comatus*) oceanic two-wing flyingfish (*Exocoetus obtusirostris*) fourwing flyingfish (*Hirundichthys affinis*) sailfin flyingfish (*Parexocoetus brachypterus*) bluntnose flyingfish (*Prognichthys occidentalis*)
Halfbeaks (Hemiramphidae)	hardhead halfbeak (*Chriodorus atherinoides*) flying halfbeak (*Euleptorhamphus velox*) balao (*Hemiramphus balao*) ballyhoo (*Hemiramphus brasiliensis*) false silverstripe halfbeak (*Hyporhamphus meeki*) Atlantic silverstripe halfbeak (*Hyporhamphus unifasciatus*) smallwing flyingfish (*Oxyporhamphus micropterus*)
Needlefishes (Belonidae)	flat needlefish (*Ablennes hians*) keeltail needlefish (*Platybelone argalus*)

Family	Species
	Atlantic needlefish (*Strongylura marina*)
	redfin needlefish (*Strongylura notata*)
	timucú (*Strongylura timucu*)
	Atlantic agujón (*Tylosurus acus*)
	houndfish (*Tylosurus crocodilus*)
New World rivulines (Rivulidae)	mangrove rivulus (*Kryptolebias marmoratus*)
Pupfishes (Cyprinodontidae)	sheepshead minnow (*Cyprinodon variegatus*)
	goldspotted killifish (*Floridichthys carpio*)
	flagfish (*Jordanella floridae*)
Topminnows (Fundulidae)	marsh killifish (*Fundulus confluentus*)
	Gulf killifish (*Fundulus grandis*)
	mummichog (*Fundulus heteroclitus*)
	saltmarsh topminnow (*Fundulus jenkinsi*)
	striped killifish (*Fundulus majalis*)
	Seminole killifish (*Fundulus seminolis*)
	longnose killifish (*Fundulus similis*)
	diamond killifish (*Fundulus xenicus*)
	rainwater killifish (*Lucania parva*)
Livebearers (Poeciliidae)	eastern mosquitofish (*Gambusia holbrooki*)
	mangrove gambusia (*Gambusia rhizophorae*)
	sailfin molly (*Poecilia latipinna*)
Squirrelfishes (Holocentridae)	spinycheek soldierfish (*Corniger spinosus*)
	squirrelfish (*Holocentrus adscensionis*)
	longspine squirrelfish (*Holocentrus rufus*)
	blackbar soldierfish (*Myripristis jacobus*)
	longjaw squirrelfish (*Neoniphon marianus*)
	bigeye soldierfish (*Ostichthys trachypoma*)
	cardinal soldierfish (*Plectrypops retrospinis*)
	deepwater squirrelfish (*Sargocentron bullisi*)
	reef squirrelfish (*Sargocentron coruscum*)
	dusky squirrelfish (*Sargocentron vexillarium*)
Seahorses and pipefishes (Syngnathidae)	pipehorse (*Acentronura dendritica*)
	fringed pipefish (*Anarchopterus criniger*)
	insular pipefish (*Anarchopterus tectus*)
	pugnose pipefish (*Bryx dunckeri*)
	whitenose pipefish *Cosmocampus albirostris*)
	crested pipefish (*Cosmocampus brachycephalus*)
	shortfin pipefish (*Cosmocampus elucens*)
	dwarf pipefish (*Cosmocampus hildebrandi*)
	banded pipefish (*Halicampus crinitus*)
	lined seahorse (*Hippocampus erectus*)
	longsnout seahorse (*Hippocampus reidi*)
	dwarf seahorse (*Hippocampus zosterae*)
	opossum pipefish (*Microphis brachyurus*)
	dusky pipefish (*Syngnathus floridae*)
	northern pipefish (*Syngnathus fuscus*)
	chain pipefish (*Syngnathus louisianae*)

Family	Species
	sargassum pipefish (*Syngnathus pelagicus*) Gulf pipefish (*Syngnathus scovelli*) bull pipefish (*Syngnathus springeri*)
Trumpetfishes (Aulostomidae)	Atlantic trumpetfish (*Aulostomus maculatus*)
Cornetfishes (Fistulariidae)	red cornetfish (*Fistularia petimba*) bluespotted cornetfish (*Fistularia tabacaria*)
Snipefishes (Macroramphosidae)	longspine snipefish (*Macroramphosus scolopax*)
Flying gurnards (Dactylopteridae)	flying gurnard (*Dactylopterus volitans*)
Scorpionfishes (Scorpaenidae)	blackbelly rosefish (*Helicolenus dactylopterus*) spinycheek scorpionfish (*Neomerinthe hemingwayi*) longsnout scorpionfish (*Pontinus castor*) longspine scorpionfish (*Pontinus longispinis*) spinythroat scorpionfish (*Pontinus nematophthalmus*) highfin scorpionfish (*Pontinus rathbuni*) devil firefish (*Pterois miles*) red lionfish (*Pterois volitans*) longfin scorpionfish (*Scorpaena agassizi*) coral scorpionfish (*Scorpaena albifimbria*) goosehead scorpionfish (*Scorpaena bergii*) shortfin scorpionfish (*Scorpaena brachyptera*) barbfish (*Scorpaena brasiliensis*) smoothhead scorpionfish (*Scorpaena calcarata*) hunchback scorpionfish (*Scorpaena dispar*) dwarf scorpionfish (*Scorpaena elachys*) plumed scorpionfish (*Scorpaena grandicornis*) mushroom scorpionfish (*Scorpaena inermis*) smoothcheek scorpionfish (*Scorpaena isthmensis*) spotted scorpionfish (*Scorpaena plumieri*) reef scorpionfish (*Scorpaenodes caribbaeus*) deepreef scorpionfish (*Scorpaenodes tredecimspinosus*)
Searobins (Triglidae)	shortfin searobin (*Bellator brachychir*) streamer searobin (*Bellator egretta*) horned searobin (*Bellator militaris*) spiny searobin (*Prionotus alatus*) northern searobin (*Prionotus carolinus*) striped searobin (*Prionotus evolans*) bigeye searobin (*Prionotus longispinosus*) barred searobin (*Prionotus martis*) bandtail searobin (*Prionotus ophryas*) Mexican searobin (*Prionotus paralatus*) bluewing searobin (*Prionotus punctatus*) bluespotted searobin (*Prionotus roseus*) blackwing searobin (*Prionotus rubio*) leopard searobin (*Prionotus scitulus*) shortwing searobin (*Prionotus stearnsi*) bighead searobin (*Prionotus tribulus*)

Family	Species
Snooks (Centropomidae)	swordspine snook (*Centropomus ensiferus*)
	smallscale fat snook (*Centropomus parallelus*)
	tarpon snook (*Centropomus pectinatus*)
	common snook (*Centropomus undecimalis*)
Wreckfishes (Polyprionidae)	wreckfish (*Polyprion americanus*)
Groupers (Epinephelidae)	mutton hamlet (*Alphestes afer*)
	graysby (*Cephalopholis cruentata*)
	coney (*Cephalopholis fulva*)
	Atlantic creolefish (*Cephalopholis furcifer*)
	marbled grouper (*Dermatolepis inermis*)
	rock hind (*Epinephelus adscensionis*)
	speckled hind (*Epinephelus drummondhayi*)
	red hind (*Epinephelus guttatus*)
	Atlantic goliath grouper (*Epinephelus itajara*)
	red grouper (*Epinephelus morio*)
	Nassau grouper (*Epinephelus striatus*)
	Spanish flag (*Gonioplectrus hispanus*)
	yellowedge grouper (*Hyporthodus flavolimbatus*)
	misty grouper (*Hyporthodus mystacinus*)
	Warsaw grouper (*Hyporthodus nigritus*)
	snowy grouper (*Hyporthodus niveatus*)
	western comb grouper (*Mycteroperca acutirostris*)
	black grouper (*Mycteroperca bonaci*)
	yellowmouth grouper (*Mycteroperca interstitialis*)
	gag (*Mycteroperca microlepis*)
	scamp (*Mycteroperca phenax*)
	tiger grouper (*Mycteroperca tigris*)
	yellowfin grouper (*Mycteroperca venenosa*)
Sea basses (Serranidae)	yellowfin bass (*Anthias nicholsi*)
	swallowtail bass (*Anthias woodsi*)
	streamer bass (*Baldwinella aureorubens*)
	red barbier (*Baldwinella vivanus*)
	yellowtail bass (*Bathyanthias mexicanus*)
	twospot sea bass (*Centropristis fuscula*)
	bank sea bass (*Centropristis ocyurus*)
	rock sea bass (*Centropristis philadelphica*)
	black sea bass (*Centropristis striata*)
	threadnose bass (*Choranthias tenuis*)
	dwarf sand perch (*Diplectrum bivittatum*)
	sand perch (*Diplectrum formosum*)
	longtail bass (*Hemanthias leptus*)
	yellowbelly hamlet (*Hypoplectrus aberrans*)
	Florida hamlet (*Hypoplectrus floridae*)
	blue hamlet (*Hypoplectrus gemma*)
	golden hamlet (*Hypoplectrus gummigutta*)
	shy hamlet (*Hypoplectrus guttavarius*)
	indigo hamlet (*Hypoplectrus indigo*)

Family	Species
	black hamlet (*Hypoplectrus nigricans*) barred hamlet (*Hypoplectrus puella*) tan hamlet (*Hypoplectrus randallorum*) butter hamlet (*Hypoplectrus unicolor*) undescribed tan hamlet (*Hypoplectrus* sp.) eyestripe basslet (*Liopropoma aberrans*) candy basslet (*Liopropoma carmabi*) wrasse basslet (*Liopropoma eukrines*) cave basslet (*Liopropoma mowbrayi*) yellow-spotted basslet (*Liopropoma olneyi*) peppermint basslet (*Liopropoma rubre*) splitfin bass (*Parasphyraenops incisus*) apricot bass (*Plectranthias garrupellus*) roughtongue bass (*Pronotogrammus martinicensis*) reef bass (*Pseudogramma gregoryi*) freckled soapfish (*Rypticus bistrispinus*) slope soapfish (*Rypticus carpenteri*) whitespotted soapfish (*Rypticus maculatus*) greater soapfish (*Rypticus saponaceus*) spotted soapfish (*Rypticus subbifrenatus*) school bass (*Schultzea beta*) pygmy sea bass (*Serraniculus pumilio*) orangeback bass (*Serranus annularis*) blackear bass (*Serranus atrobranchus*) lantern bass (*Serranus baldwini*) snow bass (*Serranus chionaraia*) saddle bass (*Serranus notospilus*) tattler (*Serranus phoebe*) belted sandfish (*Serranus subligarius*) tobaccofish (*Serranus tabacarius*) harlequin bass (*Serranus tigrinus*) chalk bass (*Serranus tortugarum*)
Basslets (Grammatidae)	fairy basslet (*Gramma loreto*) royal basslet (*Lipogramma regia*) threeline basslet (*Lipogramma trilineatum*)
Jawfishes (Opistognathidae)	swordtail jawfish (*Lonchopisthus micrognathus*) yellowhead jawfish (*Opistognathus aurifrons*) moustache jawfish (*Opistognathus lonchurus*) banded jawfish (*Opistognathus macrognathus*) mottled jawfish (*Opistognathus maxillosus*) yellowmouth jawfish (*Opistognathus nothus*) spotfin jawfish (*Opistognathus robinsi*) dusky jawfish (*Opistognathus whitehursti*)
Bigeyes (Priacanthidae)	bulleye (*Cookeolus japonicus*) glasseye snapper (*Heteropriacanthus cruentatus*) bigeye (*Priacanthus arenatus*) short bigeye (*Pristigenys alta*)

Family	Species
Cardinalfishes (Apogonidae)	bigtooth cardinalfish (*Apogon affinis*) bridle cardinalfish (*Apogon aurolineatus*) barred cardinalfish (*Apogon binotatus*) whitestar cardinalfish (*Apogon lachneri*) slendertail cardinalfish (*Apogon leptocaulus*) flamefish (*Apogon maculatus*) mimic cardinalfish (*Apogon phenax*) broadsaddle cardinalfish (*Apogon pillionatus*) pale cardinalfish (*Apogon planifrons*) twospot cardinalfish (*Apogon pseudomaculatus*) sawcheek cardinalfish (*Apogon quadrisquamatus*) belted cardinalfish (*Apogon townsendi*) bronze cardinalfish (*Astrapogon alutus*) blackfin cardinalfish (*Astrapogon puncticulatus*) conchfish (*Astrapogon stellatus*) freckled cardinalfish (*Phaeoptyx conklini*) dusky cardinalfish (*Phaeoptyx pigmentaria*) sponge cardinalfish (*Phaeoptyx xenus*)
Tilefishes (Malacanthidae)	goldface tilefish (*Caulolatilus chrysops*) blackline tilefish (*Caulolatilus cyanops*) anchor tilefish (*Caulolatilus intermedius*) blueline tilefish (*Caulolatilus microps*) tilefish (*Lopholatilus chamaeleonticeps*) sand tilefish (*Malacanthus plumieri*)
Bluefishes (Pomatomidae)	bluefish (*Pomatomus saltatrix*)
Jacks (Carangidae)	African pompano (*Alectis ciliaris*) yellow jack (*Caranx bartholomaei*) blue runner (*Caranx crysos*) crevalle jack (*Caranx hippos*) horse-eye jack (*Caranx latus*) bar jack (*Caranx ruber*) Atlantic bumper (*Chloroscombrus chrysurus*) mackerel scad (*Decapterus macarellus*) round scad (*Decapterus punctatus*) redtail scad (*Decapterus tabl*) rainbow runner (*Elagatis bipinnulata*) bluntnose jack (*Hemicaranx amblyrhynchus*) pilotfish (*Naucrates ductor*) leatherjack (*Oligoplites saurus*) bigeye scad (*Selar crumenophthalmus*) Atlantic moonfish (*Selene setapinnis*) lookdown (*Selene vomer*) greater amberjack (*Seriola dumerili*) lesser amberjack (*Seriola fasciata*) almaco jack (*Seriola rivoliana*) banded rudderfish (*Seriola zonata*) Florida pompano (*Trachinotus carolinus*)

Family	Species
	permit (*Trachinotus falcatus*) palometa (*Trachinotus goodei*) rough scad (*Trachurus lathami*) cottonmouth jack (*Uraspis secunda*)
Cobias (Rachycentridae)	cobia (*Rachycentron canadum*)
Dolphinfishes (Coryphaenidae)	pompano dolphinfish (*Coryphaena equiselis*) dolphinfish (*Coryphaena hippurus*)
Remoras (Echeneidae)	sharksucker (*Echeneis naucrates*) whitefin sharksucker (*Echeneis neucratoides*) slender suckerfish (*Phtheirichthys lineatus*) white suckerfish (*Remora albescens*) whalesucker (*Remora australis*) spearfish remora (*Remora brachyptera*) marlinsucker (*Remora osteochir*) remora (*Remora remora*)
Snappers (Lutjanidae)	black snapper (*Apsilus dentatus*) queen snapper (*Etelis oculatus*) mutton snapper (*Lutjanus analis*) schoolmaster (*Lutjanus apodus*) blackfin snapper (*Lutjanus buccanella*) red snapper (*Lutjanus campechanus*) cubera snapper (*Lutjanus cyanopterus*) gray snapper (*Lutjanus griseus*) dog snapper (*Lutjanus jocu*) mahogany snapper (*Lutjanus mahogoni*) lane snapper (*Lutjanus synagris*) silk snapper (*Lutjanus vivanus*) yellowtail snapper (*Ocyurus chrysurus*) wenchman (*Pristipomoides aquilonaris*) slender wenchman (*Pristipomoides freemani*) vermillion snapper (*Rhomboplites aurorubens*)
Tripletails (Lobotidae)	Atlantic tripletail (*Lobotes surinamensis*)
Mojarras (Gerreidae)	Irish pompano (*Diapterus auratus*) rhombic mojarra (*Diapterus rhombeus*) spotfin mojarra (*Eucinostomus argenteus*) silver jenny (*Eucinostomus gula*) tidewater mojarra (*Eucinostomus harengulus*) bigeye mojarra (*Eucinostomus havana*) slender mojarra (*Eucinostomus jonesii*) mottled mojarra (*Eucinostomus lefroyi*) flagfin mojarra (*Eucinostomus melanopterus*) striped mojarra (*Eugerres plumieri*) yellowfin mojarra (*Gerres cinereus*)
Grunts (Haemulidae)	black margate (*Anisotremus surinamensis*) porkfish (*Anisotremus virginicus*) barred grunt (*Conodon nobilis*)

Family	Species
	bonnetmouth (*Emmelichthyops atlanticus*) margate (*Haemulon album*) tomtate (*Haemulon aurolineatum*) caesar grunt (*Haemulon carbonarium*) smallmouth grunt (*Haemulon chrysargyreum*) French grunt (*Haemulon flavolineatum*) Spanish grunt (*Haemulon macrostomum*) cottonwick (*Haemulon melanurum*) sailors choice (*Haemulon parra*) white grunt (*Haemulon plumierii*) buestriped grunt (*Haemulon sciurus*) striped grunt (*Haemulon striatum*) boga (*Haemulon vittata*) pigfish (*Orthopristis chrysoptera*) burro grunt (*Pomadasys crocro*)
Porgies (Sparidae)	sheepshead (*Archosargus probatocephalus*) sea bream (*Archosargus rhomboidalis*) grass porgy (*Calamus arctifrons*) jolthead porgy (*Calamus bajonado*) saucereye porgy (*Calamus calamus*) whitebone porgy (*Calamus leucosteus*) knobbed porgy (*Calamus nodosus*) sheepshead porgy (*Calamus penna*) littlehead porgy (*Calamus proridens*) silver porgy (*Diplodus argenteus*) spottail pinfish (*Diplodus holbrooki*) pinfish (*Lagodon rhomboides*) red porgy (*Pagrus pagrus*) longspine porgy (*Stenotomus caprinus*)
Threadfins (Polynemidae)	Atlantic threadfin (*Polydactylus octonemus*) littlescale threadfin (*Polydactylus oligodon*) barbu (*Polydactylus virginicus*)
Drums and croakers (Sciaenidae)	silver perch (*Bairdiella chrysoura*) blue croaker (*Corvula batabana*) striped croaker (*Corvula sanctaeluciae*) sand seatrout (*Cynoscion arenarius*) spotted seatrout (*Cynoscion nebulosus*) silver seatrout (*Cynoscion nothus*) weakfish (*Cynoscion regalis*) jackknife-fish (*Equetus lanceolatus*) spotted drum (*Equetus punctatus*) banded drum (*Larimus fasciatus*) spot (*Leiostomus xanthurus*) southern kingfish (*Menticirrhus americanus*) Gulf kingfish (*Menticirrhus littoralis*) northern kingfish (*Menticirrhus saxatilis*) Atlantic croaker (*Micropogonias undulatus*) reef croaker (*Odontoscion dentex*)

Family	Species
	high-hat (*Pareques acuminatus*) blackbar drum (*Pareques iwamotoi*) cubbyu (*Pareques umbrosus*) black drum (*Pogonias cromis*) red drum (*Sciaenops ocellatus*) star drum (*Stellifer lanceolatus*) sand drum (*Umbrina coroides*)
Goatfishes (Mullidae)	yellow goatfish (*Mulloidichthys martinicus*) red goatfish (*Mullus auratus*) spotted goatfish (*Pseudupeneus maculatus*) dwarf goatfish (*Upeneus parvus*)
Sweepers (Pempheridae)	glassy sweeper (*Pempheris schomburgkii*)
Sea chubs (Kyphosidae)	darkfin chub (*Kyphosus bigibbus*) highfin chub (*Kyphosus cinerascens*) Bermuda chub (*Kyphosus sectatrix*) yellow chub (*Kyphosus vaigiensis*)
Butterflyfishes (Chaetodontidae)	foureye butterflyfish (*Chaetodon capistratus*) spotfin butterflyfish (*Chaetodon ocellatus*) reef butterflyfish (*Chaetodon sedentarius*) banded butterflyfish (*Chaetodon striatus*) longsnout butterflyfish (*Prognathodes aculeatus*) bank butterflyfish (*Prognathodes aya*) Guyana butterflyfish (*Prognathodes guyanensis*)
Angelfishes (Pomacanthidae)	cherubfish (*Centropyge argi*) blue angelfish (*Holacanthus bermudensis*) queen angelfish (*Holacanthus ciliaris*) rock beauty (*Holacanthus tricolor*) gray angelfish (*Pomacanthus arcuatus*) French angelfish (*Pomacanthus paru*)
Hawkfishes (Cirrhitidae)	redspotted hawkfish (*Amblycirrhitus pinos*)
Damselfishes (Pomacentridae)	sergeant major (*Abudefduf saxatilis*) night sergeant (*Abudefduf taurus*) blue chromis (*Chromis cyanea*) yellowtail reeffish (*Chromis enchrysura*) sunshinefish (*Chromis insolata*) brown chromis (*Chromis multilineata*) purple reeffish (*Chromis scotti*) yellowtail damselfish (*Microspathodon chrysurus*) dusky damselfish (*Stegastes adustus*) longfin damselfish (*Stegastes diencaeus*) beaugregory (*Stegastes leucostictus*) bicolor damselfish (*Stegastes partitus*) threespot damselfish (*Stegastes planifrons*) cocoa damselfish (*Stegastes variabilis*)
Wrasses and parrotfishes (Labridae)	spotfin hogfish (*Bodianus pulchellus*) Spanish hogfish (*Bodianus rufus*)

Family	Species
	creole wrasse (*Clepticus parrae*) bluelip parrotfish (*Cryptotomus roseus*) red hogfish (*Decodon puellaris*) dwarf wrasse (*Doratonotus megalepis*) greenband wrasse (*Halichoeres bathyphilus*) slippery dick (*Halichoeres bivittatus*) painted wrasse (*Halichoeres caudalis*) yellowcheek wrasse (*Halichoeres cyanocephalus*) yellowhead wrasse (*Halichoeres garnoti*) clown wrasse (*Halichoeres maculipinna*) rainbow wrasse (*Halichoeres pictus*) blackear wrasse (*Halichoeres poeyi*) puddingwife (*Halichoeres radiatus*) hogfish (*Lachnolaimus maximus*) emerald parrotfish (*Nicholsina usta*) midnight parrotfish (*Scarus coelestinus*) blue parrotfish (*Scarus coeruleus*) rainbow parrotfish (*Scarus guacamaia*) striped parrotfish (*Scarus iseri*) princess parrotfish (*Scarus taeniopterus*) queen parrotfish (*Scarus vetula*) greenblotch parrotfish (*Sparisoma atomarium*) redband parrotfish (*Sparisoma aurofrenatum*) redtail parrotfish (*Sparisoma chrysopterum*) bucktooth parrotfish (*Sparisoma radians*) yellowtail parrotfish (*Sparisoma rubripinne*) stoplight parrotfish (*Sparisoma viride*) bluehead (*Thalassoma bifasciatum*) rosy razorfish (*Xyrichtys martinicensis*) pearly razorfish (*Xyrichtys novacula*) green razorfish (*Xyrichtys splendens*)
Stargazers (Uranoscopidae)	northern stargazer (*Astroscopus guttatus*) southern stargazer (*Astroscopus y-graecum*) lancer stargazer (*Kathetostoma albigutta*) freckled stargazer (*Xenocephalus egregius*)
Triplefins (Tripterygiidae)	lofty triplefin (*Enneanectes altivelis*) roughhead triplefin (*Enneanectes boehlkei*) redeye triplefin (*Enneanectes pectoralis*)
Sand stargazers (Dactyloscopidae)	bigeye stargazer (*Dactyloscopus crossotus*) speckled stargazer (*Dactyloscopus moorei*) sand stargazer (*Dactyloscopus tridigitatus*) arrow stargazer (*Gillellus greyae*) masked stargazer (*Gillellus healae*) warteye stargazer (*Gillellus uranidea*) saddle stargazer (*Platygillellus rubrocinctus*)
Combtooth blennies (Blenniidae)	striped blenny (*Chasmodes bosquianus*) stretchjaw blenny (*Chasmodes longimaxilla*)

Family	Species
	Florida blenny (*Chasmodes saburrae*) pearl blenny (*Entomacrodus nigricans*) barred blenny (*Hypleurochilus bermudensis*) zebratail blenny (*Hypleurochilus caudovittatus*) crested blenny (*Hypleurochilus geminatus*) featherduster blenny (*Hypleurochilus multifilis*) oyster blenny (*Hypleurochilus pseudoaequipinnis*) orangespotted blenny (*Hypleurochilus springeri*) feather blenny (*Hypsoblennius hentz*) tessellated blenny (*Hypsoblennius invemar*) freckled blenny (*Hypsoblennius ionthas*) highfin blenny (*Lupinoblennius nicholsi*) mangrove blenny (*Lupinoblennius vinctus*) redlip blenny (*Ophioblennius macclurei*) seaweed blenny (*Parablennius marmoreus*) molly miller (*Scartella cristata*)
Labrisomid blennies (Labrisomidae)	puffcheek blenny (*Labrisomus bucciferus*) masquerader hairy blenny (*Labrisomus conditus*) mock blenny (*Labrisomus cricota*) palehead blenny (*Labrisomus gobio*) mimic blenny (*Labrisomus guppyi*) longfin blenny (*Labrisomus haitiensis*) downy blenny (*Labrisomus kalisherae*) spotcheek blenny (*Labrisomus nigricinctus*) hairy blenny (*Labrisomus nuchipinnis*) goldline blenny (*Malacoctenus aurolineatus*) rosy blenny (*Malacoctenus macropus*) saddled blenny (*Malacoctenus triangulatus*) threadfin blenny (*Nemaclinus atelestos*) coral blenny (*Paraclinus cingulatus*) banded blenny (*Paraclinus fasciatus*) horned blenny (*Paraclinus grandicomis*) bald blenny (*Paraclinus infrons*) marbled blenny (*Paraclinus marmoratus*) blackfin blenny (*Paraclinus nigripinnis*) checkered blenny (*Starksia ocellata*) Key blenny (*Starksia starcki*)
Tube blennies (Chaenopsidae)	roughhead blenny (*Acanthemblemaria aspera*) papillose blenny (*Acanthemblemaria chaplini*) spinyhead blenny (*Acanthemblemaria spinosa*) yellowface pikeblenny (*Chaenopsis limbaughi*) bluethroat pikeblenny (*Chaenopsis ocellata*) flecked pikeblenny (*Chaenopsis roseola*) banner blenny (*Emblemaria atlantica*) sailfin blenny (*Emblemaria pandionis*)` pirate blenny (*Emblemaria piratula*) blackhead blenny (*Emblemariopsis bahamensis*) glass blenny (*Emblemariopsis diaphana*)

Family	Species
	wrasse blenny (*Hemiemblemaria similis*) blackbelly blenny (*Stathmonotus hemphillii*) eelgrass blenny (*Stathmonotus stahli*)
Clingfishes (Gobiesocidae)	emerald clingfish (*Acyrtops beryllinus*) stippled clingfish (*Gobiesox punctulatus*) skilletfish (*Gobiesox strumosus*)
Dragonets (Callionymidae)	spotted dragonet (*Diplogrammus pauciradiatus*) spotfin dragonet (*Foetorepus agassizii*) palefin dragonet (*Foetorepus goodenbeani*) lancer dragonet (*Paradiplogrammus bairdi*)
Sleepers (Eleotridae)	fat sleeper (*Dormitator maculatus*) largescaled spinycheek sleeper (*Eleotris amblyopsis*) smallscaled spinycheek sleeper (*Eleotris perniger*) emerald sleeper (*Erotelis smaragdus*) bigmouth sleeper (*Gobiomorus dormitor*) guavina (*Guavina guavina*)
Gobies (Gobiidae)	river goby (*Awaous banana*) bearded goby (*Barbulifer ceuthoecus*) notchtongue goby (*Bathygobius curacao*) twinspotted frillfin (*Bathygobius geminatus*) checkerboard frillfin (*Bathygobius lacertus*) island frillfin (*Bathygobius mystacium*) frillfin goby (*Bathygobius soporator*) white-eye goby (*Bollmannia boqueronensis*) ragged goby (*Bollmannia communis*) shelf goby (*Bollmannia eigenmanni*) barfin goby (*Coryphopterus alloides*) sand-canyon goby (*Coryphopterus bol*) colon goby (*Coryphopterus dicrus*) pallid goby (*Coryphopterus eidolon*) bridled goby (*Coryphopterus glaucofraenum*) glass goby (*Coryphopterus hyalinus*) kuna goby (*Coryphopterus kuna*) peppermint goby (*Coryphopterus lipernes*) masked goby (*Coryphopterus personatus*) spotted goby (*Coryphopterus punctipectophorus*) bartail goby (*Coryphopterus thrix*) sand goby (*Coryphopterus tortugae*) darter goby (*Ctenogobius boleosoma*) slashcheek goby (*Ctenogobius pseudofasciatus*) dash goby (*Ctenogobius saepepallens*) freshwater goby (*Ctenogobius shufeldti*) emerald goby (*Ctenogobius smaragdus*) marked goby (*Ctenogobius stigmaticus*) spottail goby (*Ctenogobius stigmaturus*) yellowline goby (*Elacatinus horsti*) neon goby (*Elacatinus oceanops*)

Family	Species
	yellowprow goby (*Elacatinus xanthiprora*) sponge goby (*Evermannichthys spongicola*) lyre goby (*Evorthodus lyricus*) goldspot goby (*Gnatholepis thompsoni*) violet goby (*Gobioides broussonetii*) highfin goby (*Gobionellus oceanicus*) naked goby (*Gobiosoma bosc*) seaboard goby (*Gobiosoma ginsburgi*) rockcut goby (*Gobiosoma grosvenori*) twoscale goby (*Gobiosoma longipala*) code goby (*Gobiosoma robustum*) paleback goby (*Gobulus myersi*) crested goby (*Lophogobius cyprinoides*) dwarf goby (*Lythrypnus elasson*) island goby (*Lythrypnus nesiotes*) convict goby (*Lythrypnus phorellus*) bluegold goby (*Lythrypnus spilus*) Seminole goby (*Microgobius carri*) clown goby (*Microgobius gulosus*) banner goby (*Microgobius microlepis*) green goby (*Microgobius thalassinus*) orangespotted goby (*Nes longus*) spotfin goby (*Oxyurichthys stigmalophius*) mauve goby (*Palatogobius paradoxus*) rusty goby (*Priolepis hipoliti*) scaleless goby (*Psilotris alepis*) toadfish goby (*Psilotris batrachodes*) highspine goby (*Psilotris celsus*) tusked goby (*Risor ruber*) tiger goby (*Tigrigobius macrodon*) leopard goby (*Tigrigobius sacrus*) orangebelly goby (*Varicus marilynae*)
Wormfishes (Microdesmidae)	pugjaw wormfish (*Cerdale floridana*) pink wormfish (*Microdesmus longipinnis*)
Dartfishes (Ptereleotridae)	blue dartfish (*Ptereleotris calliura*) hovering dartfish (*Ptereleotris helenae*)
Spadefishes (Ephippidae)	Atlantic spadefish (*Chaetodipterus faber*)
Surgeonfishes (Acanthuridae)	doctorfish (*Acanthurus chirurgus*) blue tang (*Acanthurus coeruleus*) ocean surgeon (*Acanthurus tractus*)
Barracudas (Sphyraenidae)	great barracuda (*Sphyraena barracuda*) sennet (*Sphyraena borealis*) guaguanche (*Sphyraena guachancho*) southern sennet (*Sphyraena picudilla*)
Snake mackerels (Gempylidae)	snake mackerel (*Gempylus serpens*) escolar (*Lepidocybium flavobrunneum*) black snake mackerel (*Nealotus tripes*)

Family	Species
	American sackfish (*Neoepinnula americana*)
	black gemfish (*Nesiarchus nasutus*)
	roudi escolar (*Promethichthys prometheus*)
	oilfish (*Ruvettus pretiosus*)
Cutlassfishes (Trichiuridae)	Atlantic cutlassfish (*Trichiurus lepturus*)
Mackerels (Scombridae)	wahoo (*Acanthocybium solandri*)
	bullet mackerel (*Auxis rochei*)
	frigate mackerel (*Auxis thazard*)
	little tunny (*Euthynnus alletteratus*)
	skipjack tuna (*Katsuwonus pelamis*)
	Atlantic bonito (*Sarda sarda*)
	Atlantic chub mackerel (*Scomber colias*)
	king mackerel (*Scomberomorus cavalla*)
	Spanish mackerel (*Scomberomorus maculatus*)
	cero (*Scomberomorus regalis*)
	albacore (*Thunnus alalunga*)
	yellowfin tuna (*Thunnus albacarcs*)
	blackfin tuna (*Thunnus atlanticus*)
	bigeye tuna (*Thunnus obesus*)
	bluefin tuna (*Thunnus thynnus*)
Swordfishes (Xiphiidae)	swordfish (*Xiphias gladius*)
Billfishes (Istiophoridae)	sailfish (*Istiophorus platypterus*)
	white marlin (*Kajikia albida*)
	blue marlin (*Makaira nigricans*)
	roundscale spearfish (*Tetrapturus georgii*)
	longbill spearfish (*Tetrapturus pfluegeri*)
Medusafishes (Centrolophidae)	black driftfish (*Hyperoglyphe bythites*)
	barrelfish (*IIyperoglyphe perciformis*)
Driftfishes (Nomeidae)	man-of-war fish (*Nomeus gronovii*)
	freckled driftfish (*Psenes cyanophrys*)
	silver driftfish (*Psenes pellucidus*)
Ariommatids (Ariommatidae)	silver-rag (*Ariomma bondi*)
	brown driftfish (*Ariomma melanum*)
	spotted driftfish (*Ariomma regulus*)
Squaretails (Tetragonuridae)	bigeye squaretail (*Tetragonurus atlanticus*)
Butterfishes (Stromateidae)	Gulf butterfish (*Peprilus burti*)
	harvestfish (*Peprilus paru*)
	butterfish (*Peprilus triacanthus*)
Boarfishes (Caproidae)	deepbody boarfish (*Antigonia capros*)
	shortspine boarfish (*Antigonia combatia*)
Turbots (Scophthalmidae)	windowpane (*Scophthalmus aquosus*)
Sand flounders (Paralichthyidae)	three-eye flounder (*Ancylopsetta dilecta*)
	ocellated flounder (*Ancylopsetta quadrocellata*)
	Gulf Stream flounder (*Citharichthys arctifrons*)
	sand whiff (*Citharichthys arenaceus*)

Family	Species
	horned whiff (*Citharichthys cornutus*) spined whiff (*Citharichthys dinoceros*) anglefin whiff (*Citharichthys gymnorhinus*) spotted whiff (*Citharichthys macrops*) bay whiff (*Citharichthys spilopterus*) Mexican flounder (*Cyclopsetta chittendeni*) spotfin flounder (*Cyclopsetta fimbriata*) fringed flounder (*Etropus crossotus*) shelf flounder (*Etropus cyclosquamus*) smallmouth flounder (*Etropus microstomus*) gray flounder (*Etropus rimosus*) shrimp flounder (*Gastropsetta frontalis*) Gulf flounder (*Paralichthys albigutta*) summer flounder (*Paralichthys dentatus*) southern flounder (*Paralichthys lethostigma*) fourspot flounder (*Paralichthys oblongus*) broad flounder (*Paralichthys squamilentus*) shoal flounder (*Syacium gunteri*) channel flounder (*Syacium micrurum*) dusky flounder (*Syacium papillosum*)
Lefteye flounders (Bothidae)	peacock flounder (*Bothus lunatus*) eyed flounder (*Bothus ocellatus*) twospot flounder (*Bothus robinsi*) spiny flounder (*Engyophrys senta*) slim flounder (*Monolene antillarum*) deepwater flounder (*Monolene sessilicauda*) sash flounder (*Trichopsetta ventralis*)
American soles (Achiridae)	lined sole (*Achirus lineatus*) naked sole (*Gymnachirus melas*) fringed sole (*Gymnachirus texae*) scrawled sole (*Trinectes inscriptus*) hogchoker (*Trinectes maculatus*)
Tonguefishes (Cynoglossidae)	Caribbean tonguefish (*Symphurus arawak*) chocolatebanded tonguefish (*Symphurus billykrietei*) offshore tonguefish (*Symphurus civitatium*) spottedfin tonguefish (*Symphurus diomedeanus*) margined tonguefish (*Symphurus marginatus*) largescale tonguefish (*Symphurus minor*) pygmy tonguefish (*Symphurus parvus*) longtail tonguefish (*Symphurus pelicanus*) deepwater tonguefish (*Symphurus piger*) blackcheek tonguefish (*Symphurus plagiusa*) blotchfin tonguefish (*Symphurus stigmosus*) spottail tonguefish (*Symphurus urospilus*)
Spikefishes (Triacanthodidae)	spotted spikefish (*Hollardia meadi*) jambeau (*Parahollardia lineata*)
Triggerfishes (Balistidae)	gray triggerfish (*Balistes capriscus*)

Family	Species
	queen triggerfish (*Balistes vetula*) rough triggerfish (*Canthidermis maculata*) ocean triggerfish (*Canthidermis sufflamen*) black durgon (*Melichthys niger*) sargassum triggerfish (*Xanthichthys ringens*)
Filefishes (Monacanthidae)	dotterel filefish (*Aluterus heudelotii*) unicorn filefish (*Aluterus monoceros*) orange filefish (*Aluterus schoepfii*) scrawled filefish (*Aluterus scriptus*) whitespotted filefish (*Cantherhines macrocerus*) orangespotted filefish (*Cantherhines pullus*) fringed filefish (*Monacanthus ciliatus*) slender filefish (*Monacanthus tuckeri*) planehead filefish (*Stephanolepis hispidus*) pygmy filefish (*Stephanolepis setifer*)
Boxfishes (Ostraciidae)	honeycomb cowfish (*Acanthostracion polygonia*) scrawled cowfish (*Acanthostracion quadricornis*) spotted trunkfish (*Lactophrys bicaudalis*) trunkfish (*Lactophrys trigonus*) smooth trunkfish (*Lactophrys triqueter*)
Puffers (Tetraodontidae)	goldface toby (*Canthigaster jamestyleri*) sharpnose puffer (*Canthigaster rostrata*) smooth puffer (*Lagocephalus laevigatus*) oceanic puffer (*Lagocephalus lagocephalus*) marbled puffer (*Sphoeroides dorsalis*) northern puffer (*Sphoeroides maculatus*) southern puffer (*Sphoeroides nephelus*) blunthead puffer (*Sphoeroides pachygaster*) least puffer (*Sphoeroides parvus*) bandtail puffer (*Sphoeroides spengleri*) checkered puffer (*Sphoeroides testudineus*)
Porcupinefishes (Diodontidae)	bridled burrfish (*Chilomycterus antennatus*) web burrfish (*Chilomycterus antillarum*) spotfin burrfish (*Chilomycterus reticulatus*) striped burrfish (*Chilomycterus schoepfii*) balloonfish (*Diodon holocanthus*) porcupinefish (*Diodon hystrix*)
Molas (Molidae)	sharptail mola (*Mola lanceolata*) ocean sunfish (*Mola mola*) slender mola (*Ranzania laevis*)

Glossary

Abdomen. Belly.
Adipose fin. A small, fatty fin, without rays, located behind the dorsal fin on the back of some fishes.
Air bladder. A gas-filled sac located in the body cavity, below the vertebrae; also called a swim bladder.
Anal fin. The fin on the median ventral line behind the anus.
Anterior. In front of or toward the head.
Anus. The external opening of the intestine; also called the vent.
Axil. The space under a rotating fin, such as the pectoral fin.
Barbel. A slender, tactile, whisker-like projection extending from the head of some fishes.
Benthic. Living on or in the seafloor.
Branchial. Pertaining to the gills.
Buckler. A bony shield or scale.
Canine tooth. An elongated conical tooth.
Caudal. Pertaining to the tail.
Caudal fin. The tail fin.
Caudal peduncle. The slender portion of the fish's body just in front of the tail fin.
Ciguatera. A sickness caused by eating the flesh of fishes carrying toxins (ciguatoxins) derived from single-celled, dinoflagellate algae.
Cirri. Fringe-like tendrils, whiskers, or tufts of skin.
Clasper. One of a pair of elongated reproductive organs on the pelvic fins of male sharks, skates, rays, and ratfishes.
Cleaner. A fish or invertebrate that removes external parasites from other, larger fishes.
Cloaca. A combined anal-genital opening.
Demersal. Living on or near the bottom.
Dorsal fin. A fin located on the back of a fish.
Emarginate. Shallowly notched.
Finlet. A small, usually separate fin ray, located in a series behind the main dorsal or anal fin.
Gametes. Reproductive cells, where a male cell and a female cell unite to form a new organism.
Gill. A filamentous respiratory organ of aquatic animals.
Gill arch. The bony support to which gills are attached.
Gill cover. A lid or flap covering the gill; also called an opercle or operculum.
Gill filament. A thread-like structure attached to the outside of a gill arch.
Gill opening. The opening leading from the gills; also called the gill cleft.
Gill raker. A bony projection attached to the inside of gill arches and used to strain food from the water.
Gonad. A sexual organ: an ovary, a testis, or a hermaphrodite gland.
Gonopodium. A modified anal fin, functioning as a copulatory organ in certain fishes.
Incisor. A front tooth, flattened to form a cutting edge.
Jugular. Pertaining to the neck or throat.
Lanceolate. An elongate and pointed tail fin.
Lateral. Pertaining to the side.
Lateral line. The longitudinal line on each side of a fish's body, composed of pores opening into sensory organs.
Mandible. The lower jaw.
Maxillary. The second (and usually the larger) of the two bones forming the upper jaw.
Median fin. An unpaired fin along the median line of the body, such as the dorsal, anal, or caudal fin.
Midlateral stripe. A stripe on the lateral line.
Milt. The sperm of fishes.
Molar. A grinding tooth.
Nictitating membrane. A membrane that assists in keeping the eye clean in some sharks and bony fishes; sometimes called a third eyelid.
Ocellated spot. A dark spot encircled by a band of a different, lighter color.
Opercle. A gill cover; also called an operculum.
Opercular flap. A fleshy extension of the rear edge of the opercle.
Orbit. An eye socket.
Osseous. Bony.
Otolith. One of two or three small, somewhat spherical concretions (stones) found in the inner ear of fishes.
Ovary. A female reproductive gland.
Oviparous. Egg laying, with the eggs hatching outside the body.
Ovoviviparous. Producing eggs with definite shells that hatch within the body of a female, so the young are born alive.
Pectoral fin. One of a pair of fins attached to the shoulder girdle.
Pelagic. Living in open water.
Pelvic fin. One of a pair of fins below the pectoral fins; also called a ventral fin.
Pharyngeal jaws. Upper and lower gill arches in the pharynx that are connected by a joint, like jaws.
Pharyngeal tooth. A grinding tooth, located in the pharynx, on the last gill arch.
Pharynx. The section of the alimentary canal joining the mouth cavity with the esophagus.

Photophore. A luminous (light-producing) organ or spot.

Practical salinity unit (psu). A unit of measurement used to describe the amount of dissolved salts in a body of water.

Premaxillary. One of a pair of bones forming the front of the upper jaw in fishes.

Preopercle. The anterior cheekbone.

Protractile. Capable of being thrust forward.

Pseudobranchia. A small, gill-like structure on the inner side of the gill cover.

Ray. A flexible, segmented rod in a fin that may be branched near the ends.

Roe. The eggs of fishes.

Rostrum. A protruding snout.

Scute. An external, horny or bony plate or scale.

Soft ray. A flexible, jointed ray.

Spine. A stiff, unbranched rod in a fin that usually tapers to a sharp point.

Spiny ray. A very stiff, nonjointed spine, usually pointed.

Spiracle. A small respiratory opening behind the eye in sharks, skates, and rays.

Spiral valve. A corkscrew-like partition in the digestive tract of sharks.

Swim bladder. See air bladder.

Testis. A male reproductive gland.

Thoracic. Pertaining to the thorax (chest).

Truncate. An abbreviated tail fin with a straight margin.

Upper eye stripe. A dark stripe emanating from the eye of early juvenile grunts.

Vent. See anus.

Ventral fin. See pelvic fin.

Vestigial. Small and imperfectly developed (rudimentary).

Viviparous. Bearing live young.

Vomer. The median bone in the front upper part of the mouth.

Photo Locations (for Species Photographed Outside Florida)

p. 24, Sea lampreys (*Petromyzon marinus*), Deer Creek, Maryland

p. 26, Whale shark (*Rhincodon typus*), Isla Mujeres, Mexico

p. 27, Sand tiger (*Carcharias taurus*), Moorehead City, North Carolina

p. 29, Basking shark (*Cetorhinus maximus*), United Kingdom

p. 30, White shark (*Carcharodon carcharias*), Australia

p. 31, Chain dogfish (*Scyliorhinus retifer*), northeastern Gulf of Mexico

p. 32, Smooth dogfish (*Mustelus canis*), Campeche Bank, Mexico

p. 34, Tiger shark (*Galeocerdo cuvier*), with sharksuckers (*Echeneis naucrates*), Little Bahama Bank, Bahamas

p. 36, Blacktip shark (*Carcharhinus limbatus*), Little Bahama Bank, Bahamas

p. 40, Atlantic torpedo (*Torpedo nobiliana*), offshore of North Carolina

p. 45, Clearnose skate (*Raja eglanteria*), Outer Banks, North Carolina

p. 59, Bonefish (*Albula vulpes*), Berry Islands, Bahamas

p. 60, American eel (*Anguilla rostrata*), Susquehanna River, Maryland

p. 73, Redear sardines (*Harengula humeralis*), Sandy Cay, Bahamas

p. 76, Offshore lizardfish (*Synodus poeyi*), northeastern Gulf of Mexico

p. 77, Red lizardfish (*Synodus synodus*), northeastern Gulf of Mexico

p. 78, Antenna codlet (*Bregmaceros atlanticus*), northeastern Gulf of Mexico

p. 80, Southern hake (*Urophycis floridana*), northeastern Gulf of Mexico. Gulf hake (*Urophycis cirrata*), northeastern Gulf of Mexico. Carolina hake (*Urophycis earllii*), Cape Fear, North Carolina. Spotted hake (*Urophycis regia*), east of Beaufort Bar, North Carolina.

p. 83, Bank cusk-eel (*Ophidion holbrookii*), offshore of North Carolina. Polka-dot cusk-eel (*Otophidium omostigmum*), offshore of North Carolina. Stripefin brotula (*Neobythities marginatus*), Brunswick, Georgia.

p. 88, Blackfin goosefish (*Lophius gastrophysus*), northeastern Gulf of Mexico

p. 92, Longnose batfish (*Ogcocephalus corniger*), northeastern Gulf of Mexico. Shortnose batfish (*Ogcocephalus nasutus*), Walkers Cay, Bahamas.

p. 110, Spinycheek soldierfish (*Corniger spinosus*), northeastern Gulf of Mexico. Reef squirrelfish (*Sargocentron coruscum*), Little Bahama Bank, Bahamas. Dusky squirrelfish (*Sargocentron vexillarium*), Grand Bahama Island, Bahamas.

p. 111, Longjaw squirrelfish (*Neoniphon marianus*), Berry Islands, Bahamas

p. 119, Longspine snipefish (*Macroramphosus scolopax*), offshore of North Carolina

p. 130, Wreckfish (*Polyprion americanus*), Baltimore Canyon, U.S. East Coast

p. 132, Coney, bicolor phase, Abaco, Bahamas. Juvenile, Walkers Cay, Bahamas.

p. 134, Nassau grouper (*Epinephelus striatus*), Walkers Cay, Bahamas

p. 135, Red hind (*Epinephelus guttatus*), Little Bahama Bank, Bahamas

p. 136, Juvenile yellowedge grouper (*Hyporthodus flavolimbatus*), northeastern Gulf of Mexico

p. 137, Yellowmouth grouper (*Mycteroperca interstitialis*), Walkers Cay, Bahamas

p. 139, Tiger grouper (*Mycteroperca tigris*), Walkers Cay, Bahamas. Juvenile, Walkers Cay, Bahamas.

p. 144, Tobaccofish (*Serranus tabacarius*), Grand Bahama Island, Bahamas

p. 148, Shy hamlet (*Hypoplectrus guttavarius*), Walkers Cay, Bahamas. Golden hamlet (*Hypoplectrus gummigutta*), Grand Bahama Island, Bahamas.

p. 149, Peppermint basslet (*Liopropoma rubre*), Walkers Cay, Bahamas. Roughtongue bass (*Pronotogrammus martinicensis*), northeastern Gulf of Mexico. Red barbier (*Baldwinella vivanus*), northeastern Gulf of Mexico.

p. 153, Fairy basslet (*Gramma loreto*), Walkers Cay, Bahamas

p. 157, Glasseye snapper (*Heteropriacanthus cruentatus*), Grand Bahama Island, Bahamas

p. 161, Conchfish (*Astrapogon stellatus*), Grand Bahama Island, Bahamas

p. 171, Permit (*Trachinotus falcatus*), Great Bahama Bank, Bahamas. Palometa (*Trachurus goodei*), Abaco, Bahamas.

p. 177, Pompano dolphinfish (*Coryphaena equiselis*), northern Gulf of Mexico

p. 178, Juvenile dolphinfish (*Coryphaena hippurus*), northern Gulf of Mexico

p. 179, Remora (*Remora remora*), attached to a large tiger shark (*Galeocerdo cuvier*), Little Bahama Bank, Bahamas

p. 181, Queen snapper (*Etilus oculatus*), northwestern Gulf of Mexico

p. 184, Red snapper (*Lutjanus campechanus*), northeastern Gulf of Mexico. Blackfin snapper (*Lutjanus buccanella*), juvenile, Berry Islands, Bahamas

p. 186, Dog snapper (*Lutjanus jocu*), Walkers Cay, Bahamas

p. 194, Smallmouth grunt (*Haemulon chrysargyreum*), juvenile, Grand Cay, Abaco, Bahamas

p. 198, Boga (*Haemulon vittata*), Walkers Cay, Bahamas

p. 203, Jolthead porgy (*Calamus bajonado*), Walkers Cay, Bahamas

p. 217, Foureye butterflyfish (*Chaetodon capistratus*), juvenile, Grand Bahama Island, Bahamas

p. 219, Banded butterflyfish (*Chaetodon striatus*), juvenile, Little Bahama Bank, Bahamas

p. 220, Longsnout butterflyfish (*Prognathodes aculeatus*), Grand Turk, British West Indies. Bank butterflyfish (*Prognathodes aya*), northeastern Gulf of Mexico.

p. 223, Gray angelfish (*Pomacanthus arcuatus*), Little Bahama Bank, Bahamas

p. 226, Night sergeant (*Abudefduf taurus*), Seal Cay, Bahamas. Yellowtail damselfish (*Microspathodon chrysurus*), Little Bahama Bank, Bahamas.

p. 227, Blue chromis (*Chromis cyanea*), Walkers Cay, Bahamas

p. 229, Longfin damselfish (*Stegastes

diencaeus), Little Bahama Bank, Bahamas. Juvenile, Grand Bahama Island, Bahamas.

p. 230, Beaugregory (*Stegastes leucostictus*), Grand Bahama Island, Bahamas; juvenile, Grand Bahama Island, Bahamas. Threespot damselfish (*Stegastes planifrons*), Bonaire, Netherlands Antilles.

p. 233, Hogfish (*Lachnolaimus maximus*), Walkers Cay, Bahamas

p. 238, Red hogfish (*Decodon puellaris*), northeastern Gulf of Mexico

p. 239, Initial phase greenband wrasse (*Halichoeres bathyphilus*), northeastern Gulf of Mexico

p. 246, Initial phase princess parrotfish (*Scarus taeniopterus*), Walkers Cay, Bahamas

p. 255, Redlip blenny (*Ophioblennius macclurei*), Little Bahama Bank, Bahamas

p. 260, Palehead blenny (*Labrisomus gobio*), Walkers Cay, Bahamas

p. 261, Goldline blenny (*Malacoctenus aurolineatus*), Walkers Cay, Bahamas

p. 267, Wrasse blenny (*Hemiemblemaria similis*), Walkers Cay, Bahamas

p. 275, Peppermint goby (*Coryphopterus lipernes*), Little Bahama Bank, Bahamas

p. 288, Atlantic cutlassfish (*Trichiurus lepturus*), Chesapeake Bay

p. 289, King mackeral (*Scomberomorus cavalla*), North Carolina

p. 293, Swordfish (*Xiphias gladius*), Mediterranean Sea off Sardinia

p. 295, Sailfish (*Istiophorus platypterus*), Isla Mujeres, Mexico

p. 300, Harvestfish (*Peprilus paru*), Beaufort, North Carolina. Butterfish (*Peprilus triacanthus*), Beaufort, North Carolina.

p. 301, Deepbody boarfish (*Antigonia capros*), North Carolina

p. 302, Windowpane (*Scophthalmus aquosus*), North Carolina

p. 307, Peacock flounder (*Bothus lunatus*), Grand Bahama Island, Bahamas

p. 314, Black durgon (*Melichthys niger*), Little Bahama Bank, Bahamas

p. 318, Spotted phase and saddle phase whitespotted filefish (*Cantherhines macrocerus*), Little Bahama Bank, Bahamas

Index of Scientific Names

Index of Common Names